AF531340

UNDERSTANDING PALAEONTOLOGY

UNDERSTANDING PALAEONTOLOGY

By

Dr. P.R. Yadav

Lecturer

Dept. of Zoology

D.A.V. College

Muzaffarnagar (U.P.)

(India)

DISCOVERY PUBLISHING HOUSE PVT. LTD.

NEW DELHI-110 002

Published by:
Namit Wasan

DISCOVERY PUBLISHING HOUSE PVT. LTD.
4383/4B, Ansari Road, Darya Ganj
New Delhi-110 002 (India)
Phone : +91-11-23279245; 23253475; 43596065
E-mail : discoverybooksindia@gmail.com
discoverypublishinghouse@gmail.com
namitwasan9@gmail.com
web : www.discoverypublishinggroup.com

***Edition:* 2020**

ISBN: 978-81-8356-477-9

Understanding Palaeontology

Printed at:
Infinity Imaging Systems
Delhi

Preface

The present title "Understanding Palaeontology" has been written for those students interested in careers in diverse fields of biological sciences. It provides a structured approach to learning by covering all the important topics in a uniform, systematic format. The book has been comprehensively designed incorporating recent advances in this fast moving field. It also provides accessible information on palaeontology in compact form for undergraduate students in biology and related life sciences. It is intelligible to the educated layman, though it deals with some complex ideas. It is an adequate text for all the requirements of students in this area. In addition, busy lecturers who require a quick reference compendium will find it useful, particularly for tutional planning. Simple, yet hopefully clear figures and tables are provided throughout the book.

The over-riding goal of this book, and indeed of the whole *Understanding series*, is to present the essential information concering palaeontology in a compact, readily accessible form which leads itself to student learning and revision. The convergence of various approaches has generated a rich panorama of detail, the significance of which we are still attempting to unraval. The present text has been written as an introduction to this rapidly growing field.

To make the work more comprehensive and informative, the author has consulted many authoritative books, research journals, abstracts, monographs etc., so there can be no claim to originality except in the manner of treatment.

The author expresses his thanks to his friends and colleagues whose continue inspirations have initiated him to bring out this book.

The author expresses his gratitude to Mr. Wasan and staff of M/s Discovery Publishing House Pvt. Ltd. for their whole hearted co-operation in the publication of this book.

In the mean time, the author will remain sincerely responsible for any shortcomings of the book and be grateful to the readers for their suggestions and constructive criticism for the continuous betterment of the book. He takes this opportunity to appeal to the readers to send their suggestions straightaway to his Publisher.

Author

CONTENTS

1

Age of Earth

The evidence that organic evolution has occured is so overwhelming that no one who is aquainted with it has any doubt that new species are derived from previous ones by descent with modification. The extensive fossil record provides direct evidence of organic evolution and gives the details of the evolutionary relationships of many lines of descent. There are, moreover, a great many facts from all of the subdivisions of biological science which acquire significance, and make sense, only when viewed against the background of evolution.

The science of palaeontology deals with the findings cataloguing and interpretation of the abundant and diverse evidence of life in former times. The term fossil (Latin *fossilium*, something dug up) refers not only to the bones, shells, teeth and other hard parts of a plant or animal body which have been preserved, but to any impression or trace left by some previous organism. In view of the large number of fossils of plants and animals that have been found it is sobering to realize that only a small fraction of all the organisms that ever lived were preserved as fossils, and only a small fraction of these fossils have been dug up and studied to date.

Footprints or trails made in soft mud, which subsequently hardened, are a common type of fossils. From such remains one can tell something of the structure and body proportions of the animals which made them. In 1948, tracks were discovered near Pittsburgh of an amphibian from the Pennsylvanian period, some 250,000,000 years ago. That the animal moved by hopping, rather than by walking, was evident from the fact that the footprints lay opposite each other, in pairs.

Most of the vertebrate fossils are skeletal parts, from which it is possible to deduce the animal's posture and style of walking. From the bone scars, indicating muscle attachments, palaeontologists can deduce the general position and size of the muscles and, from this, the contours of the body. On the basis of such considerations, reconstructions are made of how the animals looked in life. Such qualities as the texture and colour of the fur or scales must be guessed at.

An interesting and striking type of fossil is that in which the original hard parts and, more rarely, soft tissues have been replaced by minerals—a process known as *petrifaction*. The replacing minerals may be iron pyrites, silica, calcium carbonate or other substances. The petrified muscle from a shark, more than 300,000,000 years old, was so well preserved by this process that not only individual muscle fibers but their cross striations could be observed in thin sections under the microscope. The Petrified Forest in Arizona is a famous example of the process of petrifaction.

Molds and casts are superficially similar to petrified fossils but are produced differently. Molds were formed by the hardening of the material surrounding the buried organism, followed by the decay and removal of the organism by seepage of the ground water. Sometimes the molds were subsequently filled with minerals which, in turn, hardened to form *casts* replicas of the original structures.

Occasionally, palaeontologists are fortunate enough to find organisms frozen in the soil or ice of the far North, usually in Siberia and Alaska. The remains of woolly mammoths more than 25,000 years old have been found so well preserved that their flesh was eaten by dogs. Other forms—plants, insects and spiders—have been preserved in amber, a fossil resin from pine trees. Originally the resin was a sap soft enough to engulf the fragile insect and penetrate every part; then it gradually hardened, preserving the animal intact.

The formation and preservation of a fossil require that some structure be buried. This may take place at the bottom of a body of water, or on land by the accumulation of wind blown sand, soil or volcanic ash. Many of the men and animals living in Pompeii were preserved almost perfectly by the volcanic ash from the eruption of Vesuvius. Sometimes animals were trapped and entombed in a bog, quicksand or asphalt pit, such as the famous La Brea tar pits in Los Angeles which have provide superb fossils of Pleistocene animals.

The Geologic Time Table

Geologists have found five major rock strata, each of which is subdivided into lesser strata, and on this basis the geologic time table

Table 1.1. Geological time table

Era	*Period*	*Epoch*	*Duration in Millions of Years*	*Time from Beginning of Period to Present (Millions of Years)*	*Geologic Condition*	*Plant Life*	*Animal Life*
Cenozoic (Age of	Quaternary	Recent	0.011	0.011	End of last ice age; climate warmer	Decline of wood plants; rise of herbaceous ones	Age of man
Mammals)		Pleistocene	1	1	Repeated glaciation; 4 ice age	Great extinction of species	Extinction of great mammals; first human social life
	Tertiary	Pliocene	12	13	Continued rise of mountains of western North America, volcanic activity	Decline of forests; spread of grasslands; flowering, plants, monocotyledons developed	Man evolved from man-like apes, elephants, horses, camels almost like modern species
		Miocene	13	25	Sierra and Cascade mountains formed, volcanic activity in northwest U.S.; climate cooler		Mammals at height of evolution; first manlike apes
		Oligocene	11	36	Lands lower; climate warmer	Maximum spread of forests; rise of mono-	Archaic mammals extinct; rise to

Table continue...

						cotyledons, flowering plants	anthropoids; forerunners of most living genera of mammals
		Eocene	22	58	Mountains eroded; no continental seas; climate warmer		Placental mammals diversified and specialized; hoofed mammals and carnivores established
		Paleocene	5	63			Spread of archaic mammals
	Rocky Mountain Revolution (Little Destruction of Fossils)						
Mesozoic (Age of Reptiles)	Cretaceous		72	135	Andes, Alps, Himalayas, Rockies formed late; earlier, inland seas and swamps; chalk, shale deposited	First monocotyledons; first oak and maple forest; gymnosperms declined	Dinosaurs reached peak, became extinct; toothed birds became extinct; first modern birds; archaic mammals common
	Jurassic		46	181	Continents fairly high; shallow seas over some of Europe and western U.S.	Increase of dicotyledons cycads and conifers common	First toothed birds; dinosaurs larger and specialized; insectivorous marsupials

Table continue...

	Triassic	49	230	Continents exposed; widespread desert conditions; many land deposits	Gymnosperms dominant declining toward end; extinction of seed ferns	First dinosaurs, pterosaurs and egg-laying mammals; extinction of primitive amphibians
	Appalachian Revolution (Some Loss of Fossils)					
Paleozoic (Age of Ancient Life)	Permian	50	280	Continenets rose; Appalachians formed; increasing glaciation and aridity	Decline of lycopods and horsetails	Many ancient animal died out; mammal-like reptiles, modern insects arose
	Pennsylvanian	40	320	Lands at first low; Great forests of seed great coal swamps	First reptiles; ferns and gymnosperms	insects common; spread of ancient amphibians
	Mississippian	25	345	Climate warm and humid at first, cooler later as land rose	Lycopods and horsetails dominant; gymnosperms increasingly widespread	Sea lilies at height; spread of ancient sharks
	Devonian	60	405	Smaller inland seas; land higher, more arid; glaciation	First forests; land plants well established; first gymnosperms	First amphibians; lungfishes, sharks abundant
	Silurian	20	425	Extensive continental seas; lowlands	First definite evidence of land	Marine arachnids dominant; first

Table continue...

				increasingly arid as land rose	plants; algae dominant	(wingless) insects; rise of fishes
	Ordovician	75	500	Great submergence of land; warm climates even in Arctic	Land plants probably first appeared; marine algae abundant	First fishes, probably freshwater; corals, trilobites abundant; diversified molluscs
	Cambrian	100	600	Lands low, climate mild; earliest rocks with abundant fossils	Marine algae	Trilobites, brachiopods dominant; most modern phyla established
	Second Great Revolution (Considerable Loss of Fossils)					
Proterozoic		1000	1600	Great sedimentation; volcanic activity later; extensive erosion, repeated glaciations	Primitive aquatic plants—algae, fungi	Various marine protozoa; towards end, molluscs, worms, other marine invertebrates
	First Great Revolution (Considerable Loss of Fossils)					
Archeozoic		2000	3600	Great volcanic activity; some sedimentary deposition tion; extensive erosion	No recognizable fossils; indirect evidence of living things from deposits of organic material in rock	

of eras, periods and epochs was constructed. The strata were formed by the accumulation of sediment—sand or mud—at the bottom of lakes, seas or oceans. The duration of each period or epoch can be estimated from the relative thickness of the sedimentary deposits, although, obviously, the rate of deposition varied at different times and in different places.

The layers of sedimentary rock should occur in the sequence of their deposition, with the newer, later strata on top of the older, earlier ones, but subsequent geologic events may have changed the relationship of the layers. Not all the strata occur in any one region, for some lands were exposed when others were submerged. In some regions the strata formed previously have subsequently emerged and eroded away so that relatively recent strata were then deposited directly upon very ancient ones. Moreover, certain sections of the earth's crust have undergone tremendous folding and splittings, so that early layers may now rest on top of newer ones. Sometimes the age of a rock stratum can be determined by a study of its fossil content, for some kinds of fossils were deposited in only one era or period.

Rock deposits are now dated largely by taking advantage of the fact that certain radioactive elements are transformed into other elements at rates which are slow and essentially unaffected by the temperatures and pressures to which the rock has been subjected. Half of a given sample of uranium will be converted into lead in 4.5 billion years and, by measuring the proportion of uranium and lead in a given rock, an accurate estimate of the absolute age of the rock can be made. By this method the oldest rocks of the earliest geologic period are calculated to be about 3,500,00,000 years old, and the latest Cambrian rocks to be 500,000,000 years old. These dates have been confirmed by newer methods in which the radioactive decay or rubidium-87 (half-life 47 billion years!) and potassium-40 have been utilized to measure the ages of micas and feldspars. Events in more recent times can be dated by the decay of radioactive carbon-14, which has a half-life of 5568 years.

Relatively short periods of geologic time can be determined by counting the yearly deposits of clay on the bottom of lakes and ponds. Advances in isotopic techniques have made possible some astonishing conclusions in the field of geology. For example, the proportion of the various oxygen isotopes in the calcium carbonate secreted by living organisms depends upon the temperature; consequently, by analyzing the calcium carbonate of fossil shell, it is possible to estimate the temperature of the sea in which those animals lived, hundred of millions of years ago.

Between the major eras there were widespread geologic disturbances, called *revolutions*, which raised or lowered vast regions of the earth's surface and created or eliminated shallow inland seas. These revolutions changed the distribution of a sea and land organisms and wiped out many of the previous life forms. The era known as the Paleozoic ended with the revolution that raised the Appalachian mountains and, it is thought, killed all but 3 per cent of the then existing forms of life. Similarly, the Rocky Mountain Revolution, which raised the Andes, Alps and Himalayas, as well as the Rockies, resulted in the annihilation of most of the reptiles of the Mesozoic era.

Precambrian Life

The Archeozoic Era

The rocks of the oldest geologic era are very deeply buried in most parts of the world, but are exposed at the bottom of the Grand Canyon and along the shores of Lake Superior. The oldest era does not begin with the origin of the earth but with the formation of the earth's crust, when rocks and mountains were already in existence and the processes of erosion and sedimentation had begun. This era, which lasted about two billion years, was as long as all the other eras combined. Apparently it was marked by catastrophic and widespread volcanic activity and deep-seated upheavals which climaxed in the raising of mountains. The heat, pressure and churning associated with these movements probably destroyed most of the fossils, but some evidence of life still remains. Scattered throughout the Archeozoic rocks are traces of graphite or pure carbon, which are probably the transformed remains of plant and animal bodies. If the amount of graphite in these rocks can be taken as a measure of the amount of living things (and this seems to be valid), then life must have been abundant in the Archeozoic, for there is more carbon in these rocks than in the coal beds of the Appalachians.

The Proterozoic Era

The second era, some one billion years in length, was characterized by the deposition of large quantities of sediment, and by at least one great period of glaciation, during which ice sheets stretched to within 20 degrees of the equator. The fossils found in Proterozoic rocks show not only that life was present but that evolution had proceeded quite far before the end of the era. Sponge spicules, jellyfish and the remains of fungi, algae, brachiopods, arthropods and worm tubes have been recovered from rocks of the era. Several rich deposits of Precambrian fossils have been found in South Australia, beginning in 1947. These include jellyfish, corals, segmented worms and two animals with no resemblance to any known fossil or living form. The bodies of these animals were soft, and were strengthened only by spicules of calcium carbonate.

THE PALEOZOIC ERA

Between the strata of the late Proterozoic and the earliest layers of the third major era, the Paleozoic, is a considerable gap, caused by a geologic revolution. During the 370,00,00 years of the Paleozoic, members of every phylum and class of animals appeared except the birds and mammals. Since some of these animals existed for a relatively short time, their fossils enable geologists to correlate rocks of the same era found in different localities.

The Cambrian Period

The earliest sub-division of the Paleozoic era, the Cambrian period, is represented by rocks rich in fossils, so that the reconstructions of what the world was like in those days are probably quite accurate. The forms living in this period were so varied and complex that they must have evolved from ancestors dating back to the Proterozoic era, at the latest, and possibly to the Archeozoic. All the present-day animal phyla, except the chordates, were represented, and all plants and animals lived in the sea. (The land must have been a weird, lifeless waste until the late Ordovician or Silurian when plants became established on land). There were primitive, shrimp-like crustaceans and arachnid-like forms, some of whose descendants exist almost unchanged today as the horseshoe crab. The sea floor was covered with simple sponges, corals, echinoderms growing on stalks, snails, pelecypods, primitive cephalopods, brachiopods and trilobites. *Brachiopods*, sessile, bivalved plankton feeders, flourished in the Cambrian and the rest of the Paleozoic. The trilobites were primitive arthropods with flattened, elongated bodies covered dorsally by a hard shell. The shell had two longitudinal grooves that divided the body into three lobes. There were a pair of legs of each somite but the last, and each leg had an outer gill branch and an inner walking or swimming branch. Most trilobites were 5 to 8 cm. long but a few were as large as 60 cm. There were both unicellular and multicellular algae. One of the best-preserved collections of Cambrian fossils was found in the mountains of British

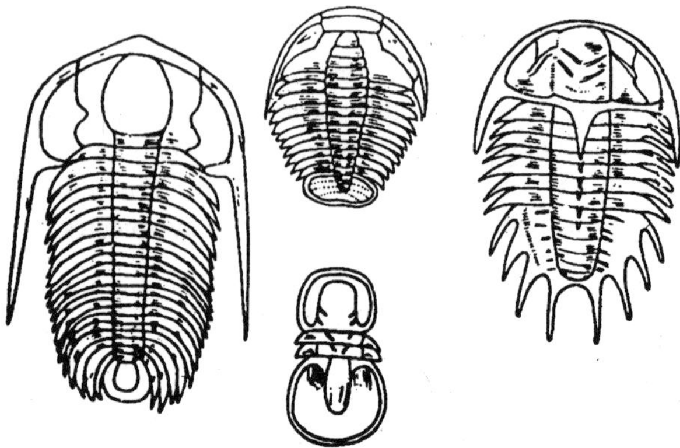

Fig. 1.1. Typical trilobites of the Cambrian seas.

Columbia; it includes annelids, crustacea, and a connecting link between annelids and arthropods, similar to the living peripatus.

Evolution since the Cambrian has not been marked by the establishment of entirely new body patterns, but the ramification of the lines already present, and by the replacement or original, primitive forms with better-adapted ones. The fact that no new phyla have originated since the early Paleozoic does not necessarily mean that no other patterns of animal organization are possible, or that mutations for new patterns did not occur. It probably indicates only that by that time the existing forms had reached a degree of adaptation to the environment which gave them a marked advantage over any new, unadapted types.

The Ordovician Period

During the Cambrian period the continents gradually had begun to be covered with water, and in the Ordovician period this submergence reached its maximum, so that much of what is now land was covered by shallow seas. Inhabiting the seas were giant cephalopods—squid or nautilus-like animals, with straight shells, 5 to 7 meters long, and 30 cm. in diameter. The Ordovician seas were apparently quite warm, for corals, which grow only in warm waters, lives as far north as Ontario and Greenland. The first traces of the vertebrates are found in Ordovician rocks; these small animals, called ostracoderms, were jawless, armored, bottom-dwelling fishes, without fins. Their armor consisted of a heavy, bony covering over the head, and thick scales over the trunk and tail; otherwise they were similar to the jawless lamprey eels of today. Apparently, they lived in fresh water, and their armor plate was a defense against the carnivorous, giant water

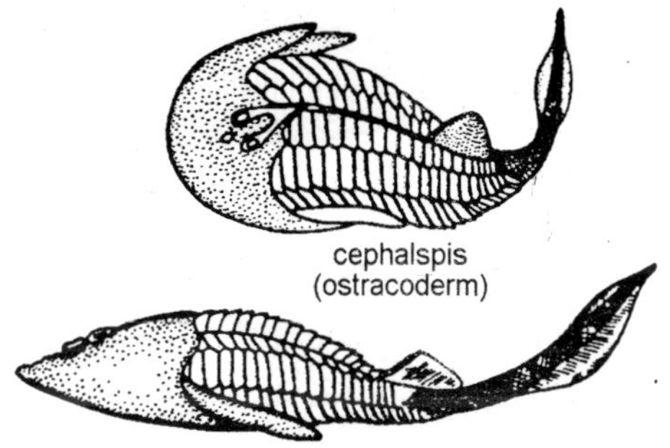

Fig. 1.2. An Ostracoderm.

scorpions—sometimes 3 meters long—called *eurypterids*, which also inhabited fresh water.

The Silurian Period

Two events of great biologic importance occurred in the Silurian period: the land plants evolved, and air-breathing animals appeared. The first land plants seem to have resembled ferns rather than mosses, and ferns were the dominant plants of the following Devonian and Mississippian periods. The first air-breathing land animals were arachnids, resembling to some extend modern scorpions. The continental area, which had been low during the Cambrian and Ordovician, rose—especially in Scotland and Northeastern North America-and the climate became much colder.

The Devonian Period

During the Devonian the original ostracoderms evolved into a great variety of fish and the period is frequently referred to as the "age of fishes." The first to evolve jaws and paired fins were the placoderms, or spiny-skinned sharks, which were small, armored, freshwater forms. These animals had a variable number of paired fins; and some had as many as five additional pairs between these two. True sharks appeared in fresh water during the Devonian, but tended to migrate to the ocean and to lose their cumbersome armor plate. The ancestors of the bony fishes also appeared in Devonian fresh water streams, and had evolved by the middle of the Devonian into three main types: lungfish, lobe-finned fish and ray-finned fish. All had lungs and an armor of bony scales. A few lungfish have survived to the present, and the ray-finned fish, after undergoing a slow evolution in the remaining Paleozoic and early Mesozoic eras, ramified greatly in the latter part of the Mesozoic to give rise to the modern bony fish, or teleosts. The lobe-finned fish, which were the ancestors of the land

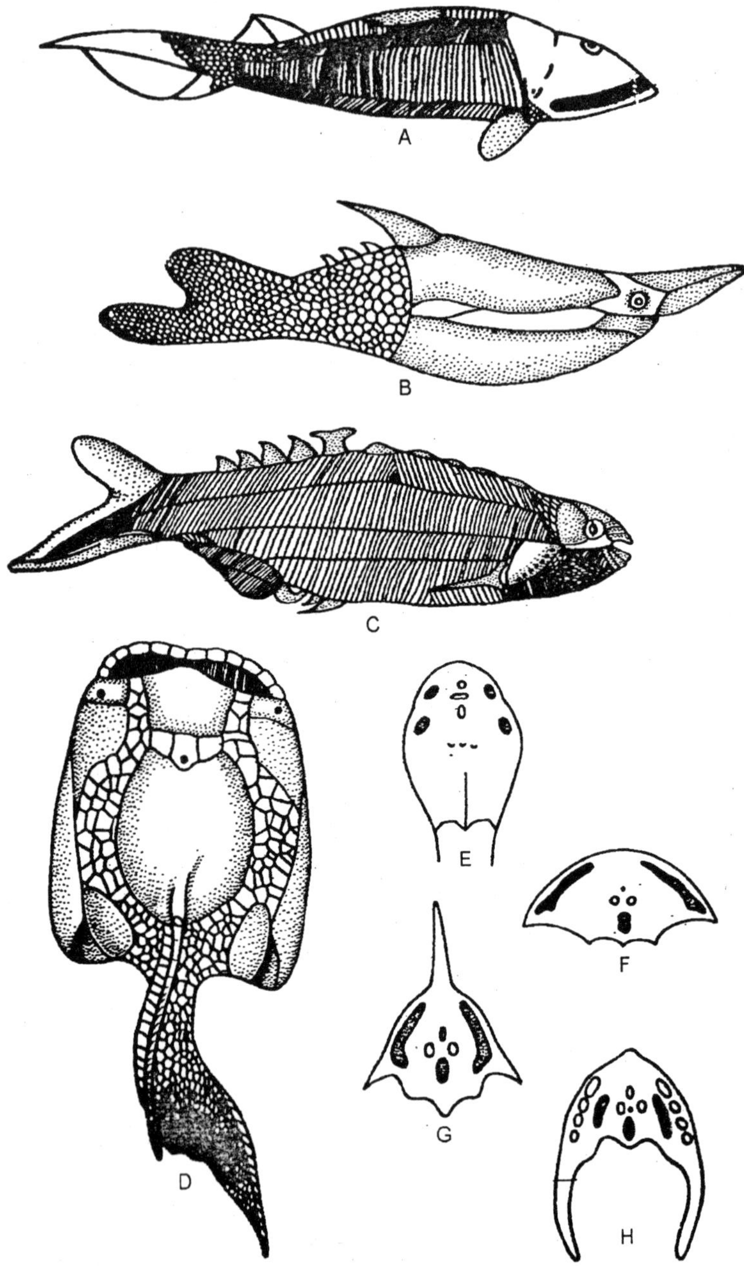

Fig. 1.3. Primitive fishes.

vertebrates, were almost extinct by the end of the Paleozoic, and it was once believed that they had vanished with the end of the Mesozoic. But since 1939 several specimens of living coelacanths nearly two meters long have been caught off the eastern coast of South Africa.

The latter part of the Devonian was marked by the appearance of the first land vertebrates—amphibians called *stegocephalians* (meaning roof-headed). These creatures, whose skulls were encased in bony armor, were similar in most respects to the lobe-fins differing chiefly in having limbs instead of fins. The Devonian was the first period characterized by true forests; ferns, club mosses, horsetails and primitive gymnosperms—the "seed ferns"—all flourished. Insects and millipedes are believed to have originated in the late Devonian.

The Carboniferous Period

The Mississippian and Pennsylvanian periods are frequently grouped together as the Carboniferous, for during this time there flourished the great swamp forests whose remains gave rise to the major coal deposits of the world. The land was covered with low swamps, filled with horsetails, ferns, seed ferns and large-leaved evergreens. The first reptiles, called *stem reptiles*, similar to their antecedent amphibians, appeared in the Pennsylvanian period, flourished in the final Paleozoic period—the Permian—and became extinct early in the Mesozoic era. Whether the most primitive reptiles known (called *Seymouria* for the

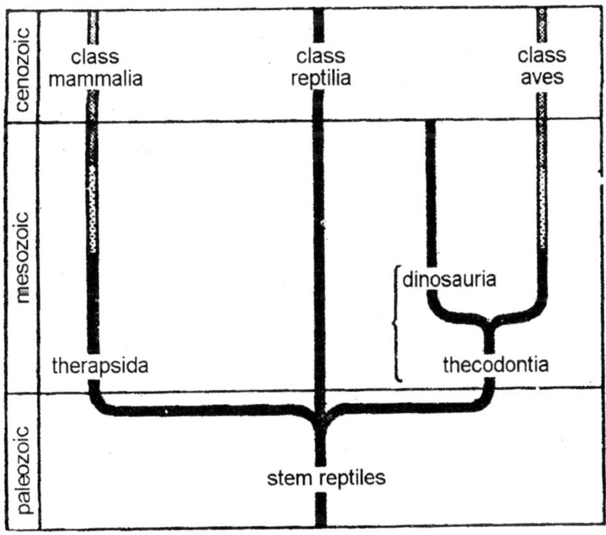

Fig. 1.4. Evolutionary representation of reptiles, birds and mammals.

town in Texas near which its fossil remains were found) should be considered an amphibian about to become a reptile, or a reptile just over the border from an amphibian is debatable. One of the main differences between reptiles and amphibians is the type of egg laid: Amphibians lay jelly-coated eggs in the water, and reptiles lay shell-covered eggs on land. Since no Seymourian eggs have survived, it may never be possible to decide to which class this animal belonged. Seymouria was a large, sluggish beast resembling a lizard. Its short, stubby legs extended laterally from the body as do a salamander's, instead of being closer together and extending directly down to form pillar-like support for the body.

Two important groups of winged insects evolved during the Carboniferous—the ancestors of the cockroaches, which reached a length of 10 cm., and ancestral dragonflies, some of which had a wingspread of 75 cm.

The Permian Period

The final period of the Paleozoic was characterized by great changes in climate and topography. The level of the continents rose all over the world, so that shallow seas which covered the region from Nebraska to Texas at the beginning of the period drained off, leaving the land a salt desert. At the end of the Permian a general folding of the earth's crust, called the *Appalachian Revolution*, raised the great mountain chain from Nova Scotia to Alabama. These mountains originally were higher than the present Rockies. Other ranges were brought into existence in Europe at this time. A great glaciation, spreading from the Antarctic, covered most of the southern hemisphere, extending almost to the equator in Brazil and Africa; North America was one of the new part of the world to escape glaciation at that time, but even its climate became much colder and drier than it had been during most of the Paleozoic. Many of the Paleozoic forms of life, apparently unable to adapt to the climatic changes, became extinct during the Appalachian Revolution. Even many of the marine forms became extinct, owing to the cooling of the water and the decrease in the amount of space available, caused by the diminishing of the shallow seas.

From the primitive stem reptiles there evolved during the late Carboniferous and early Permian the group of reptiles believed to be in the direct line which gave rise to the mammals. These were the pelycosaurs, carnivorous reptiles that were more slender and lizard-like than the stem reptiles. In the latter part of the Permian there evolved, probably form the pelycosaurs, another group of reptiles with

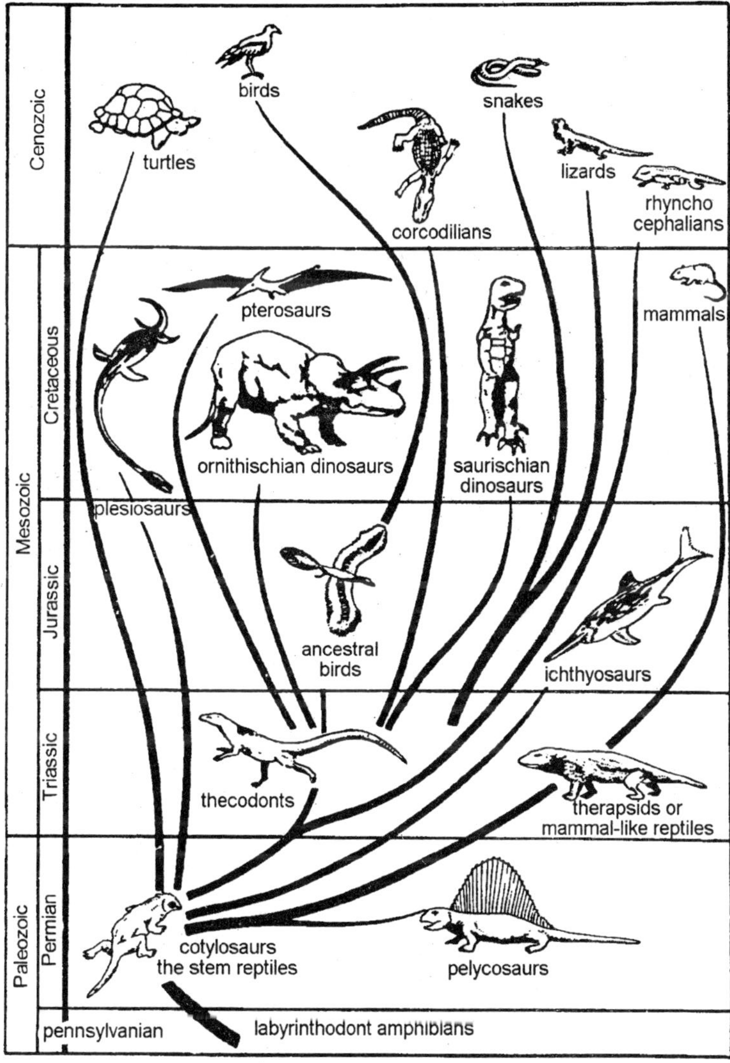

Fig. 1.5. Some vertebrates in evolutionary chart.

a few more mammalian characteristics—the therapsids. One of these *Cynognathus*, the "dog-jawed" reptile, was a slender, lightly built animal about 150 cm. long, with a skull intermediate between that of a reptile and that of a mammal. The teeth, instead of being conical and all alike, as reptilian teeth are, were differentiated into incisors, canines and molars. In the absence of information about the animal's soft parts, whether it had scales or hair, whether or not it was warm blooded, and whether it suckled its young, it is called a reptile, but if more evidence were available, it might be classified as a very early mammal. The therapsids were widespread in the late Permian, but were crowded out early in the Mesozoic by the great variety of other reptiles.

The Mesozoic Era

The Mesozoic era, which began about 230,000,000 years ago and lasted some 167,000,000 years, is subdivided into three periods. The Triassic, Jurassic and Cretaceous. During the Triassic and Jurassic most of the continental areas were above water. The Triassic climate was dry, but warmer than the Permian, and the Jurassic was warmer and moisture than the Triassic. The trees of the famous Petrified Forest in Arizona late from the Triassic period.

During the Cretaceous period the Gulf of Mexico expanded into Texas and New Mexico, and in general, the sea gradually encroached upon the continental area. There were great swamps, too, from Colorado to British Columbia. In the latter part of the Cretaceous the interior of the continent of North America was further submerged, and the bay from the Gulf of Mexico, meeting another from the Arctic, cut this continent into two parts. The Cretaceous ended in another great upheaval, called the Rocky Mountain Revolution, which raised the Rockies, Alps, Himalayas and Andes and caused much volcanic activity in western North America.

The outstanding feature of the Mesozoic era was the origin, differentiation and final extinction of a great variety of reptiles. For this reason the Mesozoic is commonly called the "Age of Reptiles."

There were six major evolutionary lines of reptiles, the most primitive of which included the ancient stem reptiles. For this reason the Mesozoic is commonly called the "Age of reptiles."

There were six major evolutionary lines of reptiles, the most primitive of which included the ancient stem reptiles and the turtles, which originated in the Permian. The turtles have evolved the most complicated armor of any land animal, consisting of scales derived from the epidermis fused to the underlying ribs and breast bone. With this protection, both marine and land forms have survived with few structural changes since before the time of the dinosaurs. There legs extended laterally, marking locomotion difficult and slow, and their skulls are unpierced behind the eye sockets, a feature essentially unchanged from the old stem reptiles.

A second group of reptiles to survive with relatively few change from the ancestral stem reptiles are the lizards—the most abundant of living reptiles—and the snakes. For the most part of the lizards have kept the primitive type of locomotion with the legs extended laterally, although many can run rapidly. Most of them are small, but the monitor lizard of the East Indies attains a length of four meters, and some fossil ones were eight meters long. The *mosasaurs* of the Cretaceous were marine lizards which reached a length of 13 meters, and had a long tail useful in swimming. During the Cretaceous, snakes evolved from lizard ancestors. The important difference between snakes and lizards is not the loss of legs (some lizards are legless), but certain changes in the skull and jaws of the snake which enable it to open its mouth wide enough to swallow an animal larger than itself. A representative of an ancient line that has managed somehow to survive in New Zealands is the lizard-like *tuatera*. It shares several traits with the ancestral cotylosaurs, one of which is the presence of a third eye on the top of its head.

The main group of Mesozoic reptiles were the *archosaurs* or "ruling reptiles," of which the only living members are the alligators and crocodiles. At an early point in their evolution from stem reptiles the ruling reptiles, which were then about one meter long, became adapted to two-legged locomotion—their front legs became short, while the hind legs became long, stout and considerably modified. These animals rested or walked on all fours, but in emergencies, they reared up and ran on. the two hind legs, assisted by their fairly long tail, which served as a balance. From the early archosaurs developed many different, specialized forms, some of which continued to use two-legged locomotion, others of which reverted to walking on all fours. These descendants include the *phytosaurs*—aquatic, alligator-like reptiles, common during the Triassic; the crocodiles, which evolved during the Jurassic and replaced the phytosaurs as aquatic forms; and the pterosaurs, or flying reptiles, which included animals the size of a robin, as well as the largest animal ever to fly—*Pteranodon*, with a wingspread of 9 meters.

There were two types of flying reptiles, one with a long tail that had a steering rudder at the end, the other with a short tail. Both these types, apparently, were fish-eaters, and they probably flew long distances over the water in search of food. Their legs were not adapted for standing, and it is believed that, like bats, they rested by clinging to some support and hanging suspended.

Of all the reptilian branches, the most famous are the dinosaurs (meaning terrible reptiles). There were divided into two main types: one with a birdlike pelvis, the other with a reptilian pelvis.

The *Saurischia* (meaning reptile pelvis) first evolved in the Triassic, and remained in existence until the Cretaceous. The early ones were fast, carnivorous, two-legged forms the size of a rooster, which probably preyed upon lizards and the archaic mammals then in existence. Throughout the Jurassic and Cretaceous there was a tendency in this group to grow larger, culminating in the gigantic carnivore of the Cretaceous, *Tyrannosaurus*. Other Saurischia beginning in the late Triassic, changed to a plant diet, reverted to a four-legged gait, and during the Jurassic and Cretaceous, evolved into tremendous amphibious forms. Among these—the largest four-footed animals that ever lived—were *Brontosaurus*, with a length of 21 meters; *Diplodocus*, which reached a length of 29 meters, and Brachiosaurus, the biggest of them all, with an estimated weight of 50 tons.

The other group of dinosaurs, the *Ornithischia* (meaning bird like pelvis), were vegetarians, probably form the beginning of their evolution. Although some of them walked upright, the majority had a four-legged gait. Having lost their front teeth, they developed a stout, horny, birdlike beak, which in some forms was broad and ducklike (hence the name "duck-billed" dinosaurs). Webbed feet were characteristic of this type; other species developed great armor plates as protection against the carnivorous saurischians. *Ankylosaursus*, which has been dubbed "the reptilian tank," had a broad, flat body, covered with an armor plate and with large laterally projecting spines. Still other ornithischians of the Cretaceous period developed bony plates around the head and neck. One of these, *triceratops*, had two horns over its eyes and another over its nose—all one meter long.

Two other groups of Mesozoic reptiles, separate from each other and from the dinosaurs, were the marine *plesiosaurs* and *ichthyosaurs*. The former were characterized by an extremely long neck, which took up over half of their total length of 15 meters. The trunk was broad, flat and rather turtle-like, the tail was small, and the animal paddled along by means of finlike arms and legs. The ichthyosaurs (Fish reptiles) had a body form superficially like that of a fish or a whale, with a short neck, a large dorsal fin, and a shark like tail. They swam by wiggling their tails, using their feet only for steering. The ichthyosaur young were apparently born alive, after having hatched from eggs within the mother, for the adults were too specialized to come out land, and

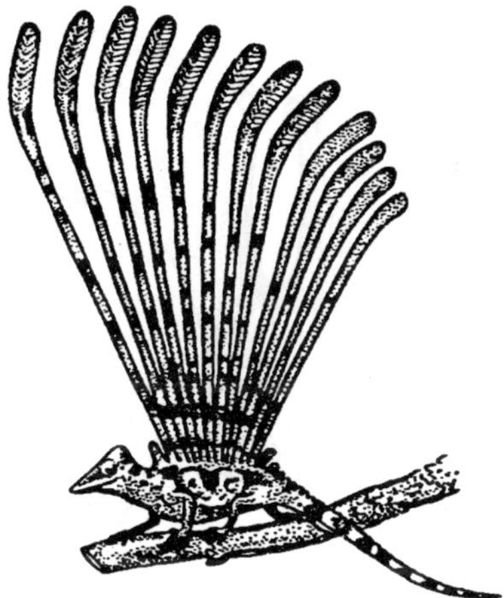

Fig. 1.6. The Longisquama.

a reptilian egg will drown in water. The presence of skeletons of the young within the body cavity of adult fossils has strengthened this theory.

At the end of the Cretaceous a great many reptiles became extinct; they were apparently unable to adapt to the marked changes brought about by the Rocky Mountain Revolution. As the climate became colder and drier many of the plants which served as food for the herbivorous reptiles disappeared. Some of the herbivorous reptiles were too large to walk about on land when the swamps dried up. The smaller, warm-blooded mammals which had appeared were beaten able to compete for food and many of them ate reptiles was probably the result of a combination of a whole host of factors, rather than any single one.

Although the reptiles were the dominant animals of the Mesozoic, many other important organisms evolved during that time; snails and bivalves increased in number and kind; sea urchins reached their peak; mammals originated in the Triassic; and teleost fishes and bird arose during the Jurassic. Most of the modern orders of insects appeared early in the Mesozoic. During the early Triassic the most abundant plants were seed ferns, cycads and conifers; but by the Cretaceous, many others, resembling present-day species, had appeared—sycamores magnolias, palms, maples and oaks.

From the Jurassic excellent fossils have been preserved, showing even the outlines of feathers, of the earliest species of bird. This creature, called *Archaeopteryx*, was about the size of a crow, had rather feeble wings, jawbones armed with teeth, and a long, reptilian

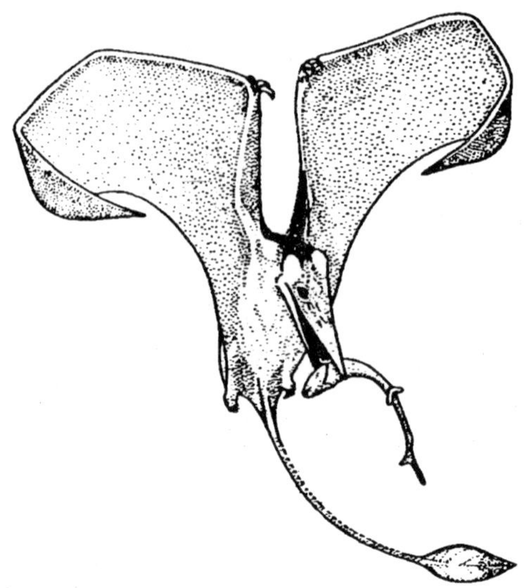

Fig. 1.7. Sorduspilosus—a pterosaur.

tail, covered with feathers. Fossils have been found in Cretaceous rocks of two other bird species—*Hesperornis*, an aquatic dividing bird that had lost the ability to fly, and Ichthyornis, a powerful flying bird, about the size of a pigeon, with reptilian teeth. Modern toothless birds evolved early in the following era.

The Cenozoic Era

With equal justice, the Cenozoic could be called the Age of Mammals, the age of birds, the Age of insects or the Age of flowering Plants, for it is marked by the evolution of all these forms. It extends from the Rocky Mountain Revolution, some 63,000,000 years ago, to the present, and is subdivided into two periods, the Tertiary, which lasted some 62,000,000 years, and the Quaternary, which includes the last million, or million and a half, years.

The Tertiary Period

The Tertiary is sub-divided into five epochs, named, from earliest to latest, Paleocene, Eocene, Oligocene, Miocene and Pliocene. The Rockies, formed at the beginning of the Tertiary, were considerably eroded by the time of the Oligocene, giving the North American continent a gently rolling topography. In the Miocene another series of uplifts raised the Sierra Nevadas and a new set of Rockies, and resulted in the formation of the western deserts. The climate was milder in the Oligocene than it is at present, and palms grew as far north as

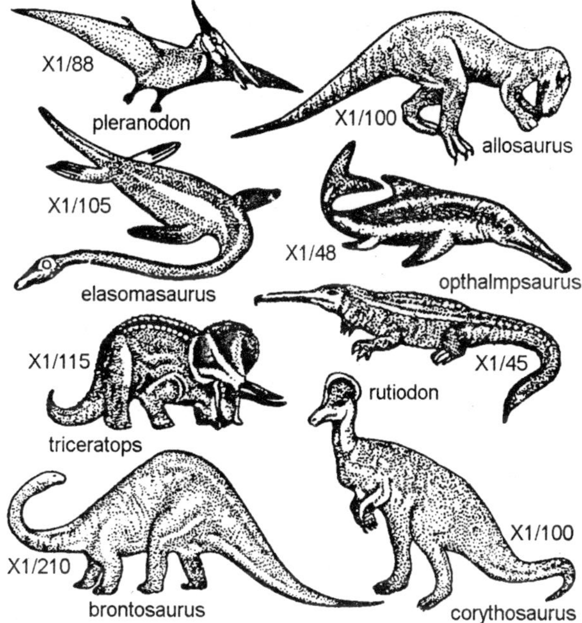

Fig. 1.8. Ruling reptiles of Mesozoic Era.

Wyoming. The uplift begun the Miocene continued in the Pliocene and, coupled with the ice ages of the Pleistocene, killed many of the mammals and other forms that had evolved. The final elevation of the Colorado Plateau, which also caused the cutting of grand Canyon, occurred almost entirely in the short Pleistocene and Recent epochs.

The earliest fossils of true mammals were deposited late in the Triassic, but the Jurassic there were four orders of mammals, all about the size of a rat or small dog. The earliest mammals (monotremes) were egg-laying animals, and their only living survivors are the duck-billed platypus and the spiny anteater of Australia. Both have fur and suckle their young, but lay eggs like turtles. The ancestral, egg-laying mammals certainly must have different from the specialized platypus and anteater, but the fossil records of those early forms are incomplete. The present-day monotremes have been able to survive this long only because they lived in Australia, which, until recently, had no placental mammals to offer competition.

By the Jurassic and Cretaceous most mammals were advanced enough to being forth their young alive, although the most primitive of them—the marsupials—gave birth to underdeveloped young which had to remain for several months in a pouch of the mother's abdomen, containing the nipples. The Australian marsupials, freed, like the monotremes, of competition from better-adapted placental mammals, which were responsible for the extinction of their cousins on other continents, evolved into a wide variety of types that superficially resemble some of the placentals; there are marsupial mice, shrews, cats, moles, bears, and one species of wolf, as well as a number of forms with no placental counterparts, such as the kangaroo, wombat and wallaby. During the Pleistocene there were giant kangaroos and wombats the size of a rhinoceros in Australia. The opossum more closely resembles the primitive, ancestral marsupial type than do any of these more specialized dorms; it is the only marsupial found outside Australia and South America.

The advanced, modern placental mammals, including man, which are distinctive in bringing forth their young alive and ready to live an independent existence, evolved from an insect-eating, tree-dwelling ancestor. Fossils of this animal in Cretaceous rocks show it to have been tiny and rather like the present-day shrew. Some of these ancestral mammals remained in the trees and gave rise, through a series of intermediate forms, to the primates—monkeys, apes and men. Others lived on or under the ground and, during the Paleocene, evolved into all the other mammals living today. The archaic mammals of the Paleocene had conical, reptilian teeth, five digits on each foot, and a small brain; they also walked on the soles of their feet instead of on their toes. During the Tertiary the evolution of grasses which served as food, and forests which afforded protection, were important factors in leading to changes in the mammalian body pattern. Concomitant with a tendency toward increased size, the mammals all displayed tendencies toward an increase in the relative size of the brain, and toward changes in the teeth and feet. As modern forms better equipped for survival arose, the archaic mammals become extinct.

Although fossils of both marsupial and placental animals have been found in Cretaceous rocks, it is rather surprising to find the remains of highly developed mammals in strata of the early Tertiary. Whether they actually arose at that time or had existed before in the highlands, and had not been preserved, is known.

In the Paleocene and Eocene epochs the first carnivores, called *creodonts*, arose from the primitive insect-eating placental mammals. They were replaced in the Eocene and Oligocene by more modern forms which eventually gave rise to the present-day carnivores, such as cats, dogs, bears and weasels, as well as to the web-footed, marine carnivores—the seals and walruses. One of the most famous fossil carnivores is the saber-toothed tiger, which became extinct only recently in the Pleistocene. These animals had tremendously elongated, knifelike upper canine teeth, and a lower jaw that could be swung down and out of the way, allowing the teeth to be used as sabers for stabbing the prey.

The larger herbivorous mammals, most of which have hooves, are sometimes referred to as the ungulates. They do not form a single, natural group, but consist of several independent lines. Although both horses and cows have hooves, they are not more closely related than either one is to a tiger. The molar teeth of ungulates are flattened and enlarged to facilitate the chewing of leaves and grass. Their legs have become elongated and adapted for the rapid movement necessary to escape predators. The earliest ungulates, called *condylarths*, appeared in the Paleocene; they had long bodies and tails, flat, grinding molars, and short legs ending in five toes, each of which bore a hoof. Corresponding to the archaic carnivores, or creodonts, were the archaic ungulates called *uintatheres*. During the Paleocene and Eocene some of these were as large as elephants, and some had three large horns projecting from the top of the head.

The fossil records of several ungulates lines—the horse, the camel and the elephant—are complete, and it is possible to trace the evolution of these animals from small, primitive, five-toed creatures. The chief evolutionary tendencies in the ungulates have been toward an increase in the over-all size of the body and a decrease in the number of toes. The ungulated were early divided into two groups, one characterized by an even number of toes, and including the cow, sheep, camel, deer, giraffe, pig and hippopotamus; the other characterized by an odd number of toes, and including the horse, zebra, tapir and rhinoceros. The elephants and their recently extinct relatives, the mammoths and mastodons, can be traced back to an Eocene ancestor the size of a hog which had no trunk. This primitive form, called Moeritherium, was close to the stem that also gave rise to such dissimilar creatures as the coney (a small, woodchuck-like animal found in Africa and Asia) and the sea cow.

The whales and porpoises descended from whalelike forms of the Eocene, called *zeuglodonts*, which in turn are believed to have evolved from the creodonts. The evolutionary history of the bats can be traced to ancestral winged types, also of the Eocene, which descended from the primitive insectivores. The evolutionary history of some of the other mammals—the rodents, rabbits and edentates (anteaters, sloths and armadillos)—is less well known.

The Quaternary Period

The Quaternary Period includes the final million or million and a half years of the earth's history and is divided into two epochs, the *Pleistocene* and the *Recent*, which began some 11,000 years ago with the recession of the last ice sheet. The Pleistocene was marked by four periods of glaciation, between which the sheets of ice retreated. At their greatest extent these ice sheets covered nearly 4,000,000 square miles of North America, extending south as for as the Ohio and Missouri rivers. The Great Lakes, which were carved out by the advancing glaciers, changed their outlines radically a number of times, and from time to time they emptied into the Mississippi. It is estimated that in the past, when the Mississippi drained lakes as far west as Duluth and as far east as Buffalo, its volume was more than sixty times as great as at present. During the Pleistocene glaciations enough water was removed from the sea and locked in the ice to lower the sea level 65 to 100 meters. This created land connections highways for the dispersal of many land forms, between Siberia and Alaska at Bearing Strait, and between England and the continent of Europe.

The plants and animals of the Pleistocene were similar to those alive today. It is sometimes difficult to distinguish between Pleistocene and Pliocene deposits, because the organisms were similar and nearly modern in form. A considerable number of mammals, including the saber-toothed tiger, mammoth and giant ground sloth, became extinct during the Pleistocene, after the appearance of primitive man. The Pleistocene was marked by the extinction of many species of plants, especially woody ones, and appearance of numerous herbaceous forms.

The paleontologic record makes it impossible to doubt that the present species arose from previously existing, different ones. The fossil record is not equally clear for all lines of evolution. Most plant tissues are too soft to leave good fossil remains and the connecting links between the animal phyla were apparently soft-bodied forms that left no fossil traces. For many lines of evolution, especially the vertebrates, the successive steps are well known; other lines have some gaps which remain to be clarified by future palaeontologists.

2

Sedimentary Rocks

The components of the Earth's atmosphere—the gases and water vapour—have two main sources; from the Earth itself by an outgassing process and from the sun. Even today the sun emits large amounts of material by solar radiation (solar wind) consisting predominantly of hydrogen nuclei (protons) and electrons. It is estimated that 1,200 million kg of material are emitted from the Sun every second at a speed of between 400-800 km/sec. This radiation bathes the planetary system and would strike the Earth were it not for its magnetic field. In addition, interplanetary material, which also contains gas, also reaches the Earth. Comets also make a small contribution. Cometary cores are composed of meteorite-like particles and volatile substances such as water (H_2O), ammonia (NH_3), methane (CH_4), carbon monoxide (CO) and carbon dioxide (CO_2), the components present in ancient and, in part, of the present atmosphere. There is evidence that meteorites have reached the Earth in past; impressive evidence of this being the Nordlinger Ries and the Steinheimer Becken, both in the Federal Republic of Germany, Meteor Crater, Arizona, and the New Quebec meteorite crater. The fall of the Tunguska "meteorite" (actually a comct) in Siberia on 30 June, 1908, caused great excitement. Soviet scientists reported that it was so bright that it was seen hundreds of kilometers away; thunder was heard up to 1,200 km away and then an enormous column of smoke was seen. The comet's fall shock the ground so violently that horses in the fields fell to their knees within a radius of 400 km from the impact site. The first investigations at the impact site showed that there was no crater but, within a radius of 70 to 80 km, trees were uprooted as though they had been caught in a large explosion and for up to 20 to 30 km, they were burnt or charred. It was calculated that the energy of the explosion was 1,000 to 2000 times more powerful than that of the first atomic bomb. The fires were caused by gas, at a temperature of several thousand degrees Celsius, following the shock wave.

Further research showed that this "meteorite" was most probably a comet head comprising ice containing innumerable solid particles.

Whereas it is readily appreciated that gases are emitted from volcanoes, it is less well known that gases are emitted from volcanoes, it is less well known that gases are constantly arriving from space.

The light gases hydrogen and helium which were abundant in the early atmosphere, have been lost from the Earth and most of the terrestrial planets. The enormous gravities of the giant planets such as Jupiter have retained these gases and consist predominantly of them. Following the earliest stage, when the Earth's atmosphere consisted of hydrogen, and helium, it may next have consisted for some time, mainly of methane (CH_4), ammonia (NH_3), and water (H_2O) together with smaller amounts of other gases, probably hydrogen sulphide (H_2S). It later changed by means of complex physicochemical reactions into an atmosphere of carbon dioxide (up to 90% CO_2), nitrogen (N_2) and various hydrocarbons, similar to that of Venus. During this early stage of atmospheric development about 4,000 to 4,500 million years ago the raw materials for the evolution of life must have been formed, and then the first development of primitive life.

The interaction between the developing biosphere and the atmosphere slowly but surely changed the composition of the latter, especially in that oxygen began to appear in the early atmosphere. This change can be traced in rocks which at first show little evidence of oxidation but later became progressively more oxidised. These important stages are reflected in the growth of the atmospheric oxygen content. About 2,700 million years ago, the oxygen content reached 1/1,000 its present level; and the 1% level, or Pasteur point, was reached development it was possible for life to make the important change to respiration. The oxygen content reached 10% of its present level about 420 million years ago, making possible the occupation of the continents. Oxygen, and particularly ozone (O_3) absorb the dangerous ultraviolet parts of the Sun's radiation and so protect living things. Before the atmospheric oxygen content reached the critical 10% level, organisms had to shelter in water but they were now able to leave its protection and colonise the land.

Since life could only develop in water, a depth of 10 m of water was necessary to afford the same protection against ultra-violet radiation as the present ozone layer which is of such importance to us. It is important to realise that the ozone layer, and the oxygen content of the atmosphere, are the result of a long process of geological evolution and need special protection.

The atmosphere also acts in other ways in protecting developing life from danger. For example, it prevents extreme temperature variation such as that of the Moon, it moderates the effect of cosmic radiation, and it prevents the loss of water from the Earth.

The atmosphere is also an important agent in geological processes; not only does it act as a weathering agent but it plays an active part in erosion by transporting rock debris, and by participating in the reaction which lead to the formation of new minerals. Without the atmosphere there would be no circulation of water, precipitation, no wind and no streams and rivers. Water and ice are important exogenetic agents which have been, and still are, active in the formation of rocks. The dynamics and thermodynamic properties of the atmosphere have been of great importance throughout geological time in determining these processes.

Because many crystalline schists are metamorphosed sedimentary rocks such as sandstone and clay, it can be concluded that atmospheric processes were at work as long as 3,000 to 4,000 million years ago.

The geological importance of the atmosphere is shown also by climatic changes, of which there have been many during the long course of geological time. There have been pronounced warm periods, and ice ages, when large areas of the Earth were covered by ice. There is also a relationship between the atmosphere and topography and the distribution of land and sea, especially with relation to the poles, is of particular importance climatically.

The behaviour of the atmosphere is influenced by changes in solar radiation, light infrared radiation, radio waves and, above all, cosmic rays. This behaviour is of a cyclic nature and in part correlates with seasonal rhythms, but it can also show cycles of longer duration as, for example, the 11 year sun spot cycle. These cyclic variations affect the circulation and condensation of water and can influence the climate as a whole. Fossil plants, for example, whose annual growth rate is unknown, sometimes show an annual variation in the rate of growth in sympathy with the activity of the Sun, just like the variation observed in the annual growth rings of modern trees. The thicknesses of growth rings of modern trees correspond to the Sun spot cycles as well as to long-term climatic changes caused by the influence of the Sun and partly by changes occurring on the Earth.

A further example of climatic change is the southward spread of continental ice across Europe during the Pleistocene ice age until it attained a state of equilibrium such that the rate of melting at the ice front was balanced by the rate at which the ice moved forwards. Clearly, the position of an ice front is not static but changes position in response to climatic change. Large amounts of sediment—called fluvio-glacial sediments—are deposited ahead of the ice front by melt water flowing from the glaciers. The amount of sediment transported by the melt water is less during the winter when the amount of water flowing from the ice is less than the during the summer. In winter, therefore, fine-grained, clayey sediment is deposited whilst during summer coarse-grained, sandy sediments are laid down. In this way altering light and dark layers of sediment, called varve clays, were deposited in lakes fed by melt-water streams. (The name *varve* derives from the Swedish word meaning circle and is used here in the sense of a climatic cycle). The layers deposited during the summer are thicker, coarser, and lighter in colour than the darker, clayey layers deposited during the winter.

Because each pair of light and dark layers represents one year, it has been possible to construct a varve chronology. Varve sequences can be correlated statistically enabling long term climatic changes, such as those caused by solar radiation, to be detected. It is possible, therefore, to establish a kind of geological calendar and to obtain information about the deposition of fluvio-glacial sediments over considerable periods of time.

Were it not for the atmosphere the Earth would not have been inhabited by life as we know it. Besides acting as a protective layer, the atmosphere, and especially the upper atmosphere, serves also to maintain a favourable temperature. The upper atmosphere may be defined as that region having a temperature far below 0°C; the stratosphere is that part of the upper atmosphere with a temperature between –50° and –60°C, at an altitude greater than about 10 to 12 km. Ice crystals forming at these altitudes fall as a result of gravity; if water were to rise higher, high energy solar radiation would cause it to dissociate into hydrogen and oxygen.

Thus the atmosphere is not only a powerful geological agent but is also essential for the evolution of life.

Minerals as Clues to Rock History

Minerals are studied by geologists not as mere constituents of the crust but as clues to the history of the rocks in which they occur. Some minerals develop exclusively in ocean water and provide the geologist with evidence that a particular sediment was deposited in the sea rather than in a freshwater lake. A thick bed of *halite* indicates aridity and evaporation so extreme that the brine had become ten times saltier than ordinary sea water. The magnetic properties of *magnetite*, an oxide of iron, can in certain situations provide clues to the position of continents relative to the earth's magnetic poles. Certain minerals form within a narrow range of conditions and can therefore be used to diagnose the pressure and temperatures involved in the formation crustal rocks and mountains. *Diamonds*, for example, form only at high temperatures and extremely high pressures. Like diamond *graphite* is a crystalline variety of carbon. However, it forms at lower temperatures and pressures. Other minerals contain radioactive isotopes that permit us to know the age of the parent rocks. By their size, crystals of feldspar give the petrologist information about now some ancient molten mass *congealed*. By their shape, grains of *quartz* contain a history of what has happened to them since they were eroded from some ancient granitic terrain. The buried mineral products of weathering may provide information about past climatic conditions in a region.

Families of Rocks

Rock Conversions

We have noted that minerals can provide information about the environment in which they formed. With study it is therefore possible to know the origin of the rocks containing those minerals. Geologists are agreed on a fundamental division of rocks into three great families according to difference or origin. *Igneous rocks* are those that have cooled from a molten state. *Sedimentary rocks* consists of materials derived from other rocks and deposited by water, ice, or wind. *Metamorphic rocks* are any that have been changed from previously existing rocks by the action of heat, pressure, and associated chemical activity. The changes may include a recrystallization of the previous minerals or growth of entirely new minerals.

In regard to the three major groupings of rocks, it is important to remember that they are not immutable. The earth's crust is dynamic and ever-changing. Any sedimentary or metamorphic rock may be partially melted to produce igneous rocks, and any previously existing rock of any category can be compressed and altered during mountain-building to produce metamorphic rocks. The weathered and eroded residue of any family of rock can be observed today being transported to the sea for deposition and conversion into sedimentary rocks. These changes can be incorporated into a schematic diagram that is designated the "rock cycle".

Although rocks are classified into groups that have had similar origin the identification of rocks is not based on origin, but on description. For identification it is necessary to know general appearance as well as chemical and physical properties. *Texture* (size, shape, and arrangement of constituent minerals) and *mineral composition* are essential for identification. Inferences regarding the origin of the rock are based on geologic observations and experimentation.

Igneous Rocks

Igneous rocks constitute over 90 per cent by volume of the earth's crust, although their great abundance may go unnoticed because they are extensively covered by sedimentary rocks. Much of our mountain scenery is sculptured in igneous rocks formed long ago. Current volcanic activity provides an often spectacular reminder of the processes that produce igneous rocks. It is appropriate name for rocks that develop from cooling masses of molten material derived from exceptionally hot parts of the earth's interior. *Magma* is the term used to describe this mixture of molten silicates and gases while it is still beneath the surface. If it should penetrate to the surface, it lose most of its gases and vapors and become *lava*.

Cooling History of Igneous Rocks

Intrusive, or *plutonic*, *igneous rocks* are those that have congealed from magma once located deep beneath the surface. Large masses of such rocks are sometimes called plutons. Their presence at the surface of the earth results from crustal uplift and erosional removal of the overlying rocks. *Extrusive*, or *volcanic*, *igneous rocks* are derived from lava and harden at the surface of the earth. The grain size of igneous rocks is an index to their history of cooling. Magmas lose heat slowly and retain water, tending to inhibit profusive formation of crystal nuclei. Thus, there is time and space for the growth of larger crystals around fewer nuclei. In typical intrusive rocks like granite, diorite, and gabbro, the intergrown crystals are large enough to be readily seen without magnification. In contrast, extrusive igneous rocks have a *finer* texture in which crystals are too small to be seen with the unaided eye. The structure of such rocks reflects sudden chilling of molten silicates as they were ejected at the surface of the earth. In lavas, water vapour is quickly lost to the atmosphere, crystal nucleation is rapid, and the melt begins to form a solid before there is sufficient time to grow larger, crystals. Such an extrusive rock is basalt, composed of ferromagnesian minerals and tiny rectangular grains of plagioclase feldspars. *Obsidian* is an extrusive rock that cooled so rapidly that there was insufficient time for crystallization; the melt therefore froze into a glass.

If coarsely crystalline igneous rocks indicated slow cooling and finely crystalline ones indicate rapid cooling, then what would be the cooling history of a rock with large crystals immersed in a very fine-grained matrix? Such rocks are said to have *porphyritic texture*. The large crystals (*phenocrysts*) were formed slowly at depth and were then swept upward and incorporated in the lava as it hardened at the surface.

Table 2.1. Common Igneous Rocks

Coarsely Crystalline Intrusive Igneous Rocks	*Approximate Extrusive Equivalents*	*Common Silicate Mineral Components*	*Average Specific Gravity*
Granite	Rhyolite	Quartz Potash feldspar Sodium plagioclase (Minor biotite, amphibole, magnetite)	2.7
Diorite	Andesite	Sodium-calcium plagioclase Amphibole, biotite (Minor pyroxene)	2.8
Gabbro	Basalt	Calcium plagioclase Pyroxene Olivine (Minor amphibole, ilmenite)	3.0

Composition of Igneous Rocks

The mineral composition of an igneous rock provides insight into the amounts and kinds of different ions in the parent magma. Altogether, in igneous rocks there are only about eight elements that are abundant: oxygen, silicon, aluminium, calcium, iron, magnesium, sodium, and potassium. These combine in specific ways to form the feldspars, ferromagnesian minerals, micas, and quartz minerals that are the constituents of igneous rocks.

For many years, geologists have recognized that intrusive igneous rocks tend to be richer in silica than are extrusive rocks. Later studies indicated that this relatively greater amount of silica served to keep the melt below the surface by increasing *viscosity* (resistance to flow). Even while still within melt, silica tetrahedra develop and tend to align themselves into chains and rings and thicken the melt and retard its upward progress.

Granite is the most abundant silica-rich intrusive rock. It is derived from melts so rich in silica that after all linkages with metallic ions are satisfied, there is still enough silicon and oxygen remaining to form quartz grains within the rock. It is true that there are silica-rich extrusive rocks as well (i.e., *rhyolite*), but they are far less abundant than extrusives containing lower percentages of silica. Most high-silica rocks are light-coloured.

Basalt is a fine-grained extrusive rock derived from low-silica melt. Basaltic lavas have been observed to flow with the approximate consistency of motor oil, a fact accounting for the frequency with which it is found at the earth's surface. Since the silica percentage is low, quartz grains are rarely in basalt.

There is one final important relationship in the chemistry of magmas and lavas. The silica-rich melts also tend to include ample quantities of potassium and sodium and therefore yield crystals of potassium feldspar, sodium plagioclase, and mica along with the ubiquitous quartz. Calcium, magnesium, and iron are rather minor constituents of silica-rich magmas, but they increase in abundance in rocks deficient in silica. The low-silica rocks utilize these elements in the formation of calcium plagioclases and ferromagnesian minerals. Because of the ferromagnesian and grayish plagioclase minerals, low-silica rocks tend to be dark gray, black, or green.

Volcanic Activity

Although deciphering the history of a granitic mass is certainly intellectually stimulating, it is unlikely to evoke the feelings of awe and excitement one experiences when viewing volcanic activity. Volcanoes are basically vents in the earth's surface through which hot gases and molten rock flow from the earth's interior. The extrusions may be quiet or explosive. Quiet eruptions are exemplified by the Hawaiian volcanoes and are frequently characterized by truly enormous outpourings of low-viscosity lava. The lava spreads widely to form the gentle slopes of a "shield volcano." Explosive eruptions are caused by the sudden release of molten rock driven upward by large pockets of

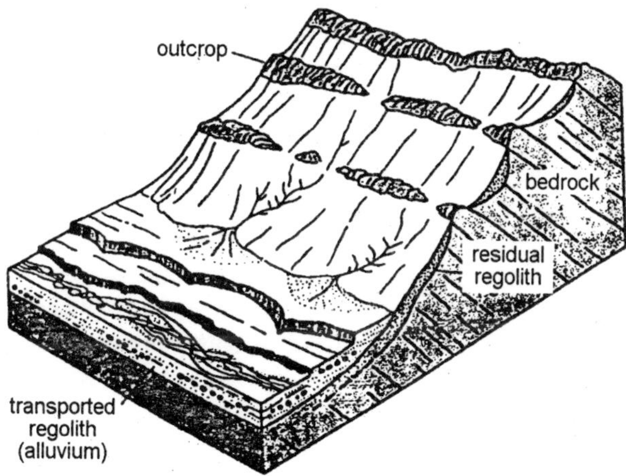

Fig. 2.1. Residual regolith on the hillside is distinguished from transported regolith in the valley bottom. Rock outcrops project through the regolith.

compressed gases. Such explosive eruptions can literally blow the volcanic cone to bits. The catastrophic eruption of Krakatoa in 1883 was heard 5000 kilometers away and was responsible for the death of 36,000 humans. Fortunately for our species, most eruptions are not so violent.

There are, of course, all stages of volcanic activity between quiet and violently explosive. Perhaps in response to changes in the composition of the parent melt, some volcanoes have even been known to shift from one type of activity to another. Volcanic activity also includes successive outpourings of lava from great fissures so as to form lava plateaus that extend over thousands of square kilometers. Such fiery floods long ago produced the Columbia and Snake River Plateau as well as the Deccan Plateau of India.

By far, the most abundant kind of volcanic rock is basalt. It underlies the ocean basins, has been built into midoceanic ridges, and has accumulated sufficiently in places like Hawaii and Ice-land to have produced substantial land area.

How and at what depth did this great volume of basalt originate? To answer this question it is necessary to refer briefly to a model of the earth's interior that has been formulated by the study of earthquake waves. The model depicts the earth's basaltic *crust* as a thin zone about 6 km thick and overlying the *mantle* of denser olivine- and pyroxene-rich rocks. The boundary between the crust and the mantle is recognized by an abrupt change in the velocity of earthquake waves as they travel downward into the earth. For many years geologists believed that basaltic lavas originated from the lower part of the basaltic crust. However, several recent lines of evidence suggest that the basaltic lavas may have come from molten pockets of upper mantle material. For example, present-day volcanic activity is closely associated with deep earthquakes that occur within the mantle far beneath the crust. It is quite likely that fractures produced by these earthquakes could serve as passages for the escape of molten material to the surface. A detailed study of earthquake shocks from particular volcanic eruptions in Hawaii indicates that the erupting lavas were derived from pockets of molten material within the upper mantle at depths of about 100 km. A weak plastic zone (called the "low velocity zone") in the mantle appears to represent the level at which the lavas originated. The mechanism by which they developed is called partial melting. *Partial melting* is that general process by which a rock subjected to high temperature and pressure is partly melted and the liquid component is moved to another location. At the new location, the separated liquid may solidify into rocks having different composition from the parent mass. The word "partial" in the expression "partial melting" refers to the fact that some minerals melt at lower temperatures than others, and so for a time the material being melted resembles a hot slush composed of liquid and still solid crystals. The molten fraction is usually less dense than the solids from which it was derived and thus tends to separate from the parent mass and work its way toward the surface. In this way melts of basaltic composition separated from denser rocks of the upper mantle and eventually made their way to the surface to form volcanoes.

Many complex and interrelated factors control where in the mantle partial melting may occur or even if it will occur at all. Generally; heat in excess at 1500°C is required, but the precise temperature for melting is also influenced by pressure and the water content of rock. As pressure increases, the temperature at which particular minerals melt also rises. Thus a rock that would melt at 1000°C higher pressure until it reaches far greater, temperatures. Water has an effect opposite to that of pressure, for its presence will allow a rock to start melting at lower temperatures and shallower depths than it would have otherwise. Laboratory experiments indicate that the melting of "dry" mantle rock can occur at depths of about 35 km but that the presence of only a little water can cause partial melting and yield basaltic liquid from depths as shallow as 100 km.

Not all lavas found at earth's surface are basaltic. Volcanoes of the more explosive type that are located at the edge of continents around the Pacific and in the Mediterranean extrude a lava called *andesite*. Andesite contains more silica than basalt, and its lava is thus more viscous. This greater resistance to flow contributes to the gas containment that precedes explosive volcanic activity. Andesites are considered to be intermediate in silica content between the rocks of the continental crust and those of the oceanic crust of the earth.

Andesitic rocks may originate in more than one way. Some emplacements result from originally basaltic magmas in which minerals

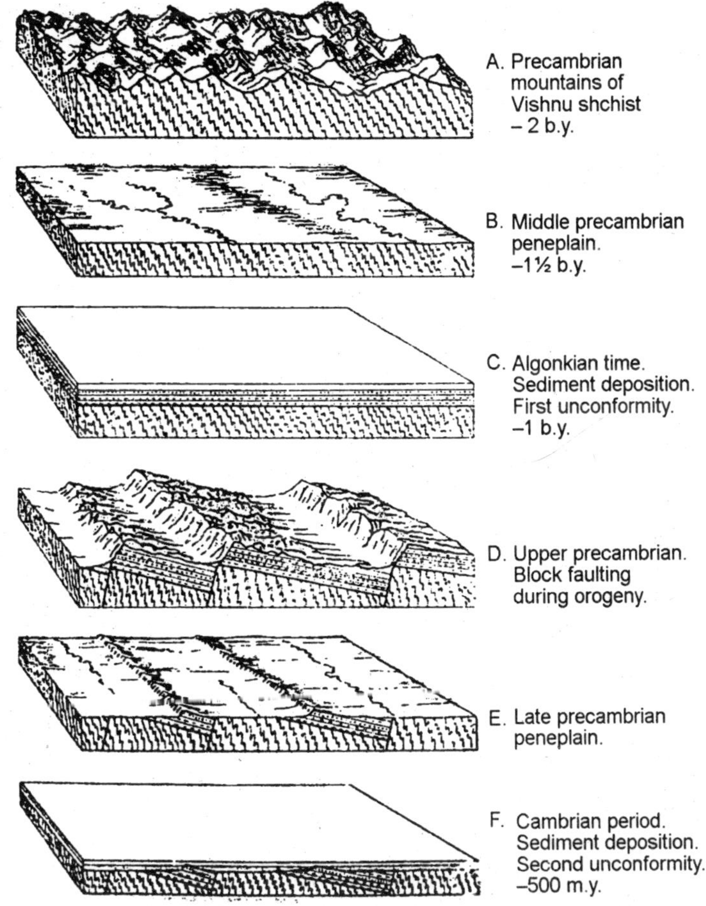

Fig. 2.2. This set of block diagrams shows the manner in which the great wedges of Algonkian strata came into existence in the lower Grand Canyon.

like olivine and pyroxene form early and settle out, thus leaving the remaining melt relatively richer in silica. This process is called *fractional crystallization*. Evidence for this mode of origin is provided by Iceland's volcanoes, Iceland is a volcanic island formed on oceanic basaltic crust. It has been observed that the longer the quiet period between eruptions of Ice-landic volcanoes, the more siliceous is the lava that is extruded. Apparently, longer periods of quiescence provide time for fractional crystallization and settling.

Andesitic melts are also believed to form by partial melting of mantle materials in the presence of some water. They may also result when a silica-rich older rock is assimilated by a basaltic magma. In addition to these theories, it has been suggested that andesites may result from the melting of oceanic crust and siliceous marine sediments as they descent into hot zones of the mantle. The wet, silica-enriched melts of andesitic composition might then rise buoyantly and erupt along volcanic island arcs.

Metamorphic Rocks

All Things Change

Sir Charles Lyell recognized that igneous or sedimentary rocks, and the chemical action of solutions and gases, can be altered to quite different kinds of rocks. Lyell borrowed the term metamorphism (from the Latin *metamorphosis*, meaning "change of form") to describe this process. It is still used today to describe alterations in rocks brought about by physical or chemical changes in the environment that are intermediate between those that result in igneous rocks and those that produce sedimentary rocks. Any previously existing rock may be converted to a metamorphic rock, and the changes primarily involve recrystallization of minerals in the rock while it remains in the solid state. In the process of recrystallization, the textural characteristics of the parent rock may be changed while at the same time new minerals develop that are stable under the new conditions of pressure and temperature. New elements need not be introduced; instead, those that are already present are incorporated into different and often denser minerals. Variations in heat and pressure may result in different kinds of metamorphic rocks, even from the same parent material.

Most metamorphic rocks exhibit a layering called *foliation*, which results from the parallel alignment of mineral grains. Whether this foliation is very fine or coarse depends upon the size and shapes of the constituent minerals. A few metamorphic rocks (*marble* is a familiar example) do not develop foliation.

Metamorphism

Alterations of rock immediately adjacent to igneous intrusions constitute *contact metamorphism*. The changes that occur in the intruded rock are largely the result of high temperatures and the emanation of chemically active vapors that accompany igneous intrusions. Such factors as the size of the magmatic body, its composition and fluidity, and the nature of the intruded rock also influence the kind and degree of contact metamorphism. Important ore deposits are commonly situated in the metamorphosed rock surrounding the intrusive. Examples of such deposits include magnetite and copper ores in metamorphic zones around granite intrusives in the Urals, central Asia, the Appalachian Mountains, Utah, and New Mexico.

Regional or *dynamothermal metamorphism* is a type of rock alteration that is areally extensive and occurs under the conditions of great confining pressures and heat accompanying deep burial and mountain building. In a subsequent chapter, we will discuss how rocks deposited in crustal troughs adjacent to continents may be compressed into mountain systems and thus be regionally metamorphosed. Metamorphic "index minerals" known to form under specific temperature and pressure conditions are used to decipher the history of growth of these

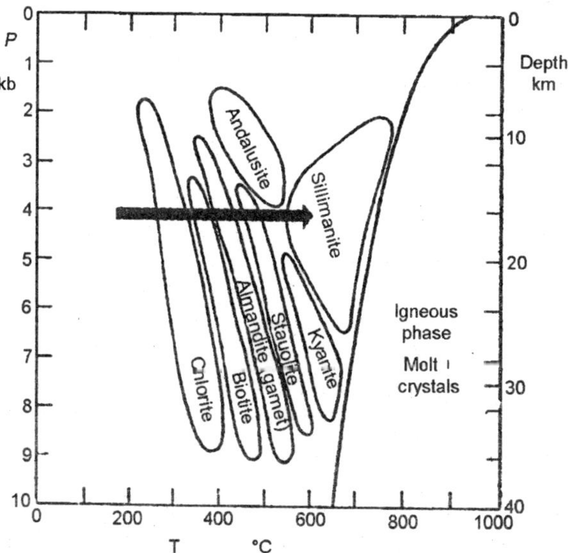

Fig. 2.3. This graph shows the grades of regional metamorphism are related to pressure and temperature. The arrow shows the typical series of changes from lower to higher grades at a given depth.

ancient mountainous regions, even when only the roots of the ranges remain.

Kinds of Metamorphic Rocks

As any rock can be metamorphosed in a number of different ways, there are hundreds of different kinds of metamorphic rocks. However, for our purposes, we need only consider several that occur extensively at the earth's surface. It is convenient to divide metamorphic rocks into two groups based on the presence or absence of foliation.

Foliated metamorphic rocks

Slate. In slate, the foliation is microscopic and caused by the parallel alignment of minute flakes of silicates with sheet structure. The planes of foliation are quite smooth, and the rock may be split along these planes of "slaty cleavage." The planes of foliation may lie at any angle to the bedding in the parent rock. Slate is derived from the regional metamorphism of shale.

Phyllite. The texture in Phyllite is also very fine, although some grains of mica, chlorite, garnet, or quartz may be visible. Phyllite

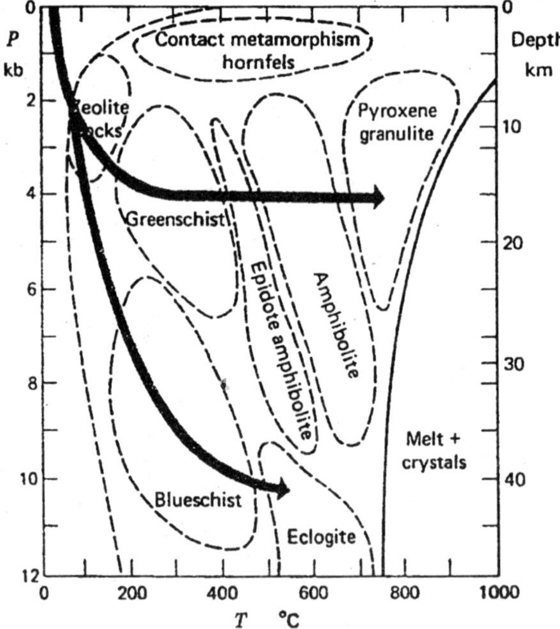

Fig. 2.4. A schematic graph of the major types of metamorphic rocks in relation to pressure (depth) and temperature. The arrow show two possible patterns of evolution in the sequence of regional metamorphic rocks.

surfaces often develop a wrinkled aspect and are more lustrous than slate. Phyllite represents an intermediate degree of metamorphism between slate and schist. The parent rocks are commonly shale or slate.

Schist. The platy or needle-like minerals in schist are sufficiently large to be visible to the unaided eye; the minerals tend to be segregated into distinct layers. Schists are named according to the most conspicuous mineral present. Thus, there are mica schists, amphibole schists, chlorite schists, and many others. Shales are the usual parent rocks for schists, although some are derived from fine-grained volcanic rocks.

Gneiss. This is a coarse-grained, evenly granular rock. Foliation results from segregation of minerals into bands rich in quartz, feldspar, biotite, or amphibole. Foliation is coarse and appears less distinct than in schist. High-silica igneous rocks and sandstones are the usual parent rocks for gneisses.

Nonfoliated metamorphic rocks

Marble. A fine to coarsely crystalline rock, marble is composed of calcite or dolomite and therefore is relatively soft. (It can be scratched with steel). Marble is derived from limestone or dolostone.

Quartzite. A fine-grained, often sugary-textured rock, quartzite is composed of intergrown quartz and therefore is very hard. Rock will break through constituent grains; it may be any colour. Quartzite is derived from quartz sandstone.

Greenstone. A dark green rock, greenstone has a texture so fine that mineral components, except for scattered larger crystals, cannot be seen without magnification. It is derived by the low-grade metamorphism of low-silica volcanic rocks.

Granitization

Before leaving the subject of metamorphic rocks, we must mention a metamorphic aspect of granite. In the previous section, we regarded granite as having formed from a silicate melt in the usual igneous manner. For a good many years, geologists were fully satisfied with that explanation. They could find many places where the contacts between the granite and the older rock were sharp, showed the effects of being baked by a hot intruding mass, and contained pieces or inclusions of the intruded rock that appeared to have "fallen into" the granite. Clearly, such granites were once molten. However, not all granite contacts were so clearly intrusive. Rock exposures were found in which true granites seemed to mere imperceptibly into rocks of metamorphic character and ultimately into sedimentary rocks. The granites seemed to have been converted from these older rocks in the course of an episode of intense regional compression. Recent studies of the isotopic composition of granites, as well as detailed field work, may provide an explanation for the two kinds of granite occurrence. One begins with a thick wedge of sediment that is compressed by mountain-building forces. Near the base of the wedge, sedimentary rocks are converted to rocks of gneissic character. Enough partial melting occurs that a film of silicate melt moves upward, surrounds older grains, and gradually converts the entire body into granite without ever completely melting it. The result is a granite with gradational contacts. This process of converting solid rocks to granites without causing them to become magma has been termed *granitization*. Higher in the deformed wedge, the films and streams of silicate liquid come together to form the classic intrusive kinds of granite with sharp contact relationships.

Historical Significance of Metamorphic Rocks

We have noted that the conditions for metamorphism are developed in regions that have been subjected to intense compressional deformation. Such regions of the earth's crust either now have, or once had, great mountain ranges. Thus, where large tracts of low-lying metamorphic terrain are exposed at the earth's surface, geologists conclude that crustal uplift and long periods of erosion leveled the mountains. Metamorphic rock exposures at many localities across the eastern half of Canada represent the truncated stumps of ancient mountain systems.

From studies of mineralogic composition of metamorphic rocks, it is often possible for geologists to reconstruct the conditions under which the rocks were altered and then to make inferences about the directions of compressional forces, pressures, temperatures, and the nature of parent rocks. Investigators are aided in these studies by the knowledge that specific metamorphic minerals form and are stable within finite limits of temperature and pressure. Maps of *metamorphic facies*, or zones of rocks that formed under specific conditions, can be constructed. Commonly, such maps delineate broad bands of metamorphic rocks, each of which formed under sequentially more intense conditions of pressure and temperature. Imagine a terrane that was once underlain by a thick sequence of calcareous shales, was subjected to compression so as to produce mountains, and then experienced loss of these mountains by erosion. One might then begin a traverse across this eroded surface on unmetamorphosed shales that were not involved in the mountain building. These shales would contain only unaltered sedimentary minerals. Progressing farther, toward the area of most intense metamorphism, one might see that the shales had given way to slates bearing the green metmorphic mineral chlorite. Still farther along the traverse, schists containing such intermediate-grade metamorphic minerals as biotite and garnet would appear. Finally, one might come upon coarsely foliated schists containing kyanite, staurolite, and sillimanite—minerals that develop under high temperature and pressure.

Metamorphic minerals do not always appear in the orderly fashion indicated in the preceding example. Depending on the nature and depth of metamorphism, temperature may increase at a faster or slower rate than does pressure, resulting in the growth of different index minerals. It is also possible for a previously metamorphosed, terrane to experience a second, less severe episode of metamorphism. In such cases, a lowering of metamorphic grade might result. Yet another factor that influences the kinds of minerals produced is the mineralogic composition of the parent rock and the amount of water present during metamorphism.

Metamorphic rocks may contain a wealth of historical information. A marble containing flakes and veins of chlorite that is dated as 1 billion years old tells the geologist many things. It records the existence of an ancient, somewhat clayey limestone that experienced a relatively low level of metamorphism. Because limestone was the parent rock, the geologist may infer that conditions on earth at that early date were suitable for the precipitation of carbonate rocks. Such conditions would include an atmosphere and hydrosphere similar to that existing today and considerably unlike the nonoxygenic environment that characterized the earth's most ancient ages.

SEDIMENTARY ROCKS

Sedimentary rocks are simply rocks composed of consolidated sediment—particles that are the product of weathering and erosion of any previously existing rock of soil. The components of sedimentary rocks may range from large boulders to the molecules dissolved in water. Sediment is deposited through such agents as wind, water, ice, or mineral-secreting organisms. The loose sediment is converted into coherent solid rock by any of several processes: precipitation of a cementing material around individual grains, compaction, or crystallization. These processes constitute *lithification*.

The most obvious feature of sedimentary rocks is their occurrence in beds or layers called *strata*. Stratification is commonly the result of changes in the conditions of deposition that cause materials of somewhat different nature to be deposited for a period of time. For example, the velocity of a stream might decrease, causing particles to settle out that might otherwise have stayed in suspension. In another situation, the kind of materials brought into a given depositional site by streams might change, and there would then be a corresponding change in composition of the accumulating layers.

Sandstone, shale, and carbonate rocks (such as limestone) constitute the most abundant sedimentary rocks. *Sandstones* are composed of grains of quartz, feldspar, and other particles that are cemented or otherwise consolidated. *Shale* consists largely of very fine particles of quartz an abundant clay. The *carbonates* are rocks formed when the carbon dioxide contained in water combines with oxides of calcium and magnesium.

Derivation of Sedimentary Materials

Sedimentary rocks must have originally come from the decomposition of older rocks. Commonly, the older rocks are igneous; indeed, these were once the only rocks on earth. It is therefore instructive to review the manner in which the common components of sedimentary rocks might be derived from an abundant kind of igneous rock. One such igneous rock is called *granodiorite*. It is an ordinary intrusive rock consisting of quartz, calcium and sodium, feldspar, potassium feldspar, biotite, and amphibole. Can such a rock, if chemically decomposed in a temperate climate, yield the materials required for the formation of sandstones, shales, and limestones?

Consider, first, the quartz in the granodiorite. Quartz will persist almost unchanged during weathering. It is one of the most chemically stable of all the common silicate minerals. As the parent rocks is gradually decomposed, quartz grains tend to be washed out and carried away to be deposited as sand that will one day become sandstones.

The feldspars decay for more readily than quartz. They are primarily aluminum silicates of potassium, sodium, and calcium. In the weathering process, the last three elements are largely dissolved and carried away by solutions as bicarbonate ions (although some may remain in soils within clay minerals). Ultimately, they reach the sea, where they may stay in solution, or they are deposited as layers of limestone. If large quantities of lake or sea water are evaporated, *evaporites* like *halite* (NaCl) or *gypsum* ($CaSO_4 \cdot 2\ H_2O$) may be formed. Of course, not all the feldspars and micas in the granodiorite necessarily decay. Some may persist as detrital grains that become incorporated into sandstones and other sediments.

The decomposition of the plagioclase feldspar in granodiorite can be expressed by the equation:

$CaAl_2Si_2O_8 \cdot 2NaAlSi_3O_8$	+	$4H_2NO_3$	+	$2(nH_2O)$
Plagioclase Feldspars		Carbonic Acid (Water plus Carbon Dioxide)		Water

yields

$Ca(HCO_3)_2$	+	$2NaHCO_3$	+	$2Al_2(OH)_2Si_4O_{10} \cdot nH_2O$
Soluble Calcium Bicarbonate		Soluble Sodium Bicarbonate		Clay Mineral

The equation shows that the principal nonsoluble product of the decomposition of feldspars (and other aluminium silicates) is clay. Here, then, is the greatest primary source for the clay of shales, claystones, and soils.

Biotite is another minerals in the granodiorite source rock. Decomposition of biotite, which is a potassium, magnesium, and iron alumino silicate, yields soluble potassium and magnesium carbonates, small amounts of soluble silica, and iron oxides. The iron oxides serve to colour many sedimentary rocks in tints of brown and red.

Variety Among Sedimentary Rocks

Sedimentary rocks are classified according to their *composition* and *texture*. The term "texture" refers to the size and shape of the individual grains and to their arrangement in the rock. A rock that has a *clastic* texture is composed of grains of sand, silt, or parts of rocks or fossils. Most sedimentary rocks have a clastic texture, but many others are formed by the intergrowth of crystals and are therefore crystalline. In very simple classifications, sedimentary rocks are divided into clastic and nonclastic categories.

Clastic rocks

Clastic (or *detrital*) *rocks* are those composed of individual fragments of mineral or rock. The fragments may range in size from huge boulders to microscopic particle. The group includes such common sedimentary rocks as conglomerate, sandstones, and shales; the materials of these rocks are derived, at least initially, from weathering and erosion of pre-existing rocks on land.

Texture is the key naming the major clastic rocks. *Conglomerate*, for example, is composed of water-worm, rounded particles larger than 2 mm in diameter. *Breccias* are composed of fragments that are angular but similar in size to conglomerates. In *sandstones*, grains range between 0.0625 and 2.0 mm. The varieties of sand stones are then subdivided largely according to compositions. *Siltstones* are finer than sandstones (0.004 to 0.0625 mm), and *shales* are composed of particles finer than 0.004 mm. Shales may contain abundant clay minerals, which are flaky minerals that align parallel to bedding planes. As a result, shales characteristically split into thin slabs parallel to bedding planes. This property is termed *fissility*. Rocks lacking fissility but composed of clay-sized particles are called *claystones* or *mudstones*.

Nonclastic rocks

Nonclastic rocks are formed from materials either directly precipitated from solution or secreted by the activities of plants or animals. In contrast to the clastic group, such sediments have undergone little if any transportation. Dissolved substances are brought to the oceans by rivers and under suitable conditions may be precipitated or extracted by organisms to form the minerals of nonclastic rocks. The most prevalent of the non-clastic group are the *carbonate rocks*, which are composed of the carbonate minerals calcite ($CaCO_3$, which crystallizes in the hexagonal crystal system), and dolomite [Ca $Mg(CO_3)_2$, which like calcite, forms hexagonal crystals]. Of these rocks, calcite is the predominant mineral in limestones, although the other carbonate minerals, as well as clay and varieties of quartz, limestones show a wide variation in texture and composition.

Limestones

The most abundant limestones are of marine origin and have formed as a result of precipitation of calcite or aragonite by organisms and the incorporation of skeletons of those organisms into sedimentary deposits. Inorganic precipitation of carbonate minerals may also from deposits of limestone. The importance of this process is questionable, however, because the precipitation is nearly always closely associated with photosynthetic and respiratory activities of organisms or with the release of tiny particles of aragonite upon the decay of green algae. Strictly speaking, it appears very few marine limestones are the result of direct chemical precipitation.

After the calcium carbonate has accumulated, it becomes recrystallized or otherwise consolidated into indurated rock that may be variously coloured—from white, through tints of brown, to gray. Limestones tend to be well stratified, frequently contain nodules and inclusions of chert, and are often highly fossiliferous (containing fossils). The rock may range in texture from coarsely granular to very fine-grained and aphanitic. In general, limestones consist of one or more of a combination of such textural components as *micrite*, *carbonate clasts*, *oolites*, or *carbonate spar*. Micrite is a uniformly fine-grained, "muddy texture in which individual particles cannot be discerned without considerable magnification. Micritic texture is apparently the result of consolidation of carbonate mud and ooze. Carbonate clasts are sand- or gravel-sized pieces of carbonate. The most common clasts are either *bioclasts* (skeletal fragments of marine invertebrates) or *oolites*, which are spherical grains formed by the precipitation of carbonate around a nucleus. Sparry carbonate is a clear crystalline carbonate that is normally deposited between the clasts as a cement or has developed by replacement of calcite. They permit classification of particular samples as micritic limestone, clastic limestone, oolitic limestone, or sparry (crystalline) limestone.

There are many varieties of limestone. Chalk is a soft, porous variety that is composed largely of extremely minute calcareous skeletal elements called *coccoliths*. Coccoliths are secreted by marine golden brown algae. *Lithographic limestone* is an etching surface in printing illustrations. Some limestones consist almost entirely of skeletal remains of reef corals and other frequently skeletonized marine invertebrates.

Dolostone

This is a nonclastic rock composed largely of the mineral *dolomite*, which is a calcium-and-magnesium carbonate. As found in exposures, dolostone is not easily distinguished from limestone. The usual field test for distinguishing dolostone from limestone is to apply cold dilute hydrochloric acid. Unlike limestone, which bubbles readily, dolostone will effervesce only slightly, if at all. In thin sections of the rock examined with the aid of a petrographic microscope, the uniform rhombic grains of dolomite are a trait useful in identification.

The origin of dolostones is somewhat problematic. The mineral dolomite is not secreted by organisms in shell-building. Direct precipitation from sea water does not normally occur today, except in a few environments where the sediment is steeped in abnormally saline water. Such an origin is not considered adequate to explain the thick sequences of dolomitic rock commonly found in the geologic record. The most widely believed theory for the origin of dolostones is that they result from partial replacement of calcium by magnesium in the original calcareous sediment. However, it is not known now long or at what time in the history of the rock this "dolomitization" occurs.

Chert

We have previously mentioned a form of microcrystalline quartz called chert (SiO_2) noting its occurrence as nodules in limestones. The origin of these, nodules is still being debated among petrologists, although the majority believe that such nodules from as replacements of carbonate sediment by silica-rich sea water trapped in the sediment. Some cherts occur in a really extensive layers and thus qualify as monomineralic rocks. These so-called bedded cherts are thought to have formed from the accumulation of the siliceous remains of diatoms and radiolaria and from subsequent reorganization of the silica into a microcrystalline quartz. Silica from the dissolution of volcanic ash is believed to enhance the process; indeed, many bedded cherts are found in association with ash beds and submarine lava flows.

Evaporites

In the previous chapter, we noted that evaporites are chemically precipitated rocks that are formed as a result of evaporation of saline water bodies. Only about 3 per cent of all sedimentary rocks consist of evaporites. Evaporite sequences of strata are composed chiefly of such minerals as gypsum $(CaSO_4) * 2H_2O)$, anhydrite $(CaSO_4)$, halite (NaCl), and associated calcite and dolomite. Extensive ancient deposits of evaporites are currently being commercially worked in Michigan, Kansas, Texas, New Mexico, Germany, and Israel. The conditions required for replication of thick sequences of evaporites include warm, relatively arid climates and a physiographic situation that would provide periodic additions of sea water to the evaporating marine basin. In the Gulf Coastal Region of the United States, as a result of the pressure of overlying rocks, deeply buried deposits of salt have flowed plastically upward to form underground domes of salt. In the process, the salt arched overlying strata, thereby producing structures in which petroleum could collect.

Coal

Coal is a carbonaceous rock resulting from the accumulation of plant matter in a swampy environment combined with alteration of that plant tissue by both biochemical and physical processes until it is converted to a consolidated carbon-rich material. The biochemical and physical changes may produce a series of products ranging from peat and lignite to bituminous and anthracite coal. For coal to form, plant tissue must be accumulated under water or be quickly buried, because vegetable matter, if left exposed to air, is really oxidized to water and carbon dioxide. With underwater accumulation or quick burial of plant material, a major part of the carbon can be retained.

Colour in Sedimentary Rocks

We have seen the colour in igneous rocks can be used to indicate the approximate amount of ferromagnesian minerals. Colour in sedimentary rock can also provide useful clues to identification. For example, varieties of chert can be identified as flint if they are gray or black, or as jasper if they are red. Colour is also useful in providing clues to the environment of deposition of sedimentary rocks. Of the sedimentary colouring agents, carbon and the oxides and hydroxides of iron are clearly the most important.

Black Colouration

Black and dark gray colouration in sedimentary rocks—especially shales—usually results from the presence of organic carbon compounds and iron sulfides. The occurrence of an amount of organic carbon sufficient to result in black colouration implies an abundance of organisms in or near the depositional areas as well as environmental circumstances that kept the remains of those organisms from being completely destroyed by oxidation or bacterial action. These circumstances are present in many marine, lake, and estuarine environments today. In a typical situation, there mains of organisms that lived in or near the depositional basin settle to the bottom and accumulate. In the quiet bottom environment, dissolved oxygen needed by aerobic bacteria to attack and break down organic matter may be lacking. There may also be insufficient oxygen for scavenging bottom-dwellers that might feed on the debris. Thus, organic decay is limited to the slow and incomplete activity of anaerobic bacteria; consequently, incompletely decomposed material rich in black carbon tends to accumulate. In such an environment, iron combines with sulfur to form finely divided iron sulfide (pyrite, FeS_2), which further contributes to the blackish colouration. Such environments of deposition are likely to yield toxic solutions of hydrogen sulfide (H_2S). The lethal solutions rise to poison other organisms and thus contribute to the process of accumulation. Black sediments do not always form in restricted basins. They may develop in relatively open areas, provided the rate of accumulation of organic matter exceeds the ability of the environment to cause its decomposition.

Red Colouration

Hues of brown, red, and green are frequently formed in sedimentary rocks as a result of their iron oxide content. Few, if any, sedimentary rocks are free of iron, and less than 0.1 per cent of this metal can colour a sediment a deep red. The iron pigments not only are ubiquitous in sediments but also are difficult to remove in most natural solutions.

Iron forms two sorts of ions; *ferrous* iron has two positive charges, whereas *ferric* iron has three positive charges. Thus, iron may form two oxides; FeO (ferrous oxide) and Fe_2O_3 (ferric oxide). In air, ferrous iron is slowly oxidized to ferric iron. When oxygen is in short supply, ferric iron may be similarly reduced to ferrous iron. Ferric minerals like *hematite* tend to colour the rock red, brown, or purple, whereas the ferrous compounds impact hues of gray and green. Hydrous ferric oxide (*limonite*) is often yellowish in colour.

Red Beds

Strata coloured in shades of red, brown, or purple by ferric iron are designated *red beds* by geologists. Oxidizing conditions required for the development of ferric compounds are more typical of nonmarine than marine environments; most red beds are flood plain alluvial fan, or deltaic deposits. Some, however, are originally reddish sediment carried into the open sea. Electron microscope studies of red beds forming today in Baja, California, indicate that the red colouration developed long after the sediment was deposited. After burial, the decay of clastic ferromagnesian minerals released iron that was oxidized by the oxygen is underground water circulating through the pore spaces. Thus, red colouration may be imparted in the subsurface and may be independent of climate. The paleoenvironmental interpretations one can draw from red beds should be based to a large degree on the associated rocks and sedimentary structures. Red beds interspersed with evaporite layers indicate warm and arid conditions.

Although red beds are more likely to represent nonmarine than marine deposition, occasionally the reddish strata are interbedded with fossiliferous marine limestones. In such cases, the colour may be inherited from red soils of nearby continental areas. Lands located in warm, humid climates often develop such reddish soils. When the soil particles arrive at the marine depositional site, they will retain their red colouration if there is insufficient organic matter present to reduce the ferric iron to the ferrous state. Otherwise, they will be converted to the gray or green colourations of ferrous compounds.

In summary, sedimentary rocks of red colouration may be a product of the source materials, may have developed after burial as a result of a lengthy period of subsurface alteration, or may be the result of subaerial oxidation. Geologists are suspicious of the last possibility, because most modern desert sediments are not red unless composed of materials from nearby outcroppings of older red beds.

3

Detecting Age

Rocks are the essential building materials of which the earth is constructed. The architecture of our planet results from the kinds of rocks that are present, the positions or attitudes they assume, and the processes acting upon them.

The geologist's definition of rock is broader than that of everyday language. Ice and even water may be considered by the geologist—as *rocks*. The engineering geologist regards loose - superficial material, whether soil or sediment, as *rock* because it has been derived from once compact material. The intimate mixtures of minerals characteristic of ore deposits form another type of rock.

Most rocks are aggregates of two or more individual minerals. The preponderance of the earth's crust, however, consists of only the 25 minerals - out of almost 2,000 known species - which are to any extent abundant.

The geologic significance of a rock specimen lies in the faithfulness with which it represents the larger body from which it came.

Kinds of Rock Bodies

According to the basic nature of their mineral content rocks fall into four different categories, as follows:

1. *Monomineralic rocks,* consisting essentially of a single mineral, which occurs on a large enough scale so that it can be considered an integral part of the structure of the earth. Other minerals may be incorporated as impurities, though many such rocks are remarkably simple in their uniformity. Some limestones and marbles belong to this category.

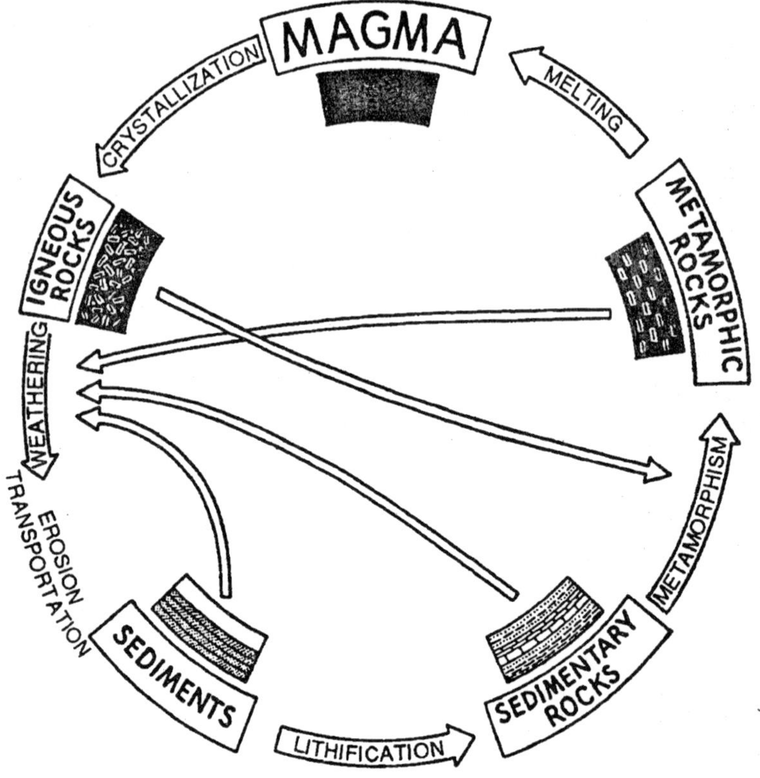

Fig. 3.1. The rock cycle.

2. *Natural glass,* often nearly homogeneous but not having a composition that can be expressed by a chemical formula, because it varies from place to place in the same mass.
3. *Organic matter,* an animal or vegetable product.
4. *An aggregate of two or more minerals,* with or without a groundmass of natural glass. Many such rocks contain dozen different minerals, mostly observable only under high magnification. The great majority of rocks belong to this group.

Types of Rocks

All rocks may be divided into three large groups based on their modes of origin. Certain members of each of these groups grade into one another (unlike minerals rocks are not classified by sharp boundaries). But, in general, practically all rocks can be classified without difficulty, as follows:

1. *Igneous Rocks*: Rocks formed by the solidification of cooling molten material.
2. *Sedimentary Rocks*: Rocks formed at the surface of the earth from accumulations of mud, sand, and gravel derived from the weathering and transport of pre-existing rocks (cemented together by deposits of mineral matter once held in solution in underground water). Other sedimentary rocks, such as *limestone* and *gypsum*, consist almost entirely of material deposited from solution. Sedimentary rocks cover three quarters of the earth's Land area.
3. *Metamorphic Rocks*: Rocks formed at depth, under great heat and pressure, by the alteration of either igneous or sedimentary rock.

The one conspicuous exception to this classification is, granite, which seems to originate both from a molten state and by the transformation of sedimentary rocks under conditions of extreme metamorphism. This latter process, called *granitization*, is now an accepted mechanism for producing granite, but its quantitative importance in nature is not known and is still a subject of heated debate among geologists. In the field, rocks are classified *megascopically*, by examining them in the outcrop or in the hand. In the laboratory a much more elaborate classification is possible because thin sections (slices of rock ground to a standard thickness of nearly .03 millimeter, mounted on a glass slide and protected by a cover glass) can be examined *microscopically*. In the study of elementary geology, a modified field classification is used for examination of rocks because it requires only a hand lens, a bottle of acid to test the carbonate content, a streak plate, a magnet, and a knife blade or glass plate to identify the minerals by their most easily determined properties. Characteristic specimens of all the important kinds of rocks should be studied and, if possible, compared with their occurrence in the field.

Igneous Rocks

Igneous rocks are classified on a two-fold basis: (1) chemical (mineral) and (2) textural. Texture refers to size, shape and pattern of the mineral grains, and is indicated by the position of the rock name in the vertical columns. The position of a rock in the horizontal row indicates its chemical or mineral composition (its position in the sequence commonly referred to as acid to basic).

Texture

The igneous rocks, both *intrusive*, and *extrusive*, have different textures which usually indicate the condition under which they cooled. Although the chemical composition of the magma playa a part, the size of the mineral grains depends chiefly on the rate of cooling, as determined by temperature and pressure and the presence of volatiles. Coarse interlocking textures (called *phaneritic*, *granitoid*, or *granitic*) result from slow cooling, aided greatly by large amounts of water and other volatile substances. Fine textures (called *aphanitic* or *felsitic* if the grains are indistinguishable) result from rapid cooling, which indeed may take place so fast that only a glass is formed. Mixed textures (called *porphyritic*) are generally explained as representing two stages of solidification; they consist of larger crystals called *phenocrysts* embedded in a *groundmass* of finer aphanitic crystals, which solidified about the phenocrysts. Broken igneous rocks, shattered by volcanic explosions and reassembled afterward, have a *fragmental*, or *pyroclastic*, texture.

Mineral Content

The kinds and amounts of the various minerals in an igneous rock depend principally on the chemical composition of the *magma* or *lava*. *Acidic rocks* (also called *silicic* or *persilicic* rocks) have a high content of silica; quartz and feldspar predominate in them, and they are typically light in colour and low in specific gravity. Examples are granite and rhyolite. *Basic rocks* (*subsilicic rocks*) have a lower content of silica but more iron and magnesium, which yield *ferromagenesian* (also called *mafic* or *femag*) minerals, such as pyroxene, amphibole, biotite, and olivine. These minerals make basic rocks darker and heavier, even though some feldspar is frequently present. Examples are gabbro, dolerite, and basalt. Extremely bask rocks, with almost no feldspar, are known as *ultra basic*. Examples are dunite, peridotite and pyroxenite.

The distinction between acidic and basic rocks is arbitrary, since there is complete gradation from one extreme to the other. It is therefore convenient to consider a group of *intermediate rocks*, which lie between acidic and basic rocks in composition.

Kinds of Igneous Rocks

Granite rocks

Granitic rocks include true granite and other rocks, commonly caned "*granite*" to which different names more properly apply; much of the granite used in buildings and monuments is of the latter class.

The essential minerals necessary to the classification of true granite are potash feldspar and quartz. Plagioclase feldspar and biotite mica or amphibole (hornblende) are usually present. Many granites contain scattered grains of muscovite mica, as well as minor *accessory minerals* (such as magnetite, apatite, and zircon) which do not influence the naming of the rock. The colour of granite - white, gray, pink, or red is largely determined by the colour of the feldspar.

Exceptionally large amounts of certain minerals give rise to varieties such as biotite granite, hornblende granite, and so forth. When some of the minerals grow to abnormal sizes, the rock is known as *pegmatite*, other kinds of igneous rocks sometimes possess a pegmatitic mixture. Rock of a uniformly fine-grained texture and granitic composition is referred to as *aplite*.

Syenite rocks

Similar in texture to granite, butwith less silica and little or no quartz is syenite; it is much less common than granite. The other minerals are rather similar to those in granite. *Nepheline syenite* is an important though uncommon rock containing, in addition to feldspar, the mineral nepheline, which belongs to the feldspathold group of minerals.

Monzonite rocks

A granitoid intrusive in which both potash and plagioclase feldspar are present in about equal proportions (they have somewhat different colours) is called *monzonite*. The dark minerals are mainly biotite mica, amphibole (hornblende), and pyroxene (augite). If quartz is also present: as in granite, the rock is *quartz monzonite* or *granodiorite* (the difference between them is a technical one, based upon the ratio of potash to plagioclase feldspar).

Diorite rocks

Plagioclase is the dominant feldspar in diorite, another intrusive igneous rock less common than granite. Diorite contains abundant dark minerals, resembling monzonite in this respect. When quartz is present, the rock is termed *quartz diorite* or *tonalite*.

Felsite rocks

The acidic and intermediate igneous rocks that are so fine-grained that the minerals can scarcely be recognized without a microscope are grouped together under the name *felsite*. Banding due to the flow of congealing lava, inclusions, and gas cavities (*vesicles*) are common features of felsites. Each of the main types of felsites which is of extrusive origin) corresponds in quantity and kind of feldspar to an jntrusive rock of similar chemical composition. However, the minerals in felsite may differ in some respects and certain minerals are apt to stand out as they do in porphyries.

Rhyolite is the extrusive equivalent of granite; its most prominent grains are quartz, though potash feldspar and biotite mica are common. *Trachyte* is the equivalent of syenite. *Phonolite* (with nepheline) is the equivalent of nepheline syenite, and often emits a ringing sound when struck. *Latite* is the equivalent of monzonite, and *quartz latite* of quartz monzonite or granodiorite. *Andesite*, the equivalent of diorite, shows mostly feldspar grains. *Dacite* is the equivalent of quartz diorite. Rhyolite and andesite are very abundant and are found in various colours Rhyolite is usually buff, cream, or purplish; andesite is darker, often gray or greenish.

Gabbro rocks

Its chief minerals are pyroxene (augite) and plagioclase feldspar (usually *labradorite*). Gabbro may also contain hornblende and olivine and is a typical basic igneous rock. An important variety of gabbro consisting almost exclusively of labradorite is called *anorthosite*; another kind Containing hypersthene (a pyroxene) is named *norite*. *Dolerite* or *diabase* is a medium-grained gabbro.

Peridotite rocks

The intrusive igneous rock consisting principally of olivine and pyroxene is named *peridotite*. When the rock consists almost entirely of olivine, *itisdunite*; when pyroxene is the sole essential mineral, it is *pyroxenite*. A special variety of peridotite is the diamond be anngkmoer life of South Africa and Arkansas. These rocks often contain, or are associated with, the heavy metals such as nickel, chromium and platinum. Extrusive equivalents are extremely rare.

Basalt rocks

The aphanitic equivalent of gabbro is basalt, the most abundant of all lavas. Its vesicles when filled by minerals such as quartz or calcite are known as *amygdales*. Basalts and similar dark, fine-grained, igneous rocks often go under the field name of traprock.

Porphyry rocks

An appreciable number of phenocrysts in the ground mass makes any igueous rock *aporyhyry*. Thus, among those described above are granite porphyry, syenite porphyry, monzonite porphyry, diorite porphyry, felsite porphyry, basalt porphyry, and related rocks. Many of the ordinaryaphanitic rocks, no matter what name they bear, tend generally to be porphyritic.

Obsidian rocks

The glassy equivalent of the acidic and intermediate rocks is called *obsidian*. In spite of its dark colour (usually black; but sometimes gray, brown, or red), obsidian is not a basic rock. When magnified, obsidian shows incipient crystallization which accounts for the dark colour.

If appreciable water is present, natural glass is called pitchstone or perlite,according to the luster.Fronthy glass, consisting of winding tubes filled with air so that it floats on water is pumice.

The basic equivalent of obsidian is known as *basaltic glass* or *tachylyte*; the corresponding cellular rock is *scoria*.

Pyroclastic rocks

Volcanic ash from an eruption becomes tuff when consolidated; some becomes welded tuff when fused together in a fiery cloud of hot gases. Coarser fragments produce *volcanic breccia*, which may be of either explosive or flow origin.

Sedimentary Rocks

Sedimentary rocks are composed of material ultimately derived from the disintegration, by weathering and erosion, of older igneous,sedimentary, or metamorphic rocks. Sedimentary material falls into two categories: (1) dissolved mineral matter which is precipitated by inorganic or organic agents and (2) solid fragments, or sediment, which accumulates to become a body of rock. Dissolved mineral matter-forms the precipitated sedimentary rocks, and the solid fragments accumulate to form the *fragmental* or *clastic sedimentary rocks*. The precipitated mineral matter also contributes to the cementation of the clastic material. These sedimentary materials become compact rocks by the following processes:

1. *Compaction*. In this water is squeezed out (most sediments are deposited in water, though someare carried by the wind or ice) and the individual particles are pressed together by the weight of the overlying, sediments. Older sediments are usually more closely consolidated than younger ones, because they are apt to have been more deeply buried or subjected to earth movements, but degree of compaction is no proof of age.
2. *Cementation*. Here the mineral matter held in solution by underground water is deposited between the grains to bind them together. The many substances cementing sedimentary rocks include calcium carbonate (*calcareous cement*) and silica (*siliceous cement*),

as well as lesser amounts of iron oxide (*ferruginous cement*), clay, and gypsum.

3. *Recrystallization.* It enables small grains to grow into larger and stronger ones, or new minerals to form in the open spaces between others. Replacement may later substitute new and more stable minerals for earlier formed ones.
4. *Chemical alterations.* It includes reduction, especially of iron compounds by organic matter; destructive distillation of organic matter, and the activities of bacteria and bottom-dwelling animals.

Fragmental Rocks

Sedimentary rocks composed of particles of sediment from a previous source constitute the group known as the *fragmental clastic*, *detrital* (from *detritus*, debris); or *mechanical* rocks. The loose material is classified on the basis of the size of the fragments.

Sedimentary rocks can also be classified according to the agent of deposition (e.g., wind-blown dune sand), but unfortunately the geologic history is not always known. The name of the resulting sedimentary rock corresponds not only to the size of the fragments but also to their *texture* (shape and arrangement) and to the composition-of both grains and cement. Besides the actual name of the sedimentary rock, a combination of terms may be needed to describe it adequately according to its various aspects. Typical rock specimens should be studied indoors and out, to learn more about the variations which exist in nature.

Precipitated Rocks

Mineral matter that is dissolved in water may be removed from it in two principal ways:

1. *By inorganic chemical processes*. Becoming a *chemical* or *inorganic precipitate*. Among the solutions leached from the land that are particularly abundant are sodium chloride, calcium sulphate, silica, carbonates of calcium and magnesium, and compounds of phosphorus, barium,manganese, and iron.
2. *By the action of plants and animals*. Becoming an *organic* or *biogenic precipitate*. Living things extract such chemical as silica, calcium carbonate, and phosphates from fresh and sea water for the development of their supporting and protective hard structures such as bones, shells, and teeth. Other organisms cause chemical reactions to occur in what seems to be the ordinary way of inorganic processes. It is often extremely difficult to determine in which manner a given sedimentary rock was actually created.

Kinds of Sedimentary Rocks

Mixtures of clastic and non-clastic sedimentary rocks are very common, for the precipitated matter that serves as a cement assumes an increasing proportion of the total volume of a normally fragmented rock. Even an organically produced rock can be broken up and its pieces recemented together as fragments.

Conglomerate rocks

Cemented gravel is termed *conglomerate.* Any kind of rock material may constitute the fragments, but quartz and *chalcedony* are especially abundant. The size of the pieces may vary widely, and usually sand grains fill the interstices. Glacial deposits form rough, coarse conglomerates called *tillites.* When the gravel is relatively unworn rubble, with sharp edges and pointed comers, the rock is called a *sedimentary breccia. Rudite* is a general term for conglomerate and breccia.

Sandstone rocks

Grains the size of sand become lithified to *sandstone* or *arenite.* Quartz is the typical mineral, making a *quartzose sandstone*, but interesting sandstones may be composed largely of gypsum or coral. Heavy placer minerals such as magnetite, rutile, and zircon may be sufficiently abundant to produce so-called *black sands* or *yellow sands.* Placer minerals are those very resistant to weathering and heavy, which therefore become concentrated in running water or along beaches. Gold placers are produced in this way. *Greensand* or glauconite sandstone contains a high percentage of the mineral glauconite. The colour of sandstone is usually, however, determined by the nature of the cement-red, brown, and green being due to the presence of iron. *Arkose* is a variety of sandstone in which feldspar is prominent in addition to quartz. *Graywacke* is a sandstone consisting largely, of dark rock fragments, usually slates or fine-grained basic igneous rock. *Calcarenite* consists of sand-sized grains of calcite and hence is a detrital limestone.

Shale rocks

The most frequently occurring sedimentary rock on all the continents is shale; composed of mud (silt and clay), the finest particles of sediments. Hence it appears homogeneous to the eye, though mica and quartz are generally present to-some extent. Sandy shales are termed arenaceous;organic matter makes a black, carbonaceous shale. Calcilutite consists of mud-sized grains of calcite.

Limestone rocks

Of the dominantly nonclastic sedimentary rocks, limestone is the most common. Some limestone has undoubtedly been formed by direct chemical precipitation, but most has evidently been built up in large part by the accumulation of shells and skeletons of organisms which remove calcium carbonate from sea water inorder to make their resistant parts. The open spaces are then filled by deposits of the same material crushed to powder by the waves or precipitated from the water. *Calcite*, the chief mineral component of limestone effervesces in acid.

Chalk is a soft, porous limestone. *Coquina* is a limestone consisting largely of shell fragments that are clearly visible. *Marl* is a very finegrained limy material often mixed with day (the word has other meanings, especially among foreign geologists).

Dolomite rocks

Resembling limestone in most ways, *dolomite* or *dolostone* is formed when magnesium replaces part of the calcium in limestone. The mineral dolomite is somewhat harder, heavier, and less soluble in acid than calcite; it effervesces in cold acid only when scratched or powdered.

Gypsum rocks

Thick beds of the mineral gypsum constitute one of the common sedimentary rock, which goes by the same name, and is interlayered with other sedimentary rocks that also are produced by the evaporation of sea water.

Anhydrite rocks

Composed of the mineral anhydrite, the rock of this name can change to *gypsum* in the presence- of moisture, and the reverse change can take place when gypsum is heated or strongly compressed to drive off its water content.

Rock salt rocks

The enormous quantity of *halite* or rock salt dissolved in the oceans is ample to explain the thicklayers of this material that have been deposited throughout geologic history. When reasonbly pure, this rock is easy to identify by its taste. Other *evaporites* occur in various parts of the earth and -may be commercially valuable. Among the useful evaporites are potash salts, nitrates, phosphates, and borates.

Coal rocks

Coal is regarded as a sedimentary rock because it is found in layers. However, it originated neither as fragments nor by chemical precipitation.

METAMORPHIC ROCKS

Igneous and sedimentary rocks can, under suitable conditions, be materially changed, without melting, into the third great group, the *metamorphic rocks*. The transformation or metamorphism, is the result of a changed geologic environment, in which the stability of the rocks can be maintained only by a corresponding change in their make-up. *Metamorphism* is characterized by the development of new textures, new minerals, or both, and these are often so unlike the former ones that it is frequently difficult to determine the nature of the original rock

New textures are produced by *recrystallization*, whereby the minerals grow into larger crystals having a different orientation. A characteristic structure of metamorphic rocks is *foliated* (leaf-like), wavy, or banded due to a parallel arrangement of platy or elongated minerals. Such rocks tend to separate along the foliation as *rock cleavage*. New minerals are created by *recombination*; the chemical constituents form new partnerships, in which minor impurities take on a more significant role. Although there are a considerable number of minerals of strictly metamorphic origin, most metamorphic rocks are similar in composition to the rocks from which they were derived.

Factors in Metamorphism

The rather drastic changes involved in metamorphism are the effects- of heat, pressure, and fluids, usually acting together. Heat from within the earth and molten bodies of rock as well as from pressure and friction, speeds up chemical activity. Pressure may come about by simple burial, but movements of the crust are more effective in altering textures. Water and gas supply the mobility for the required changes to take place, and they may carry elements from a nearby *magma* to facilitate the chemical changes.

Kinds of Metamorphism

According to the factors involved, four types of metamorphism may be distinguished, although, as is true of virtually all geological processes, they merge and interact with one another.

1. *Geothermal metamorphism*. It results from the deep burial of rocks, such as that which occurs when sediments are buried in geosyncline The pressure of overlying material, together with the heat thereby generated, produces a transformation of the rock.
2. *Dynamic*, or *kinetic, metamorphism*. It refers to the change produced by crystal folding, usually at shallow depths. The crushing force

frequently gives the rock involved a broken or *cataclastic* (down-breaking) structure, associated with faults. The confining pressure at greater depths is sufficient, under these conditions, to produce a finely pulverized, recrystallized rock called *mylonite*.

3. *Hydrothermal metamorphism*. It covers the transformations sometimes very profound that accompany the action of hot magmatic solutions and gases. *Replticement* and deposition of ore minerals are common phenomena.
4. *Contact metamorphism*. It embraces the complex effects resulting from the intrusion of a magma; high temperature and high pressure, often combined with mountain-making forces, are powerful agents. These effects reach their maximum intensity around the upper margins of batholiths, especially in adjacent limestone, where the intruding fluids are most corrosive. From some large igneous bodies the highly fluid magma is injected intimately into the surrounding "*country rock*" to produce *migmatite* or *injection gneiss*. Under extreme conditions,various rocks are converted into a product hardly distinguishable from ordinary granite; this converting process is called *granitization*.

Kinds of Metamorphic Rocks

The specific product that will result from metamorphism depends upon the character of the original rock, the types of metamorphic processes involved and the intensity with which they have operated.

Gneiss rocks

The most coarsely banded metamorphic rock is *gneiss*, usually consisting of alternating bands or lenses of unlike appearance. Bands rich in feldspar and quartz (the dominant constituents of most gneiss) are more granular and lighter in colour than those rich in biotite mica, hornblende (amphibole), or garmet.

Schist rocks

When the platy or micaceous constituents dominate, *gneiss* grades into *schist*. The visible minerals, moreover, are likely to be much more uniform in appearance and composition, having little feldspar, and adjacent layers generally consist of the same minerals. The extreme foliation of schist causes it to split readily so that this kind of separation is known as *schistosity*. On the basis of the most prominent mineral present, the varieties called *micaschist*, *hornblende* or *amphiboleschist*, and *chlorite schist*, etc., are recognized.

Phyllite rocks

Phyllite is intermediate in texture between schist and slate, and tends to break into slabs, the surfaces of which show minute crumpling.

Slate rocks

A uniformly fine-grained rock, slate splits easily into smooth lustrous plates (*slaty clecwage*). It often contains black carbon in the form of graphite, as well as iron and manganese minerals, which give it various colours such as red or green.

Marble rocks

The metamorphism of either limestone or dolomite produces the massive (nonfoliated) crystalline rock called *marble*. Impurities tend to be segregated into knots or spread out in the striking patterns so familiar in this rock. The principal minerals are *calcite* or *dolomite*.

Quartzite rocks

Sandstone thoroughly altered by metamorphism becomes *quartzite*. It has a glassy appearance on a fractured surface, and the fractures pass indiscriminately through the grains and the cement around them.

Other metamorphic rocks consisting almost exclusively of a single mineral include *serpentine* (composed of the mineral of the same name), *soapstone* (mostly talc), and *amphibolite* (mostly hornblende).

Dating of the Rocks

Although several methods exist to determine the age of the earth, yet none of them appears to be accurate to estimate the age of our planet, earth, because its date of birth is not known. But all the methods we know reveal that the antiquity of the earth lies between 3 and 5 billion years or more. A few methods of estimating geologic time have been described in the following paragraphs:

Salt content of the ocean: Many geologists believe that the present-day oceans originally composed of freshwater and the salts present in them must have been derived from the earth's original crust, for rain water does not contain any. Of course a lot of time must have been consumed in converting such large fresh-water bodies into the present salty oceans. However, if we known the total quantity of salt in the oceans and the amount of salt discharged annually by rivers into them, the time elapsed in acquiring the present salt content by oceans can be calculated by simple division. The total oceanic salt content has been estimated to be about 16,000,000,000,000,000 tonnes of sodium. Similarly the quantity of sodium annually discharged into the oceans by the river waters has been estimated to be about 158,000,000 tonnes. So the time elapsed since the deposition first began is:

16,000,000,000,000,000 / 158,000,000 = 101,265,000 years

However, this figure does not agree with the estimated figures by other methods.The main objections to this method are:

1. Much of the oceanic salt has undergone several cycles of deposition.
2. Much of the original oceanic salt is trapped and buried deeply in the marine sedimentary rocks.
3. Age of the earth is much greater than the age of the oceans.

Deposition of the sediments or the Geological method his is the simplest and earlist method employed by the geologists to determine the age of the rocks. In the method if we were able to know the average rate of deposition of the sediments per year, and if we safely assume that the rate of deposition was the sameinthe past geological times also, then the total thickness of these sediments will give us an indication as to the age of the earth. Calculations made by this method range from 100 to 600 million years. This method has several limitations. Firstly, it is applicable to the aquatic sedimentsonly. Secondly different types of sediments accumulate at different rates. Thirdly, frequent braks or unconfimities are present in the deposition of the sediments, and lastly, due to the fact thickness of the ancient must have been subjected to the earth movements several in the past. They have also undergone the processes like weathering and denudations, etc.

Rate of Erosion

In contrast to the preceding method, this method tries to determine that since how long the erosion of the rock began on the earth. If we are able to estimate the number of years required to erode one meter of rock and know the total number of meters eroded, by simpled division, we can estimate the age of the earth. Estimates made by this method indicate that the earth is very old.

Like other methods, this method has the following drawbacks:

1. The total amount of the sedimentary rocks is not known.
2. The sediments of the sedimentary rocks have several times been eroded and redeposited.
3. The present rate of erosion is not necessarily the standard rate of erosion throughout the geological past.

Stratigraphy

This subject deals with the nature and origin of stratified rocks; with their sequence in the earth's crust; and with their correlation, i.e. the identication of isolated outcrops of rocks with those of similar age else where. The first practicalstep is stratigraphy is the grouping of stratara into lithilogical units, e.g. limestones, clays, etc. The second step is to establish the sequence of these units in time. For this purpose the basis of all stratigraphical work is the principal of supervposition of stratasimply/ that in any normal undisturbed sequence of sedimentary rocks a stratum is younger than on which it rests. This principal has been used to establish the correct sequence of strata in the stratigraphi showed that the fossils which occur in any one part of the stratigraphical colmn are distinctive and different from those which occur at other levels.This made possible the delineation of groups of strata containing particular characterstics fossils(zonal fossils)and known as biozones. Once the order of these biozones was established it became possible to refer any set of sedimentary rocks to its correct position in the stratigraphical column by examination of the fossil content.

Stratigraphical Colomn

The complete succession of stratified rocks which contain fossils is summrized in the stratigraphical colomn in which the strata, combined into major groups and systems, are arranged in sequence with the oldest at the base and the youngest at the top. The earliest rocks in which fossils occur in appreciable numbers, the cambran, overlie still older rocks, the Precambrian, in which fossils are rare.

4

STRATIGRAPHY

Sedimentary rocks have been built up by layer upon layer of sediment, which sometimes has been much the same for long periods of time but at other times has changed its character rapidly. The individual layers within sedimentary rocks are separated by bedding planes. These bedding planes are time horizons, and the history of a rock sequence is reflected in its layering. Each layer in the rock sequence must have been laid down on a pre-existing layer, so that the oldest rocks are at the bottom of an exposed sequence, the youngest at the top, unless the succession has been tectonically inverted. This is the *principle of superposition*, recognised as long ago as 1669 by the Italian scientist Nicolaus Steno.

Stratigraphy is concerned with the study of stratified rocks, their classification into ordered units and their historical interpretation. It bears not only upon past geological events but also upon the history of life and is perhaps the most basic part of geology.

Much of stratigraphy is concerned with chronology; the geological record has to be divided up into time periods, standardised, as far as possible, all over the world. One of the primary aims of stratigraphy has been to produce an accurate chronology in which not only the order of events but also their dates are known. Stratigraphical classification is basic to all of this.

There are three principal categories of stratigraphical classification, *lithostratigraphy*, *biostratigraphy* and *chronostratigraphy*, all of which are ways of ordering rock strata into meaningful units.

LITHOSTRATIGRAPHY

Lithostratigraphy is concerned with the erection of units based upon the characters of the rocks and differentiated on types of rock, e.g. siltstone, limestone, clay, etc. It is useful in local areas and essential in geological mapping, but there is always the danger that even in a small area rock units cut across time planes. For instance, if a shoreline has been, direction a articular suite of sediments, probably of the same general kind, will be left in its wake, Though this bed will appear in the rock record as a single uniform layer, it will not have been deposited at the one time; since it cuts across time planes it is said to be diachronous. Such diachronism is common in the geological record. Furthermore many suites of sediments are laterally impersistent; different sedimentary facies may have existed at the same time within a small space–a sandstone layer, for instance, passing into a shale some distance away. Lithostratigraphy is thus only of real value within a relatively small region.

The division erected in lithostratigraphy are arranged in a hierarchial system; *group*, *formation*, *member* and *bed*. A bed is distinct layer in a rock sequence. A member is a group of beds united by certain common characters. A formation is a group of members, again united by characters with features in common. It is the primary unit of lithostratigraphy and is most useful in geological mapping. Hence it is formations that are normally represented by different colours on geological maps and cross sections, and a formation is normally defined for its mapping applications. Finally a group ranks above a formation; it is composed of two or more formations and is often used for simplifying stratigraphy on a small scale map.

BIOSTRATIGRAPHY

In biostratigraphy the fossil contents of the beds are used in interpreting the historical sequence. It is based upon the principle of the irreversibility of evolution. This means that at any one moment in the Earth's history there was living a unique and special assemblage of animals, characteristic of that period and of no other. As time went on these were replaced by others; each successive fossil assemblage is a pale reflection of the life at the time the enclosing sediments were deposited. Thus during the early Palaeozoic trilobites and brachiopods were the most common fossils; by the Mesozoic the most abundant preservable invertebrates were the ammonites; they too become extinct, and snails and bivalves are the commonest relics of Cenozoic time. This is how it appears on a broad scale. But when the time ranges of individual fossil species are examined it is evident that some of these lasted for only a fraction of geological time, characterising very precisely a particular brief historical period.

In any local area, once the sequence of fossil faunas has been precisely established through assiduous collection and documentation from exposed sections this known succession can be used for correlation with other areas. Certain fossil species have been found to be particularly good stratigraphical markers. They characterise short sections of the geological succession known as zones. To take an example, ammonites are particularly good zone fossils for Mesozoic stratigraphy. The Jurassic period lasted some 55 million years, and in the standard British succession there are over 60 ammonite zones by which it is subdivided, so the zones are defined historical periods which have an average duration of less than a million years each.

The practical problems in biostratigraphy are, however, very complex, and some parts of the geological succession are much more closely zoned than others. The main problems are as follows:

(a) Many kinds of fossils, especially those of bottom-dwelling invertebrates, are facies-controlled. They lived in particular environments only, e.g., lime-mud sea floor, reef, sand or silty sea floor. They were often highly adapted for particular conditions of temperature, salinity or substrate and are not found preserved outside this environment. This means that they can only be used for correlating particular environments and thus are not universally applicable.

(b) Some kinds of fossils are very long-ranged. Their rates of evolutionary change were very slow. They can only be used in a broad and general sense for long period correlation and are of very little use for establishing close subdivisions.

(c) Such good fossils as the graptolites are delicate and only preserved in quiet environments, being destroyed by more tubulent conditions.

(d) Since fossil species could migrate following their own environment through time, there is always a possibility of diachronous faunas. The zone as defined in one area may not therefore be exactly time-equivalent to that in another region.

In the example of a graptolite, therefore, for the reasons outlined in (c) and (d), the total range or biozone of a species is not likely to be preserved in any one area, and it is therefore hard to draw ideal isochronous boundaries or time lines.

Ideally, zone fossils should have a particular combination of characters to make them fully suitable for biostratigraphy. These would be

(a) a wide horizontal distribution preferably inter-continental.

(b) a short vertical range so that they could be used to define a very precise part of the geological column.

(c) enough morphological characters to enable them to be identified and distinguished easily.

(d) strong hard shells to enable them to be commonly preserved

(e) independence of facies, as would be expected from a free-swimming animal.

All of these conditions are seldom fulfilled in fossils used for zonation; perhaps the neritic ammonites came closest to it and it is not surprising that the principles of really precise stratigraphical correlation were first worked out fully with these fossils, notably by the German paleontologist A. Oppel in the 1850s.

It was Oppel too who first recognised that there are various ways of using fossils in stratigraphy, which partially circumvent the difficulties mentioned, and hence different types of biozones. There are four main kinds of biozones generally used (Hedberg 1976). Assemblage zones are beds or groups of beds with a natural assemblage of fossils. They many be based on all the fossils preserved therein or on only certain kinds. They are usually very much environmentally controlled and therefore of use only in local correlation. Range zones are perhaps of more general application. A range zone usually represents the total range of a particularly useful selected element in the fauna. One may therefore refer to the Psiloceras planorbis zone, based upon the eponymous ammonite that defines the lowest zone of the European Jurassic, above which is the Scholorezmia angulata zone. Each range zone is always named after a particular species which occurs within it. Where there are a number of zonally useful species, or where the ranges of individual special are long, a more precise time definition may be given by the use of overlapping stratigraphical ranges. Such zones are therefore called concurrent range zones. Acme or peak zones are useful locally. An acme zone is a body of strata in which the maximum abundance of a particular species is found, though not its total range. Such acme zones may be narrow but are often useful as marker horizons in geological mapping. Finally an interval zone is an interval between two distinct biostratigraphical horizons. It may not have an distinctive fossils, or indeed any fossils at all, being simply a convenient way of referring to a group of strata bracketed between two named biostratigraphically defined zones.

Biostratigraphical units, unlike litho and chromo-stratigraphical units, are not hierarchially arranged, apart from in the case of subzones, which are local divisions where a zone can be divided more finely in a particular region than elsewhere.

A different kind of stratigraphic concept, the biomere was defined as regional biostratigraphic unit bounded by abrupt, non-evolutionary changes in the dominant element of a single phylum. These changes are not necessarily related to physical discontinuities in the sedimentary record and they may be diachronous. The biomere concept has proved most useful in studies of late Cambrian trilobite faunas, where a repeated pattern of events is evident from the fossil record. In each biomere the shelf sediments contain an initial fauna of low diversity and short stratigraphical range (one or two species only). However, later faunas within the biomere become much more diversified and of longer stratigraphic range and suggest, by this stage, 'sound adaptive plans' and the zenith of the trilobite fauna. At the top of the biomere there is often a rather specialised fauna of short -lived trilobites, then all the groups become extinct abruptly.

The succeeding biomere begins in the same way as its predecessor, often with trilobites of similar appearance to those at the base of the first one. They may have migrated in from a stock of more slowly evolving trilobites in an outlying, possibly deeper water area. The later development of the new biomere is as before: expansion and diversification followed by extinction. Several such biomeres have been defined in the intensively studied Upper Cambrian of North America. The pattern is invariably similar and could probably be discerned in other parts of the geological column as well.

Stratigraphic Units

Stratigraphy uses three complementary methods of subdividing the record of Earth history. *Lithostratigraphy*, whose units are the bed and the formation, based on examination of the lithological sequence. *Biostratigraphy* matches the fossil record to the lithostratigraphic framework. Its smallest unit is normally the zone whose boundaries are determined by the geographical and temporal range of the characteristic species. Hence it is possible for several independent zonal succession to co-exist side by side, based on different groups to organisms. *Chronostratigraphy* is the outcome of the best available synthesis of biostratigraphic and lithostratigraphic data. It is therefore an interpretation of the observed relationships. Again the standard unit is the zone or 'chronozone' and since this, like the 'biozone', is named after a fossil, confusion between the two is common. Zones are grouped into stages that are generally named after localities. Several stages constitute a system, while systems in their turn are grouped into eras. There are two opposed points of view concerning the delimitation of chronostratigraphic units. The first tries to site the boundaries at natural breaks, whose geographical extent is however, often debateable. The second view regards chronostratigraphy as a convenient but arbitrary framework into which events in the Earth's history can be slotted.

Absolute dating based on the natural decay of radioactive isotopes with a suitable half-life is another possibility.

CHRONOSTRATIGRAPHY

Chronostratigraphy is more far reaching than either bio-or lithostratigraphy but has its roots in both of them. Its purpose is to organise the sequence of rocks on a global scale into chronostratigraphical units, so that all local as well as worldwide events can be related to a single standard scale. Hence it is concerned with the age of strata and their time relations. To do this a hierarchical classification of time-equivalent units must be employed. The conventional hierarchical system used in shown in Table 4.1

Table 4.1. The conventional hierarchical correlation between chronostratigraphical and geochronological units.

Chronostratigraphical units	*Geochronological units*
eonothem	aeon
erathem	era
system	period
series	epoch
stage	age
chronozone	choron

Chronostratigraphical units relate quite simply to geochrono-logical units; thus the rocks of the Cambrian system were all deposited during the Cambrian period. Most of these terms are self-explanatory, but it should be recognised that they are all, at least in theory, worldwide in extent.

The *Psiloceras planorbis* chronozone is a time unit equivalent to the time in which the said ammonite was in existence, even if it was confined to certain parts of the world only. It is hard indeed, however, to be able to delimit chronozones accurately, since most fossils were confined to certain geographical regions or provinces, as are most of the animals living today. There are relatively few well established chronozones or 'world instants' as they have been called, and so

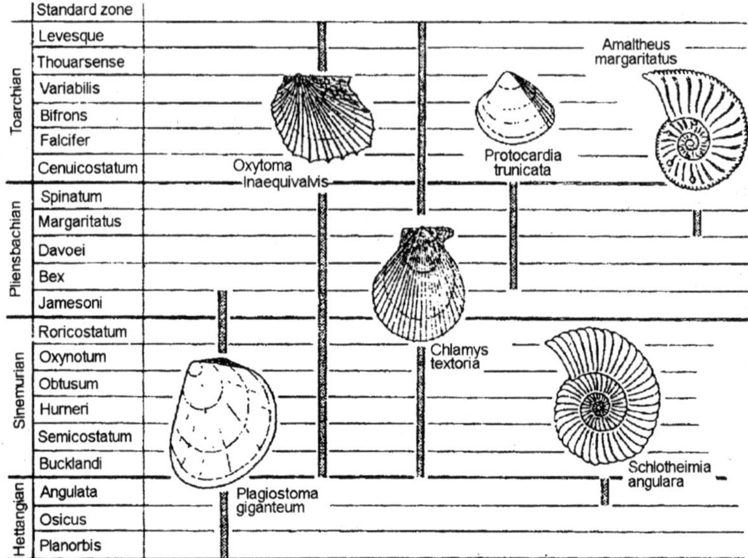

Fig. 4.1. Different stratigraphic value of bivalves and ammnites in the Lias.

'chronozone', though it has a real meaning, is not a term applicable to most practical stratigraphy. A stage, on the other hand, is a group of successive zones having great practical use, especially since it is normally the basic working time unit of chronostratigraphy, the narrowest that can actually be used on regional scale.

It is usually at the stage level that rocks of widely different facies can be correlated. As an example there are some difficulties in making precise zonal correlations between. Ordovician trilobite-brachiopod faunas and time equivalent faunas with graptolites. Graptolite are rarely preserved in the siltstones and limestones favoured by the shelly fossils, and the latter being benthic could not inhabit the stagnant muds in which the graptolites were best preserved. In some areas, of course, the faunas to alternate in vertical sequence since the sites of deposition of these two facies fluctuated with oscillating shorelines, but though precise zone-to-zone correlations are possible at some levels it is found in practice that Ordovician graptolite zones correlate best with stages defined on shelly fossils.

Fossils give a relative chronology which can be used as the primary basis of chronostratigraphy. But it is often hard to correlate precisely beds of equivalent age in widely separate areas. The fossil sequences, though well documented with any one area, may contain

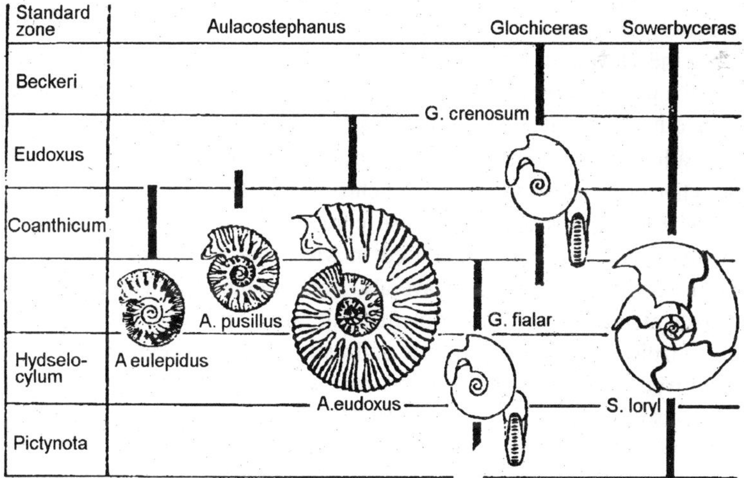

Fig. 4.2. Different stratigraphic value of upper Jurassic ammonites.

very few elements in common, if indeed any at all, for they belong to different faunal provinces which are hard to correlate. Sometimes, however, the boundaries of such provinces may have oscillated to and fro. There may thus appear elements of adjacent faunal provinces in vertical succession thus facilitating stratigraphical correlation. And at most stratigraphical horizons there are usually some ubiquitous worldwide fossils, so intercontinental correlation is not impossible.

In chronostratigraphy the relative sequence give by the fossils is supplemented and enhanced by absolute dates which can be affixed at certain points wherever appropriate rocks occur. These are usually lavas bracketed between fossiliferous sediments, and their occurrence is not too common. It is most unlikely, therefore that radiometric dating will supersede palaeontological correlation; the two are entirely complementary, and the great success of chronostratigraphy, in spite of its limitations, owes much to both.

Zone Fossils

Various fossil groups have been used as zone fossils to subdivide the last 600 million years of Earth history since the beginning of the Cambrian period.

Groups of Zone Fossils

The trilobites are the most stratigraphically useful fossils in marine sediments of Cambrian age whereas in the Ordovician and Silurian systems this role is taken by the graptolites, Cephalopods, or more precisely the ammonoidea, are used to subdivide the time interval from the Devonian to the end of the Cretaceous. The standard zones based on these fossils from our ortho-chronology whose precedence over other possible systems is supported by decades of usage.

Where these zone fossils are absent, the stratigrapher has to rely on other organisms to erect a para-chronology. Here the most important Palaeozoic groups are the rugose corals and the brachiopods, while bivalves and gastropods are used in Mesozoic and Cainozoic strata. More recently, microfossils have also come to the fore. The conodonts, whose affinities remain problematical, are extremely useful in the Palaeozoic while foraminiferids and ostracods have been successfully employed in Mesozoic and cainozoic sediments. Spores and pollen have also proved to be of great stratigraphic value because like other microfossils they are very abundant and can be found in most rock samples. Hence microfossils have great practical usefulness in the dating of core samples, for example, form oil wells.

Completely different organisms serve as zone fossils in terrestrial or fresh-water environments. Plant remains are indispensible in the case of Carboniferous and Permian strata while mammalian remains are also very useful in Tertiary and Quarternary sediments. Human artifacts (tools, utensils, ornaments) can be used to date late Quaternary rocks.

Requirements of Zone Fossils

The stratigraphic value of any fossil depends on a number of factors. A good zone fossil should be abundant and easily identified. It should also have a wide geographical distribution, a short stratigraphic range and should be as independent as possible of environment or lithology. Only a few groups of fossils fulfil all these requirements.

Abundance is important not only in relation to ease of finding the species in question but also because we cannot be assure of the true range of a rare species. When numbers are small the level of maximum abundance may more readily be diachronous between one locality and another. Hence rare species are not often used as zone fossils.

For several groups of organisms reliable identification of genera and species demands time-consuming this section or serial section studies. This is true, for example, of some corals and brachiopods, where forms with a similar external morphology can be distinguished only by their internal structure. Similar external appearance develops at different times. Because such organisms are difficult or impossible to identify in the field, the beds containing them cannot be dated quickly. Species with more readily recognisable features are easier, to use stratigraphically.

All organisms are dependent to some extent on the environment and since this also affects lithology, some facies control of organisms is usually observed. There are, however, some species whose controlling factors are not reflected in the host sediment.

The more a living organism is tied to its habitat, the clearer is its facies dependence. The inhabitants of coral reefs afford a classic example. Their habitat currently depends on a particular combination of clearly defined environmental parameters and it would appear that their requirements remained fairly constant throughout Earth's history. Their morphology is strongly influenced by their habitat. Hence the overall faunal character of coral reefs has remained very similar since

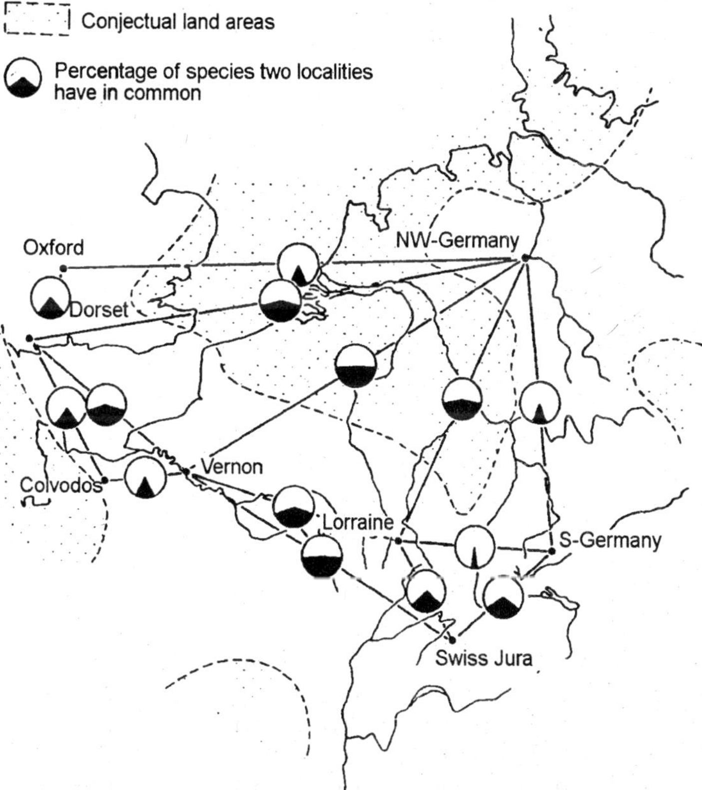

Fig. 4.3. Degree of difference or similarity between ostracod faunas of the same age from the upper Jurassic of Europe.

the Palaeozoic. Such organisms are termed 'facies fossils'. They are confined to limited areas where suitable living conditions exist. Their spatial distribution is therefore somewhat limited.

Plankton and the active swimmers of the oceans tend naturally to be the most widespread organisms, and can therefore be expected to provide the best zone fossils although even they experience some limits to their distribution. Moreover, many of them lack skeletons and are thus incapable of fossilisation, and useless as zone fossils despite their geographic range. Planktonic foraminifera, radiolaria, and grapolites are among the exceptions. Other groups of marine index fossils, for example the ammonoidea are facies-dependent to a certain degree but strike a balance between wide distribution and ease of fossilisation.

A species may have a short vertical range either because of its dependence on certain environmental conditions or because of rapid evolution of the group. In the first case the range at different localities may be markedly diachronous. The thinner the beds containing the fossil, the more likely is this explanation to be true. Rapidly evolving organisms on the other hand are among the best zone fossils. Whereas the average span of a typical Lower Jurassic bivalve species is about 5-10 million years, the contemporaneous ammonite species, like the graptolite ones of the Silurian, last only about 1 million years. The rate of evolution does, however, differ even among closely related organisms. Some of the Neogene and Pleistocene mammals (for example, elephants, hominids) evolved even faster. Even relatively slowly evolving groups are not without stratigraphic value but they are not suitable for very precise dating.

Evolutionary Series

Zone fossils that evolve in a particular direction are especially useful. Within such an evolutionary series is more or less synchronous from a geological point of view in difference localities. If we know the first and last members of the series, the intermediate stages can be arranged in morphological order. Moreover if we know when the series started and finished we can date the intermediate forms by assuming a constant rate of evolution.

Working with evolutionary series can, however, be fraught with dangers. Hardly any series develops at one single locality over an appreciable time span. Immigration and emigration are always possible. Reconstruction of the series is therefore never completely reliable because suppose intermediate forms remote localities may in fact be side lines or vice versa. The assumption of a constant rate of change

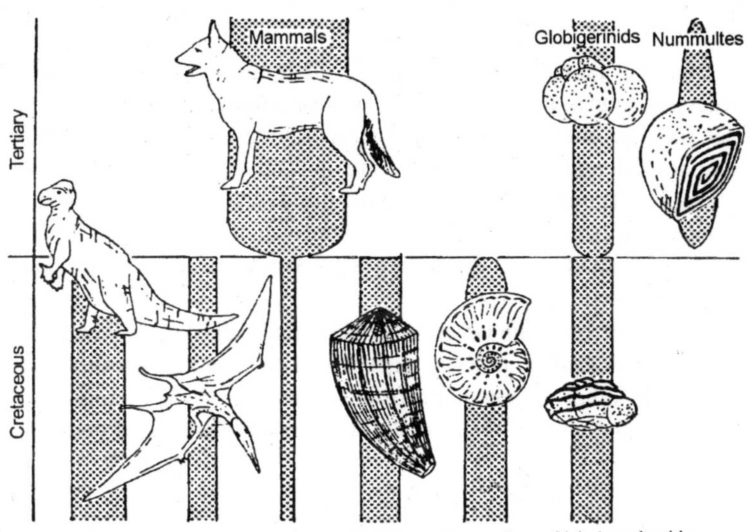

Fig. 4.4. Extinction and flowerings at the Cretaceous/Tertiary boundary.

is also unlikely to be valid. Phases of slow change usually alternate with bursts of rapid transformation. Furthermore some 'directed' evolutionary changes are reversible.

Stratigraphic dating can be based on the degree of evolution of whole floras or faunas as well as on the morphological development of particular species. Among the Tertiary and Quarternary molluscs, for example, the number of extant species increases from 1-5% in the Eocene, through 20-40% in the Miocene, to 90-100% in the Pleistocene. When such studies are undertaken, we must always remember that environmental factors may produce great differences between faunas and floras of the same age and that faunas of different ages may display greater similarity. For example, because of climatic conditions, the Arctic Eocene flora bears stronger resemblance to the Miocene flora of central Europe that to the Oligocene one.

Ecostratigraphy

Within a limited area, we can exploit the ecological dependence of organisms to make stratigraphic correlations. In the Baltic area, for example, three species which elsewhere are contemporaneous can be used to recognise successive marine and limnic phases in the postglacial sediments. Pollen analysis, which is particularly useful for Quaternary correlations, is based on similar foundations since the changes in pollen assemblages mirror the climatic history of the area.

Faunal Breaks

Distinct faunal breaks, that is, obvious differences between successive faunas might appear to be ideal stratigraphic markers but since they are often related to ecological factors, and occur at different times in different localities, they are not suitable for inter-regional correlation. The environmental change that alters the flora or fauna is usually accompanied by a change in sedimentation. Stratigraphic breaks that are not evident in the sedimentary sequence can also produce faunal breaks.

There are also, however short time intervals during which the organic realm changes unusually quickly with numerous groups becoming extinct and other taking their place. Such world-wide breaks are especially subdivisions of the stratigraphic column are based on them.

5

REMAINS OF PAST

In common usage the word "fossil" carries a distinctly derogatory implication but, as has been seen already, it referred originally to anything dug out of the ground. The root Latin words is *fossilis* meaning dug up, and therefore minerals as well as the remains of animals and plants were called *fossils* but the term soon came to be restricted to the remains of animals and plants found in rocks. These used also to be called petrifactions. Curiosities of all kinds were, in the past, also regarded as fossils and professor J.B. A. Beringer of Wurzburg, in 1767 described in good faith and illustrated 200 "fossils" in his *Lithographia wirceburgensis* but which were fakes manufactured as a practical joke possibly by his students and placed where he would find them. He realised his mistake just after the publication of his book, and, to save his good name, he bought the stock and hid it but the copies were found after his death. Such fakes were later known in Germany as *Lugensteine*—literally "deceptive stones".

In 1713 Johann Jakob Scheuchzer described a giant reptile as *Homo tristis* because the supposed it to be a human skeleton, and there are many other instances of such mistakes and misinterpretations. More important, however, are the important discoveries which resulted in the understanding of the development of life on Earth. It has been noted already that William Smith recognised that certain fossils were characteristic of the strata in which they occur, and, indicated a definite age; these are known as characteristic fossils.

It is necessary now to consider how the remains of living things came to be preserved for such a long time. Immediately after death organism begin to decay. The slow process of oxidation takes place in

Fig. 5.1. Figure showing the eruption of Volcanio.

the presence of oxygen, and simple compounds of oxygen with hydrogen, carbon, nitrogen, sulphur and phosphorus are the final products. In the absence of oxygen, however, fermentation takes place, resulting in the formation of carbon and nitrogen. The channels through which gas has escaped from decaying organisms are sometimes preserved in sediments.

In addition decomposition can be caused by bacteria and other organisms that live on dead material.

The hard parts of organisms resist decomposition more effectively than the soft parts and become buried in and protected by sediment. Under favourable condition either the complete animal, or parts of it, may be preserved in their original state. This has happened in permafrost areas such as Siberia, where bodies of ice age mammoths have been preserved. Permeation by salt or complete impregnation of the tissues by oil or wax has a similar effect. A fairly well preserved woolly rhinoceros found at Starunia in the south of Poland is one such example.

If organic remains are transported for any distance, they are usually broken or eroded so as to become almost unrecognisable, but in favourable circumstances, large numbers of dead animals can be deposited and entombed in sediments. Typical examples are shallow water, inshore shell banks and many fossil shell beds and "mussel bands" have been preserved. In many instances the fossils are effectively part of the sediment in which they are embedded. The covering of the fossils by successive layers of sediment and the resulting increase in pressure causes the sediments to become compacted and lithified, and also affects the fossils themselves. The most important change is a reduction, with increasing pressure, in the size of the pore spaces of the rocks and the consequent loss of water. This is accompanied by a series of chemical reactions which are not discussed further here, but which are together known as diagenesis. Such reactions frequently involve the remains of animals and plants. The processes of fossilisation take place over long periods of time.

Clearly fossils are only preserved under the most favourable circumstances; for all practical purposes they are restricted to the sedimentary environment and do not occur in igneous rocks which are produced by the crystallisation of magma. Sedimentary rocks are formed mainly in sea water. Terrestrial animals decompose quickly after death and are rarely preserved as fossils.

Most fossils are the hard parts of organisms which have been embedded in sediments and altered by chemical reaction. They may, for example, become calcified or silicified and thereby resistant to attack. The organic material itself may be replaced, molecule by molecule, by minerals. Alternatively the hard parts may be dissolved to leave in the compacted, surrounding rock a cavity which is an imprint of the former organism. The shells of molluscs or snails for example are sometimes filled with mud, clay or sand, which preserves an impression of the interior surface of the shell in the form of an internal cast.

Carbonisation plays a special role in the preservation of plant fossils, the cellulose being reduced to carbon in the absence of air. Coal was formed in this way. Coal-bearing strata contain the remains of numerous plants, particularly in those of Carboniferous (late Palaeozoic) age and in the Tertiary (early Cainozoic) brown coals. Impressions of leaves, fruits, branches and stems of plants are frequently preserved in the shale and seat earth above and below a coal seam.

Some fossils are so small that they can only be seen with a microscope; these are known as microfossils and are of great value in the search for oil, for they are delicate indicators of the age of sediments, particularly of Mesozoic and later age. The unidirectional development of life on Earth provides a calendar of the passage of time and there is, therefore, a close connection between stratigraphy and the development of life through time, termed biostratigraphy.

Fossils are not merely interesting or beautiful structures which occur in the sediments in which they lived or in which they were buried, but are rather the means whereby the process of evolution can be traced and the clues to the interpretation of past events are recorded cryptically in sedimentary rocks to form a diary of Earth history. Palaeontology has come a long way from the days of the medieval cabinet of curiosities. For example, it is possible, by means of the mass spectrometer, to measure the amounts of two oxygen isotopes in the shells of certain fossils. Isotopes have the same number so that they differ from one another in mass number. The proportion of light to heavy isotopes in an organic system depends on temperature; therefore it is possible, from the isotopic ratio, to determine the temperature of the water in which these animals lived.

The earth is about 4.6 billion years old. Life has been present on it for about 3.2 billion years. The science that seeks to understand all aspects of the succession of plants and animals over the great span of time is called *palaeontology*. It is a science based on the study of fossils, the remains or traces of ancient life. The work of *palaeontologists* is far more complicated (and exceedingly more interesting) than merely describing and cataloguing ancient creatures. They must determine how these animals and plants of long ago lived and how they grew. They seek out the ancestors and descendants of these life forms, attempt to know their true "fleshed-out" appearances, and try to determine the

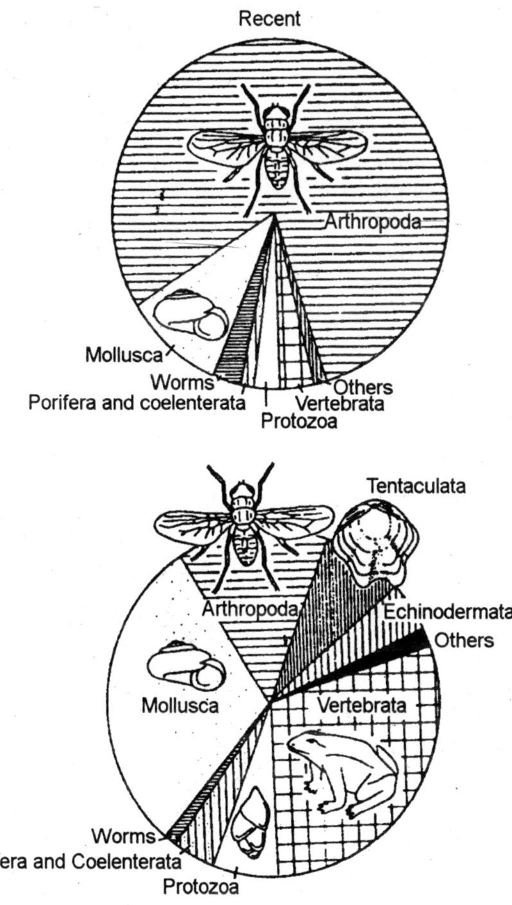

Fig. 5.2. The relative abundance of species of the different groups in the present-day fauna and a comparison with the number of described fossil genera.

extremes of temperature, salinity, or moisture they could tolerate. Paleontologists must determine how the past inhabitants of the earth fit into the total web of life, how they interacted with each other and the environment, and how they can now be utilized in deciphering the age of strata.

PRESERVATION

Types of Fossilization

When one considers the many ways by which organisms are completely destroyed after death, it is remarkable that fossils are as common as they are. Attack by scavengers and bacteria, chemical decay, and destruction by erosion and other geologic agencies make the odds against preservation very high. However, the chances of escaping complete destruction are vastly improved if the organism happens to have a mineralized skeleton and dies in a place where it can be quickly buried by sediment. Both of these conditions are met especially well in the sea, where shelled invertebrates flourish and are covered by the continuous rain of sedimentary particles. Although most fossils are found in marine sedimentary rocks, they also are found in terrestrial deposits left by streams and lakes. On occasion, animals and plants have been preserved after becoming immersed in tar or quicksand, trapped in ice or lava, or engulfed by rapid falls of volcanic ash.

The term "fossil" often implies *petrification*—literally, a transformation into stone. After the death of an organism, the soft tissue is ordinarily consumed by scavengers and bacteria. The empty shell may be left behind, and it is sufficiently durable and resistant to dissolution, it may remain basically unchanged for a long period of time. Indeed, unaltered shells of marine invertebrates are known from deposits over 100 million years old. In many marine creatures, however, the skeleton is composed of a kind of crystalline calcium carbonate called aragonite. Although aragonite has the same composition as the more familiar mineral known as calcite, its ions are packed differently, and it crystallizes in a different crystal system. Aragonite is also relatively unstable and in time changes by recrystallization to the more stable calcite.

Many other processes may alter the shell of a clam or snail and enhance its chances for preservation. Water containing dissolved silica, calcium carbonate, or iron may circulate through the enclosing sediment and be deposited in cavities once occupied by veins, canals, nerves, or tissues. In such cases, the original composition of the bone or shell remains, but the fossil is made harder and more durable. This addition of a chemically precipitated substance into pore spaces is termed *permineralization*.

Petrification may also involve a molecular exchange of the original substitution of mineral matter of a different composition. This process is termed *replacement*, because as each original molecule is removed, it is replaced by a molecule of silica or calcium carbonate, or less frequently by sulfides and oxides of iron. Replacement can be marvelously precise, so that even the growth lines on shells and structures of wood may be perfectly preserved.

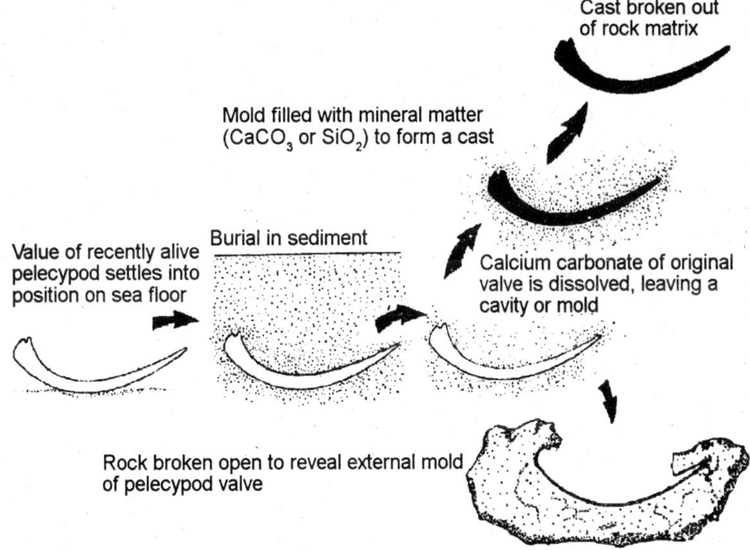

Fig. 5.3. Origin of fossil molds and casts.

Another type of fossilization, known as *carbonization*, occurs when soft tissues are preserved as thin films of carbon. Leaves and tissue of soft-bodied organisms such as jellyfish or worms may accumulate, become buried and compressed, and lose the volatile constituents. The carbon often remains behind as a blackened silhouette.

Fossils may also take the form of molds, imprints, and casts. Any organic structure may leave an impression of itself it if is pressed into a soft material and if that material is capable of retaining the imprint. Commonly, among shell-bearing in vertebrates, the shell is dissolved after burial and lithification, leaving a vacant *mold* bearing the surface features of the original shell. In many instances, molds are later filled, forming *casts* that faithfully show the original form of the shell.

Although it is certainly true that the possession of hard parts enhances the prospects of preservation, organisms having soft tissues and organs are also occasionally preserved. Insects have been found preserved in the hardened resins of coniferous trees. X-ray examination of thin slabs of rock sometimes reveals the ghostly outlines of tentacles, digestive tracts, and visual organs of a variety of marine creatures. Soft parts, such as skin, hair, and viscera of ice age mammoths have been preserved in frozen soil or have been "pickled" in the oozing tar of oil seeps. The delicate remains of the world's oldest unicellular organisms are found embedded in chert that hardened over them 3 billion years ago.

Evidence of ancient life does not consist solely of petrifications, molds, and casts. Sometimes the palaeontologist is able to obtain clues to an animal's appearance and how it lived by examining tracks, burrows, and borings. Such markings are called *trace fossils*, and the study of trace fossils is termed *ichnology*. The tracks of ancient vertebrate animals may indicate whether the animals that made them were bipedal (walked on two legs) or quadrupedal (walked on four legs), digitigrade (walked on toes) or plantigrade (walked on the flat of the foot), whether it had an elongate or a short body, if it was aquatic (with webbed toes) or possibly a flesh-eating predator with sharp claws).

Trace fossils of invertebrate animals are more frequently found than the tracks of vertebrates, and they are also useful indicators of the habits of ancient creatures. One can sometimes infer if the trace-making invertebrate was crawling, resting, grazing, feeding, or simply living within a relatively permanent dwelling. For example, crawling traces, as might be expected, are linear and show directed movement. Shallow depressions that more or less reflect the shape of the animal may be resting traces. Simple or U-shaped structure more or less perpendicular to bedding are often dwelling traces. Grazing traces occur along bedding planes and are characterized by a systematic meandering or concentric and parallel patterns that represent the animal's effort to cover the area containing food in an efficient manner. The three-dimensional counterparts of grazing traces are called feeding races. Feeding traces consist of systems of branched or unbranched burrows.

The Incomplete Record of Life

The record of life on earth is incomplete. If it were a finished record, if would have to include information on all past forms of life for every increment of time as well as documentation of their abundances and geographic distribution. Clearly, this goal is unattainable. Only a limited number of animals and plants have survived the hazards of preservation. Many that were preserved have never been exposed to our view by erosion or drilling. Others have simply not yet been discovered. The record is rather more complete for marine life with external skeletons—and less complete for land life lacking bone or shell. Because of erosion, there are many gaps in the record; these are not unexpected.

Figure 5.4, from an article by Joseph Barrell, illustrates the intermittent nature of the rock record and, hence, the contained fossil record as well. The figure depicts the accumulation of a sequence of strata during a period of general marine transgression (advance of the sea over the land) at a particular locality. As one might expect, the major transgressions have minor regressions superimposed upon them. These lesser comings and goings of the sea may result from a variety of factors, including the release or entrapment of glacial ice on continents, changes in shape of ocean basins, or movements of the earth's crust. Because the lands are exposed to erosion during the regression phases, there is a loss of previously deposited sediment or at least a lack of deposition. The rock record is built up only during times when the sea level rises again. Thus, the rock column shows on the side of the figure represents only a small fraction of time—namely, the parts of the time graph.

Of course, in nature there will be other factors, including those causing changes in the supply of sediment, that would affect the sedimentary sequence. The importance facts, however, are that the

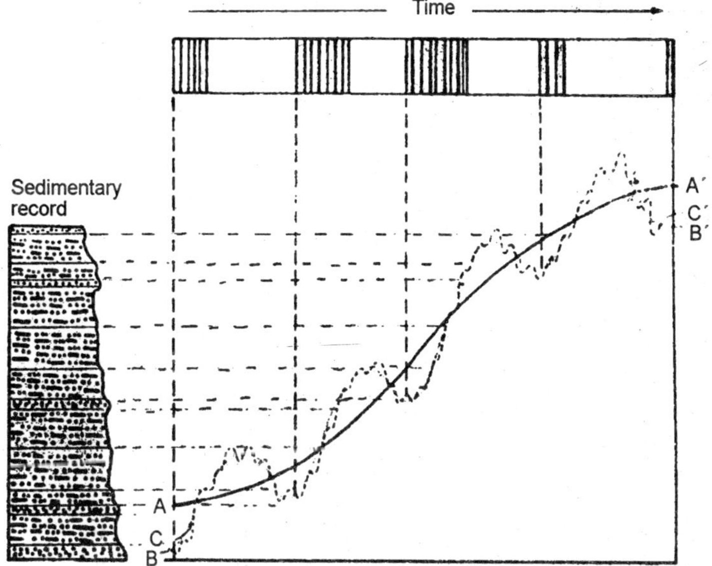

Fig. 5.4. Diagram showing the sedimentary record resulting from oscillations in sea level and the time value of the sedimentary record. Curve A-A′ represent a major trend of rising sea level with resulting marine transgression. The two smaller curves (B-B′ and C-C′) represent shorter frequency oscillation (minor transgressions and regressions).

rock record is incomplete and that where there are no rocks of a particular age there can be no fossils for that period. Even when the sedimentary sequence is relatively representative of a given time span, there may be numerous rock intervals that are barren of fossils. In spite of these difficulties, by piecing together the data from many localities around the world, paleontologists have managed to uncover a history of life that merits most serious attention.

Nature of Fossils

To many the term fossil implies a petrification—literally a turning to stone—and while in many instances the gradual addition to or replacement of the organic material by some mineral substance has occurred, that is not always the case, so the student has come to recognize several sorts of fossils and to group them under the following heads.

Actual Preservation Intact

In nature's cold storage ware house, notably in the arctic tundras of Siberia, frozen either in the paleocrystic ice or in the soil itself, are found animal remains with more or less of the original substance intact. How many such remains have been found is doubtless unrecorded, as they were in more than one instance swept away by the waters as the ice broke up or were devoured by dogs and wolves, possibly by the natives.

Petrification

As a rule, however, when more or less of the original material is preserved, it has undergone a certain mineralization to which the term *petrifaction* is applied. Naturally the degree of mineralization varies, but is usually greater the older the fossil in time. The parts thus preserved are almost invariably woody tissue or the hard parts of the animals's anatomy—bones, teeth, or shell. Petrifaction implies interstitial addition or an extremely gradual replacement, molecule for molecule, as the original substance is lost through disintegration. The resultant fossil retains, therefore, not only the external form but the histologic characters of the original structure as well. Next to the rare preservals by cold or amber, the petrifaction is the most valuable in recompense for careful study. The elaborate exposition of the structure of the fossil cycads by Wieland is based upon this type of preservation, for while in almost every instance in a very large series of specimens the trunk only is preserved, it has been possible by drilling out a cylindrical portion where the buds occur and by cutting it into thin microsections in several planes to reconstruct with entire accuracy the foliage and flowers of the plant.

While histometabasis usually occurs with plant tissues, the more perishable soft parts of the vertebrates are, however, occasionally preserved. In one striking instance described by Bash ford Dean the muscle fibers and kidney structure are in a state of admirable conservation in an ancient Devonian shark (*Cladoselache*) from the Cleveland shale of Ohio. Microsections of the muscle tissue magnified one thousand diameters show very clearly not only the clean-cut character of the individual muscle fibres, but the cross striation of the skeletal muscles, and in one or two places the delicate membranous sheath or sarcolemma by which the fibres were enclosed.

The replacing substances may be iron pyrites, iron oxide, sulphur, malachite, magnesite, silica, or carbon. Woody tissue, the limy shells of molluscs and the calcareous skeletons of corals and certain sponges are apt to be replaced by silica, giving in some instances a perfect replica of the original in form, though not in minute structure, in a totally different substance, thus forming what is technically called a *pseudomorph*. There is evidence that certain hexactinellid sponges the original substance of which was silica have been replaced by lime, the reverse of what may happen in the calcareous types.

As time goes on, the molecules of the mineralizing substance tend to rearrange themselves according to the laws of crystallography, so that first the minute structure becomes impaired, the external form modified and obscured, and ultimately after an inconceivable lapse of time all trace of the organic original may be lost.

Natural Moulds and Casts

Another group of fossils are the natural moulds in which neither the material nor the minute structure is preserved. These are formed by the hardening of the surrounding material in which the organism was buried, followed by the decay and subsequent removal of the organic material by percolating waters, thus leaving a cavity which retains the exact form of the original. In Pompeii, which may in a sense be considered a fossil city, at least two thousand people perished in the eruption of Vesuvius in 79 A.D. The city was covered to a depth of many feet by volcanic ash, finely divided rock which drifted in through the various openings of the houses and buried man and beast as well as the result of man's handiwork. At first when human remains were discovered, the bones were merely dug out and thus preserved; later it was found that if the cavity wherein they lay were filled with liquid plaster of paris the latter would, when hardened, give an admirable replica of the form and features of the victim. The remains of several people, men and women, European and Ethiopian, have been thus reproduced, together with a dog and the vanished doors and wooden portions of the household furniture.

Some of the fossil vertebrates of the Connecticut valley, which antedate those of Pompeii by millions of years, have lost all trace of the original bone through the percolation of dissolving waters. The impression, however, still remains, by means of which a fairly perfect restoration of much of the skeleton may be made.

Often these moulds are filled in with other material, so that a natural cast of the object is formed differing from the petrifaction in that it retains the form of the organism but not its structure. By this means such evanescent things as jellyfishes have been preserved. Again, interior cavities, as of shells of the brain chamber of a vertebrate skull, may be filled with subsequently hardening sediment, so that a perfect cast of long vanished soft parts is produced. This in the case of shells is often deceptive and has led to some confusion and duplication of names because of the striking dissimilarity of the outer and inner surfaces of the same shell. In the vertebrates, however, the brain replica is of the utmost importance, for form, size, and proportions of parts are all preserved with absolute fidelity. It is however, a cast of the dura mater or outer membrane of the brain, and while blood-vessels and nerve roots are often clearly indicated, the depth and complexity of the minor convolutions are not recorded.

Footprints and Trails

A certain group of phenomena should be considered in this connection—the footprints and traits of vertebrates and invertebrates with their attendant meteorological records of rain-prints, ripple-marks, and mud-cracks caused by the drying of surface mud after showers. The footprints are of double interest, for not only are they often times so well preserved as to enable the student to trace much of the entire animal, especially when to the impressions of the hind feet are added those of the hands and tail. Furthermore, the footprints bring before the observer more clearly than any other records of the past the individuality of the creature, for they are fossils of *living beings*, while all of the other relics are those of the dead.

Coprolites

A clue to feeding habits is generally gained from the study of the mouth armament in comparison to that of living allies, but conclusions are sometimes verified and supplemented by finding the fossil rejectamenta in association with bones or foot prints. To such relics the term *coprolite* is applied. Sometimes the unvoided intestinal contents preserves the outline and extent of the alimentary canal. This is distinctly seen in several of the salamandrine forms preserved in Carboniferous nodules from Mazon Creek, Illinois.

Conditions of Fossilization

Fossilization is the sum of the phenomena by which the remains of animals plants are preserved in the earth's crust. The two common but not prerequisites fossilization, are the possession of hard parts and immediate burial.

Possession of Hard Parts

Forms having hard parts are much more likely to become fossilized than are those lacking a skeleton. These hard body parts consists of bones, teeth, shell, chitin or the woody tissues of plants. However, even the most delicate organisms may also be preserved as impressions, carbon residues or petrifactions. It occurs when very favourable conditions are present.

Escape from Immediate Destruction

Out of the millions or more species of living animals, only a very small percentage will be preserved as fossils. The remains of most of these organisms are destroyed by the work of the atmosphere, mechanical forces such as wave action or crushing and biological agents such as predators, scavengers and bacterias. The possibility that any given animal will leave a fossil record is very small.

Immediate Burial in a Medium Capable of Retarding Decomposition

It is the first prerequisite for fossilization and it should be such that no oxidation of the material should take place. The burial is done by water-carried sediment, deposited in the seas and oceans, especially in the shallower regions and at the river deltas.

The burial is also done by the lava, volcanic ash and dust which has yielded fossils of land-living beings. Land animals living in very dry regions may be preserved by *desication* (extreme drying). Such remains are usually called *natural mummies*. Many extinct insects and spiders were trapped in the sticky antiseptic resin exuding from coniferous trees that once occupied the Baltic region The ancient forests are now burried beneath the Baltic sea, but the amber (resin) is cast upon the beach by the waves.

Miring (Plunging into Mud)

It is also a notable mode of preservation. Death and immediate burial take place by mining in bogs (swamps) and quick sand. Its examples are the remains of ancient *mastodons* found in New York and the adjacent states, the remains of the great *Irish deer*, found in the peat bogs of Ireland.

Larger animals have been trapped in tarpits and oil seeps. The most remarkable death traps in the world are found in Rancho La Brea in Los Angeles, California, where remains of the largest mammals have been found, such as *Elephants*, *Mastodons* and *Paramylodons*. Here an ancient oil seep, through loss of its volatile materials has been changed into a pool of asphalt. The above prehistoric animals were trapped in the asphalt. Their accumulated bones have been removed from the pit in an excellent state of preservation.

Ice or Frozen Soil

Animals having in periods of gland hold were frozen and incorporated into ice or frozen ground. Their remains are found in a very fine state of preservation that even the flesh, hair and contents of the stomach have been perfectly preserved.

Subsequent Vicissitudes (Changes)

After all the conditions for fossilization have been fulfilled, the resultant fossil is subjected to various vicissitudes (changes). As time goes on , the changes take place due to pressure, elevation, folding and subsequent erosion of the strata and due to the show circulation of acidulated and other waters through the rock, either from above or below. The acidulated water dissolves away the shell or bone and the mould is formed or the fossil becomes mineralized (petrifactions).

SIGNIFICANCE OF FOSSILS

Geographical History of the Earth and Distribution of Plants and Animals

Fossils indicate the geographic history of the earth, i.e., the distribution of seas and shores and the mountains or the continental areas of the past. They also tell, though incomplete, about the animals and plants which inhabited those continents and seas from the beginning of the life on earth over a billion years ago to the present.

Index Fossils

Fossils indicate the age of the rock in which they are found. It is relic of the life which inhabited the earth when that sediment, which now forms the enclosing rock, was being deposited. It was early observed that succeeding rocks contain different fossils when we follow from the lower (older) to the higher (more recent) beds.

Ancient Geography

Fossils tells us the presence of land barriers which are now not present at that place, of land bridges which formerly united lands but now they are long separated. For example, *Trinucleus* (a species of trilobite) is found in Ordovician deposits of both Europe and America, which shows that there must have been a continuous beach line, a land bridge between the two continents at that time. Presence of the same species of elephant in the Pleistocene strata of both Europe and America indicates a land bridge at that time. Fossils provide much information as to the type of environment in which these prehistoric organisms lived.

Fossils as Climatic Indicators

Fossils plants also indicate the variation of temperature and degree of moisture and it is also supported by the fossil animals. If we find the remains of tropical plants or animals in a region that has a temperate or cold climate today, we may assume that a tropical climate prevailed in that area at one time. For example, fossil ferns found in Antarctica and the fossil magnolias reported from Greenland indicate a much warmer climate for these areas in that times. The increasing aridity in the West during Tertiary shows no remains of the flora. It is also proved by the rapid diminution in the number and kind of browsing animals and increase of grazing forms after the beginning of Miocene.

Fossils as Evidence or Organic Evolution

Palaeontology is the chief contribution for evolution. It indicates definite lines of ancestry for living plants and animals.

Palaenotology is the science of the past life of the earth. It aids the botanist and zoologist in the gaining knowledge of the ancestors of the present day plants and animals. It aids the geologist in determining the age and origin of the rocks in which the fossil remains occur.

One of the best known records of evolutionary change is found in the series of fossil horses from Coenozoic rocks. This record begins with the small, primitive, four-toed *Hyracotheriun* (*Eohippus*) of early Eocene time and continues for some sixty millions years through the Oligocene, Miocene, Pliocene and Pleistocene, and upto the present time. This series ends in our modern horse, *Equus*.

Fossils as Economic Tools

Since many of our important resources are associated with sedimentary rocks, fossils, when present, may be of help in locating ores, coal, oil and gas.

Fossiliferous rocks may also provide evidence as the where these ore-bearing rocks may be found. Coal deposits, which are commonly associated with fossil plants, have been found in much the same way. Valuable deposits of radioactive minerals have been discovered in sedimentary rocks. In some of these rock strata are found the ore concentrations which are closely associated with fossils. Uranium was found in some specimens of fossil wood and bones of certain fossil reptiles and mammals.

Some fossiliferous limestones and sandstones are useful as building stones. The Trigonia stone of the Lower Cretaceous of Texas is an example of such stone. It contains a large number of casts and moulds of marine clams and snails, the most abundant of which is *Trigonia*, a calm.

Geological formations also contain accumulations of oil and gas and for finding out the oil-bearing rocks, knowledge of fossils is basic.

Imperfection of the Geological Record

The vast majority of animal and plant remains could not become preserved as fossils due to their partial or complete disintegration. Such organisms will remain foreover unknown to man. Most of the animals of the remote past had no hard parts to be preserved as fossils. For example, most of the Protozoa, though some of them (foraminiferans) had calcareous shells. These shells in the aggregate have produced thick deposits of lime stone. Most of the coelenterates, except corals, and worms had also failed to leave fossil remains. Skeletal calcareous supports of corals have also produced thick deposits of limestone. Besides these, probably all the animals possessing hard parts could not be fossilized.

The chances of fossilization vary greatly. Animals living in the ocean have the best chance of being preserved as fossils. Animals inhabiting fresh-water have the next best chance, while the terrestrial animals have the least chance. In order to be preserved as a fossil the body must be prevented from complete destruction. Not only the softer portions of the body but also the skeleton will disintegrate in a few years if exposed to action of predatory animals, scavangers, insects larvae, bacteria and the weather. A dry climate favours preservation of bones, but even dry bones disintegrate in time. Hence it is necessary that bones be protected by being covered. Wind-blown soil, such as that which produced thick deposits of clay known as loess, may provide the protective covering. Or if the animal becomes mired in a bog or in quicksand bones may gradually sink and be covered. Fossils in the Rancho La Brea asphalt pits are a special case of this procedure. Or if bones happen to lie in the flood plain of a river they may be covered by a deposit of soil left by the river when it overflows it bank in time of flood. Or the river, in flood stage may sweep bones into a lake or into sea, where they will be mingled with remains of aquatic animals. Fossils of Tertiary mammals of White River Bad Lands of South Dakota and Nebraska were preserved by this procedure.

Deposits which remain deeply buried under younger strata of metamorphic rocks of under sea will remain largely unknown. Some exception to the above statement is afforded by mines, in the walls of which fossils may be found.

—

6

PROCESS OF FOSSIL FORMATION

Fossils owe their existence to a break in the natural cycle. Normally all of the material used in living organisms is recycled, but fossilisation represents a transfer of material from the biosphere to the lithosphere. In the long term, of course, weathering of rocks may return material to the biosphere. The three stages of the process of fossilisation are mortality, biostratinomy, and diagenesis. The whole field is sometimes called taphonomy.

MORTALITY

Here we are concerned with the cause and consequences of death. In nature very few organisms die simply of old age. Far more die prematurely as a result of suffocation, thirst, inadequate light, severe temperature or pressure changes, poisoning, illness, parasites, injury, or predation. The cause of death can only rarely be established in fossils Examples include encasement in resin or amber, drowning in asphalt or pitch and predation, when the material is preserved in coprolites. In other cases death may have resulted from a number of different causes, for example, an organisms buried by sediment might die of asphyxiation, hunger, or pressure of overburden.

An abnormally high number of fossils is often attributed to mass death. At present such events occur where water is polluted by phytoplankton toxins or becomes chilled or oxygen deficient. Care should be taken, however, when attempting to apply such modern observations to fossil occurrences.

Traces left by death-throes are very scarce but we can sometimes observe the attitude of death and the effects of rigor mortis where the

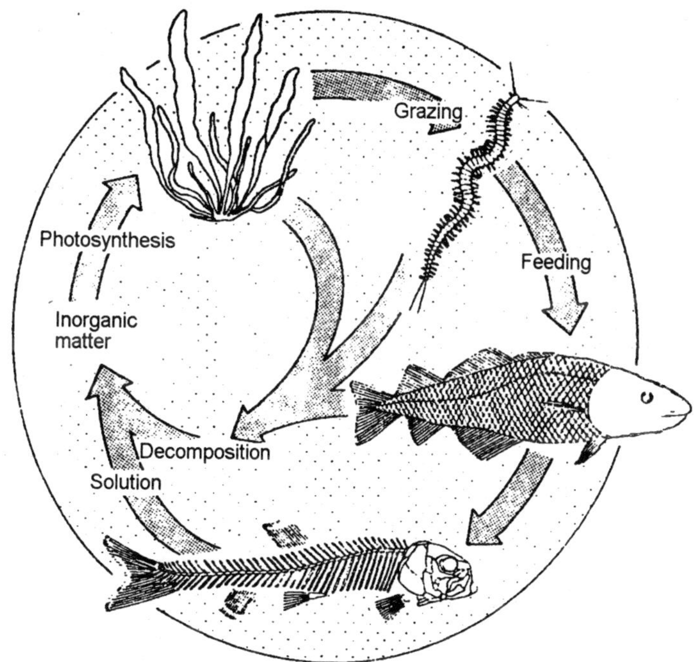

Fig. 6.1. The natural cycle. Decomposition of organic matter replaces the stock of inorganic materials.

contraction of ligaments and tendons may cause the head to be thrown back in the vertebrates.

Biostratinomy

Biostratinomy is concerned with the rate of an organism's body from the time of death until it is finally buried by sediment.

Soft Tissues

Among the soft tissue the most important organic components are proteins, fats, and carbohydrates. These begin to break down both chemically and bacterially immediately after death. The cause of this decomposition is controlled by environmental factors. In the presence of water and oxygen and organic substances break down into the simplest inorganic components such as CO_2 and H_2O under the attack of aerobic bacteria. Where oxygen is absent however, in stagnent environments, anaerobic bacteria break down the original molecules using some for their metabolism and converting the remainder into hydrocarbon mixtures of high molecular weight. This process may produce bituminous muds whose organic matter may subsequently convert to bitumens, oil,

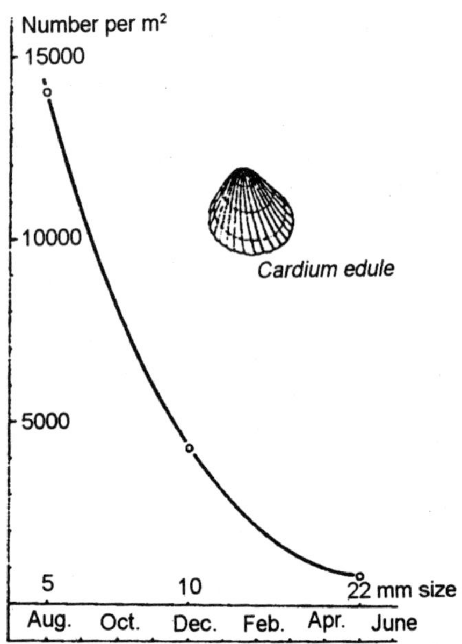

Fig. 6.2. Mortality of juvenile stages as shown by bivalve Cardium edule.

and natural gas, under the influence of higher temperatures and pressures. Stagnation also permits the preservation of organic matter. True preservation of soft tissue may occur in the absence of water in arid regions or where hygroscopic materials are present. It also results from freezing or incorporation of the body in resin, asphalt, syngenetic silica concretions or other media containing neither water nor oxygen. Often during slow decomposition of soft parts, the infiltration of mineral substances or relatively stable organic materials pseudomorphs the original structure.

Skeletal Material

Relatively few materials are available to the organic world war the formulation of protective and supporting hard parts. The most readily available are calcium carbonate (calcite and aragonite), calcium phosphate, opal, chitin and cellulose. Agglutinating organism produce hard parts by cementing together foreign particles with an organic or inorganic cement.

Calcite is the preferred skeletal material of the archaeocyathids, octocorals, bryozoa, brachiopods, some annelids, polychaetes, many bivalves, pectinids, limids, ostreids, some cephalopod structures several crustacea, echinoderms and red algae. Calcite may include isomorphous

$MgCO_3$ whose abundance increases with temperature. Precipitation of calcite is facilitated by war, shallow water but by no means confined to that environment. Its solubility in pure water or in alkaline solutions is small (14 mg/l in pure H_2O), but in CO_2-saturated water its solubility increases to 1 gm/l; in acid media calcite is not preserved. In the modern oceans solution of calcite is particularly widespread in the cold depths of the oceans. In sediments, calcite is often leached from permeable strata but more readily preserved in fine grained and impermeable rocks.

Aragonite occurs in the hydrozoa, hexacorals, gastropods, most bivalve and cephalopod shells and the green algae. It is metastable in shallow marine environments but in fresh water it gradually alters to calcite at normal temperature and pressure. Its solubility is some 10% higher than that of calcite.

Calcium phosphate ($Ca_5(PO_4)_3$ (OH) is important to the vertebrates (bone material) and is also found in some brachiopods and arthropods (trilobites, crustacea). It is almost insoluble in neutral solutions (0.01 gm/l) and is well preserved in weakly acid or alkaline environments. In more strongly acid environments or where it remains in contact with acid for a long period calcium phosphate dissolves leaving bones fragile and flexible.

Opal forms the test of radiolaria, diatoms, siliceous sponges, and many flagellates. It is weakly soluble in water (river water contains about 10 mg/l of S_iO_2 in temperate regions and 30 mg/l in the tropics), stable in acids other than HF and slowly soluble in alkalis.

Chitin, a nitrificated polysaccharide occurs in algae, fungi, lichens, cnidarians, priapulids, bivalves, annelids, onychophores, pentastomids, arthropods and tentaculates. It forms the skeleton or arthropods (crustaceans, tribolites, insects, etc.), graptolites, and some brachiopods. It is subject to decomposition like other organic substances but is more stable that most. Other organic substances that are involved in the building of skeletal material include proteins spongin (in some sponges) and keratin (horn, hair, scales, feathers, and claws of vertebrates).

Cellulose forms the cell wall of plants and thereby gives them a certain rigidity. It also appears in tunicates. In the presence of air cellulose breaks down to CO_2 and H_2O. Lignin (a benzol derivative) is an important structural constituent in the wood of pteridophytes and spermatophytes.

Decomposition of the soft tissue affects the preservation of the skeletal material since the gases evolved (CO_2, H_2S, NH_3 etc.) vary depending on environmental conditions and later the pH by producing acids (H_2CO_3, H_2SO_4) or bases (NH_4OH) which may corrode or even dissolve calcium carbonate, calcium phosphate, or opal.

Disintegration of Skeletal Material

Articulated skeletons whose components are held together by organic tissue tend to disintegrate into their individual parts following death of the organism, the pattern depending on the type of skeleton.

Invertebrates

In bivalves, post-mortal relaxation of the adductor muscles allow the valves to spring apart. The ligament decomposes more slowly than the rest of the soft tissue but when it has done so the valves separate. The remainder of the fossilisation process may precede in various ways. Many ammonites posses an aptychus that was probably a mandible aperture. If the organism is in an upright position during decomposition of the soft parts, the structure tends to slip down into the bottom of the body chamber where it is often preserved. When the aptychus consists of two plates these separate along the symphysis only at a late stage. Like most arthropods, the trilobites had a segmented carapace which disintegrated after death. However, since the resulting fragments are identical to those produced by sloughing of the carapace during growth of the trilobite it is difficult to distinguish between the two. In many echinoderm species the test comprises individual elements that are rarely fused together. They are instead held together mainly by ligaments and cartilagenous material. Disintegration begins where this material is most readily accessible, that is, in distal parts of the body such as the tips of the arms.

Fish

Whether a dead fish floats to the surface or sinks to the bottom depends on the amount of gas in its swim bladder at the time of death. Where a fish sinks onto an oxygen-deficient bed, its skeleton may remain intact as it is entombed in sediment. The same thing may happen to fish stranded on the beach. Gas produced by putrefaction accumulates mainly in the central cavity of the body, the corpse therefore floats belly up with head and tail dangling. The belly later ruptures and as the gas gradually escapes the fish sinks, lands on its head or tail and then usually rolls over onto its side.

Fish with a weak axial skeleton may begin to disintegrate while still afloat. In the ganoids with their overlapping armoured scales the otoliths are dropped first from the fishes ears. The scale-armour then

Fig. 6.3. Post-mortal decay of fish. A–floating and sinking of the corps; B and C–stages in the breakdown of a floating corpse (B only with heavy scale-armour).

separates from the skull and backbone which also fall apart fairly quickly. Disintegration of the skull begins with solution of the lower jaw. The scale-armour may remain intact for along period - in Recent ganoids often for months — but it then starts to corrode around the periphery and finally dissolves completely. In modern fish (without a scale-armour) disintegration again begins with the otoliths. They also soon lose their scales. Once the abdominal wall ruptures, the link between head and body weakens and they may separate completely. As in the ganoids, decomposition of the skull begins with the lower jaw whose components separate along their symphyses. The jaw joints come apart later as do the upper elements of the skull. The linkage of the tail-fin to the backbone also loosens or fails and finally the individual ribs and vertebrae fall apart. When a fish starts to decompose during drift, large sections of the corpse (head, torso, tail) tend to sink to the bottom where they disintegrate further. Since most parts of a fishes skeleton decompose unusually easily, it is often only the otoliths, teeth, scales, and spines that survive.

Reptiles and mammals

Whereas fish posses a weak integument that can hold the corpse together for only a short time, the marine reptiles and mammals have a much stronger one. Moreover they have a large abdominal cavity and the considerable evolution of gas that results from decomposition of the soft tissue can keep them floating for a long time. While it is drifting the *corpse* loses individual skeletal elements in a regular order. These fragments therefore sink to the bottom in different places. Disintegration begins with the lower jaw, the epidermis and the tip of the tail. The coherence of the skeleton is then gradually lost in the abdominal area, the skull falls away and finally the spinal column and rib-cage disintegrate into individual bones and vertebrae. Complete skeleton are preserved in some littoral or oxygen-deficient environments. Where the rate of disintegration of the soft tissue is greatly retarded, the skeleton may be preserved completely but in a rather tangled jumble. In completely anaerobic environments the skeleton remains intact during burial. Terrestrial reptiles and mammals decompose on the same sequence and if they are covered by fluvial or lacustrine sediments the results are similar to those of marine environments. It is only if they die on dry land that aridity and other subaerial processes may produce a different pattern of decay.

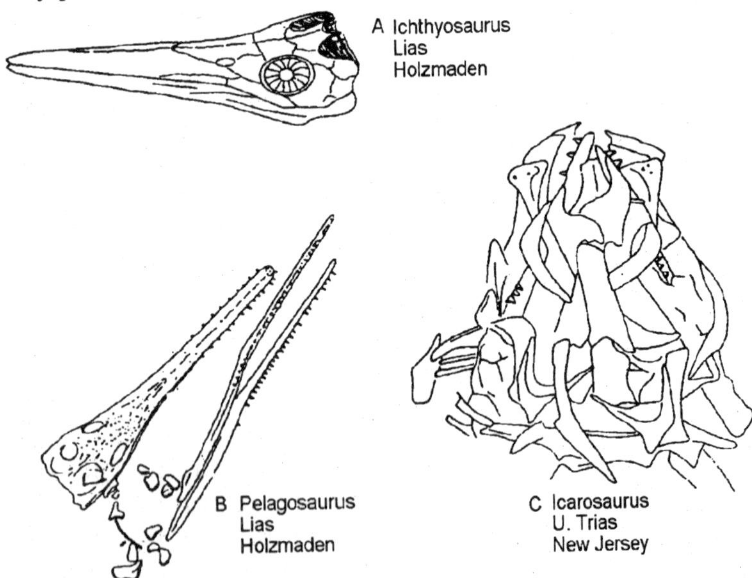

Fig. 6.4. Stages in the breakdown of reptile skulls.

Birds

Because their feathers trap plenty of air and because their bones have air spaces, dead birds float on water for a long time. During this period, their skeletal elements are scattered over the sea floor. The down goes first, then the foot bones, then the head separates from the body although it may remain linked for a time by means of the tough trachea. Later, the pinions and tail feathers become detached from the decomposing periostreum and finally the torso disintegrates. From observations of modern gulls it would appear that complete skeletons are preserved only on land or on the beach, but whether these findings apply to all other Recent and fossil birds remains uncertain.

Burial

The chances of *in situ* preservation depends on mode of life and whereabouts at death. Many benthonic organisms live under the sediment-water interface and are readily preserved in their growth position. Others living on the surface appear simply to have fallen over. In most cases, however, the organism did not live where we find it. A phase of transport intervened between death and burial, and during this period the organic remains were damaged, current-sorted and aligned.

When organic remains undergo transport either alone or with sediment they may be damaged or destroyed. Shattering is the dominant process affecting large particles in gravel deposits whereas abrasion predominates in sandy sediments with a grain size between about 0.05 and 0.5 mm. Abrasion is produced by other small shell fragments as well as by the sand grains themselves. The silt and clay fractions have little mechanical effect. Where the fossil is being transported with a mass of shifting sand it is equally abraded on all sides whereas facetting is likely to result if the shell is lying on the surface of the sediment or anchored to a bed over which a uni-directional current is flowing. Bowl-shaped objects develop first a round and then a horseshoe shaped notch whose opening faces up-current. If the shell changes position or the current direction fluctuates several facets are developed. The results are similar to those seen in ventifacts. In areas of stronger currents or tidal flow where the fossils tip over and roll, all the projecting parts of the shell are abraded to produce shapes similar to those of shells abraded within moving sediment. Circular and horseshoe shaped pits are again developed. Glide facets are produced when the shell slides along an abrasive bed and has its base eroded.

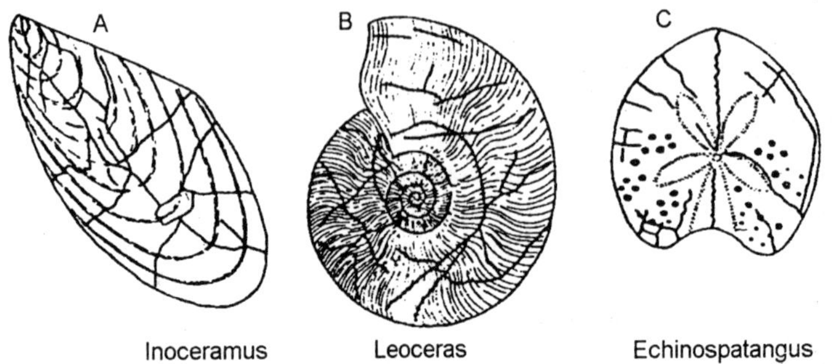

Fig. 6.5. Fracture patterns caused by mechanical stress and overburden pressure.

Shattering and dislocation of skeletal material results not only from purely mechanical processes but also from organic activity – boring micro-organisms such as algae and fungi, bivalves, sponges, and other animals all weaken the skeletal structure. Scavenging also plays a considerable part in biogenic disintegration.

The fate of bodies that lack any inherent buoyancy depends greatly on the ease with which they can be carried in suspension. Suspendibility can be calculated from the formula S -= 0/4G (where O = surface area in mm^2 and G = weight in gms). The higher the value of S, the more easily is the material transported. Organic debris which is easily suspended is carried largely in the turbulent zone above the bed whereas less easily suspended material is carried in the bed load. These two different modes of transport produce different types of wear. Buoyant bodies experience quite different conditions. They float until they are either washed shore or lose whatever feature (gas, air chambers, fat droplets, etc.) confers that bouyancy.

During transport, organic debris is sorted according to its size, weight, shape, and resistance to abrasion. The strongest skeletal elements survive longest. The strongest skeletal elements survive longest. The greater the range of suspendibility the more widespread the area over which the material is deposited. From their sorting curve we may determine whether the remains have suffered long or short transport.

The suspendibility of fossil material can be greatly affected by the presence of projections that act as anchors. Flat, disc-shaped bodies such as some gastropod opercula lie tight against the bed. They present only a small surface area to the current and may therefore remain unmoved despite their high suspendibility while other larger objects are carried away. If the area dries out and is then recovered by water, such objects may be held up by surface tension and transported further than other material.

Current sorting should not mistaken for other kinds of selective, post-mortal processes. All juvenile forms of a species may be killed off while the adults resist the attack and survive. The resulting fossil population would then only appear to have undergone size sorting. Selective diagenetic processes such as the solution of aragonitic tests and preservation of calcitic material may also be confused with current sorting.

Flow conditions in any medium (water, air, ice, sediment) affect objects within or at the surface of that medium. They may result in orientation of shell material. Movement about a horizontal axis may produce over turning or imbrication while rotation about a vertical axis produces alignment.

Moveable objects within a fluid are readily overturned when they rise or fall under the influence of gravity, buoyancy, and drag. Curved objects turn convex side down. This happens, for example, when individual valves of bivalves settle through still water, or when they are reworked from bottom sediment then allowed to settle again. When conical objects such as gastropods are free to move within a bottom mud or other sediment whose consistency enables objects to be held in any position, they tend to come to rest point down where buoyant forces predominate but right way up where gravitational forces predominate, provided that a reasonable amount of vertical motion can take place.

Where the body being reoriented lies at the boundary between two media, its freedom of movement is partly inhibited by one of the media; the sediment in the case of a sediment-water interface, the ground in the case of a sediment-water interface, the ground in the case of a subaerial land surface. The restrictions imposed have considerable biostratonomic and palaeogeographic significance, because shell or other organic detritus tend to become overturned into the attitude where they offer least resistance to the current flow. This is particularly apparent in the case of cup-shaped shells. Here the rule is that in subaqueous environments with a sufficient current flow velocity, cup-shaped objects will normally come to rest and be buried convex side up. This fact affords an important indicator of the way up of a sedimentary sequence.

Exceptions to the rule are common when shells are able to sink into the sediment or are so closely packed that they mutually interfere.

In the former case the shells often come to rest concave side up, while the latter situation produces random orientation. Irregularities in shape of the shell may also cause inversion or random orientation.

Where alignment occurs within one medium it may be described as 'free' whereas at the interface between two media it is termed 'limited'. It is best seen in elongate bodies, and most information is available when such bodies have differently shaped end, for example, if they are elongate cones. When bodies of this shape are lying on a substrate over which a uni-directional current is flowing they roll into a position where they are offering least resistance to the flow. This position occurs when the long axis of the body is parallel to the current and its pointed end is usually directed downstream although with open-ended shell such as turriform gastropods the pointed end may face up current. When such bodies are being rolled around by an oscillating current, for example, waves they lie with their long axis perpendicular to the flow.

Spindle-shaped or cylindrical objects roll with their long axes perpendicular to the current. Preferred orientation of the most easily aligned shells requires current velocities in excess of 15 cm/s while a flow rate of over 40 cm/s may be needed in the case of large, heavy, or strongly ornamented shells. When the object has a projection that can serve as an anchor, the body pivots about it until the pivot point is directed up-current. A shell without such a projection can be pivoted in the same way if it is partly projecting out of the water or lying on a cohesive substrate.

Preferred orientation of elongate bodies can be analysed by means of a current rose. Symmetrical, bimodal distributions (propellor-shaped) suggest orientation perpendicular to the flow. If the shells are conical we can then deduce that the current was an oscillating one. Such a deduction cannot however be made in the case of cylindrical objects. Uni-modal distributions indicate orientation parallel to a uni-directional current with, in the case of pivotable objects, the pivot end up-current. Non-anchored cones may, however, lie with their points upstream.

In addition to these mechanisms, we also find preferred orientation of biogenetic origin caused by the living organism aligning itself in the current in such a way that will allow its various organs to function most efficiently.

'Sole marks' are imprints on the sediment surface produced physically by the movement of pebbles, or, less commonly, by current-drifted hard parts of dead organisms.

Fossil Communities

A 'biocoenosis' is a community of living organisms. It cannot possibly remain unchanged in a fossil state. An assemblage of dead organisms that lived together and have been preserved in situ is known as a 'taphocoenosis' is a fossil assemblage whose members are derived from different habitats. It may comprise either a mixture of autochthonous and allochthonous individuals or a purely allochthonous assemblage. Most fossil communities are taphocoenoses.

Fossils assemblages are always preserved in sediment whose structures and texture enable us to distinguish various biofacies types. In the vital-astratal biofacies, a single biocoenosis flourished there for a long period. It forms massive unbedded sediments, for example, reefs. The vital-pantostratal biofacies is characterised by the presence of an autochthonous thanatocoenosis in well bedded sediment. Remains of animals and plants foreign to the community may also be present. The vital-lipostratal biofacies displays many indications of reworking and sedimentary breaks. In it the remains of thanatocoenoses are

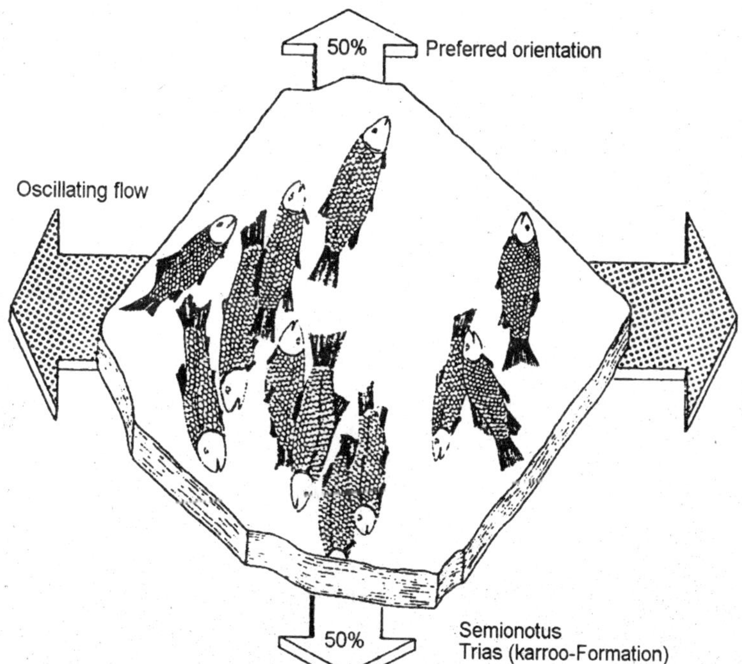

Fig. 6.6. Alignment of fish corpses with their long axes perpendicular to the flow directions of an oscillating current.

supplemented by allochthonous taphocoenoses. The same sedimentary characteristics reappear in the lethal-lipostratal biofacies whose fossil assemblage is an allochthonous taphocoenois without any autochthonous organisms. Autochthonous life forms are again missing from the lethal-pantostratal biofacies whose assemblages again consists entirely of allochthonous forms. In this case, however, the bedding is undisturbed as the sediment has not been reworked.

Thanatocoenoses and taphocoenoses turn up in a number of frequently recurring guises the most important of which are shell banks, pavements, sorted pavements and bonebeds. Shell banks (lumachelles) are accumulations of skeletal remains (mostly shells) whose state of preservation range from intact through broken to shattered. Most shell banks are allochthonous taphocoenoses. They occur widely on shorelines and beaches, in estuaries, ahead of river deltas, on submarine ridges and on submarine dunes. Shells can also be piled up on the surface of reworked sediment as a result of the activity of burrowing organisms. In this case the assemblage is autochthonous. Shell pavements are bedding planes enriched in shell material. The density of shell cover may vary from sparse to almost complete. Pavements form in the surf zone either as allochthonous taphocoenoses or by winnowing of the autochthonous material. Most of the shells always lie convex side up: *Lesedecken* are pavements on which the shells have undergone size- and shape-sorting as a result of current activity. The long axes of the shells are commonly oriented parallel to the flow, for example, in belemite 'graveyards'. Bonebeds are tabular enrichments of bones and

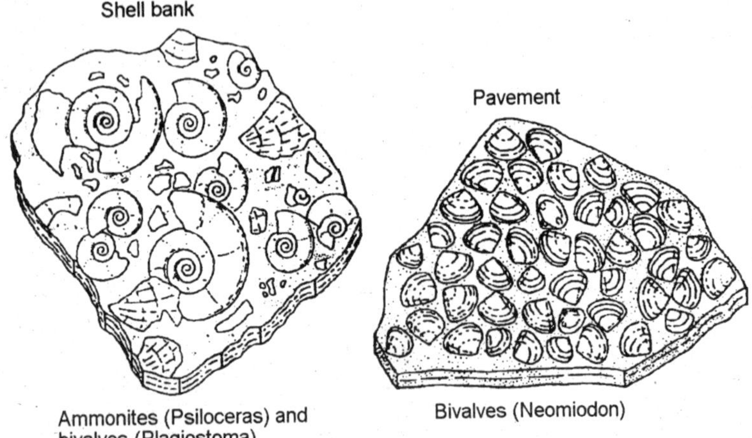

Fig. 6.7. Shell bank and pavement.

teeth. Their components are always allochthonous and have often suffered considerable abrasion. Apart from the fissure fillings in karstified areas which also contain allochthonous taphocoenoses, bone beds are the most widespread sites of rich vertebrate assemblages. The famous localities in the lethal-pantostratal biofacies, for example, Mansfeld, Monte San Georgio, Holzmaden, and Solenhofen are exceptional cases.

Concentration of fossil material on a considerable scale can arise as a result of condensed sequences that form when sedimentation ceases for a time or when sediment is winnowed. The organic remains continue to accumulate at their usual rate during these periods giving enriched sequences that can be recognised by the differing stratigraphic age of the various constituents.

Concentrations may also arise by selective sorting (for example, placer deposits or bone beds) or by being washed into a cavity. Unusually well-preserved material occurs where decomposition of soft tissue is prevented or greatly retarded. Articulated skeletons whose elements are held together by soft tissue may then survive intact. Such conservation may occur in stagnant oxygen-deficient environments, for example, in bituminous shales, or in preserving media such as resin or peat, or in early diagenetic concretions, or in rapidly deposited sediments.

Diagenesis of Fossils

Diagenesis of fossils covers the fate of organic remains after they are buried in sediment. It is controlled largely by conditions of sedimentation and by the petrography of the sediment.

Preservation of Material

In the simplest case, the chemistry and structure of the organic remains does not change. Such fossil remains are fairly common in Recent sediments but with increasing age of the material the likelihood of some kind of charge increases. Original material is rarely preserved in Palaeozoic rocks. A sediment in which the circulation of pore water is reduced to a minimum is the main pre-requisite for preservation of such material.

Solution

Organic remains may be dissolved away by circulating ground water especially in course clastic sediments. Solution is assisted by acids and bases in that water. Aragonite is usually the first mineral to be destroyed; calcitic, phosphatic and siliceous materials are most resistant. Thus we get a selective destruction of fossils depending on the mineralogy of their skeleton.

Recrystallisation

Recrystallisation without change in the chemical composition of skeletal material can arise in a number of ways. Metastable phases of a polymorphic substance are transformed into stable phases in the course of time, for example, aragonite transforms to calcite. The shell structure may survive unaltered during this inversion although the finer details are usually lost. Grain growth involves growth of the largely crystals in the skeleton at the expense of the smaller ones and thus produces a completely new texture. The diagenesis of echinoid tests provides a fine example of this process. Here, the individual plates which in life are permeated by a network of cavities are each transformed after death into a single crystal of calcite with almost complete loss of the original structure. Occasionally, crystal growth does not stop at the shell margin but extend out into the sediment. Diagenesis of skeletal opal produces fine-grained quartz with the loss of water of crystallisation. Grain growth may or may not involve mineralogical transformations.

Micritisation involves the early diagenetic breakdown of skeletal material into a cryptocrystalline and usually structureless aggregate. Algae are usually responsible and aragonite is the mineral most affected.

In many cases, the original skeletal material undergoes chemical change (metasomatism). Silification, calcification, and pyritisation are particularly common. Silicification, which involves the removal of the original shell material and its replacement by silica, often begins at particular points then spreads outward to produce concentric rings (beekite structure). It may occur during early or late diagenesis. Calcification affects mainly siliceous tests incorporated in alkaline, calcareous sediments. While pyritisation (involving the replacement of skeletal material by the iron sulphides pyrite and marcasite) is found particularly in the clastic sediments deposited in reducing environments with stagnant pore water.

Diagenesis of plant matter such as lignin or cellulose can go in a number of directions, but, in the absence of free oxygen, coalification involving a loss of O, N and H and a relative increase in O occurs. Methane, water, and carbon dioxide are released

$$4C_6H_{10}O_5 \rightarrow 2\ C_9H_6O + 2\ CH_4 + 4\ CO_2 + 10\ H_2O$$

Under higher temperature and pressure the reaction proceeds towards the production of almost pure carbon (anthracite and graphite).

Impregnation

Minerals are often precipitated in the pore spaces of skeletal material and thereby impregnate them. The minerals involved naturally depend on the chemistry of the ground water. Calcite, silica, and barytes are common.

Encrustation

Skeletal material may become surrounded by crusts, especially of calcite, produced either by algal activity by chemical precipitation from supersaturated water (of ooliths). Such material is sometimes incorrectly as 'mummified'.

Internal Moulds

An internal mould (sometimes called a cast) is replica of the inner surface of shells such as foraminifera, brachiopods, bivalves, gastropods, cephalopods, and echinoids. The material involved is usually derived from the surrounding sediment, although in the case of tightly closed shells that prevent the ingress of sediment, a crystalline fill precipitated from aqueous solution may be present. When the shell is only partly filled with sediment with the remainder either empty or containing crystalline precipitates the geopetal texture indicates which way up it should be like a 'fossil spirit level'. When no deposition occurs within tightly shut shells, and when the shell material itself is later dissolved after lithification, hollow moulds are produced. These may subsequently have a skin of crystals developed on their walls in the position of the original shell.

Internal moulds consisting of pyrite or marcasite often originate in anaerobic clays. They form in the same way as concretions and often infill only the most inaccessible parts of shell, for example, the innermost chambers of cephalopods or the tip of gastropod shells and so produce an illusion of stunted growth.

In thin-shelled organisms the inner and outer surfaces usually have similar shapes whereas in thick-shelled animals they are often very different. If such a shell subsequently dissolves in still unconsolidated sediment the imprint of the external surface of the shell may be compaction become impressed on the internal cast to provide a 'sculptured cast'.

In ammonites, sediment can only enter the body chamber because the phragmocone chambers are hermetically sealed by septae. For this reason these chambers are often preserved empty or filled with later crystalline calcite. Since many other ammonite shells do, however,

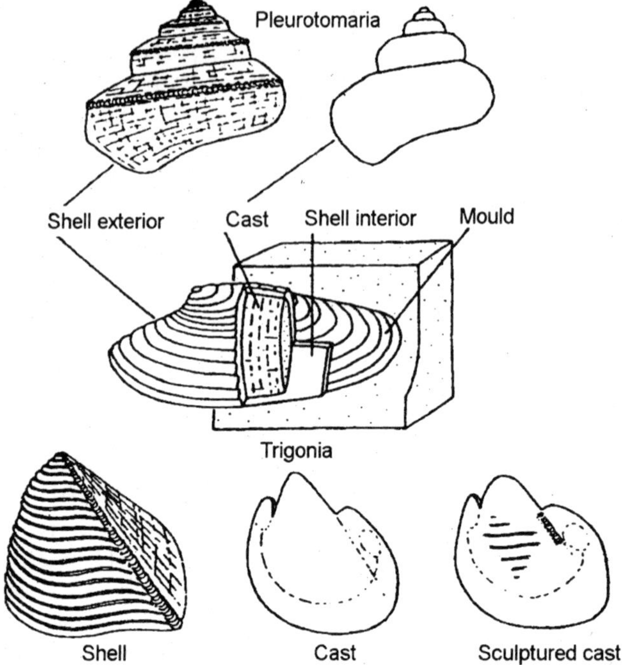

Fig. 6.8. The relationship of casts and moulds to shell and the difference in ornament of shell and cast in thick-shelled molluscs.

display internal moulds of sediment, the coarser-grained sediment must have entered through fractures in the shell while fine-grained material appears to have passed in through the siphuncle. Where structural elements such as keels are separated from the interior of the shell by a partition (septate keels) they may remain empty when the internal moulds forms to produce hollow keels that contrast with the filled keels produced where such partitions are absent.

Internal moulds are mostly formed during early diagenesis. Moulds on which epifaunas are present must have been reworked and have remained on the sea floor long enough for the epifauna to develop.

Concretions

Concretions result from the localised segregation of originally dispersed material. They may consist of calcite, siderite, silica, pyrite, marcasite, phosphate or other minerals and often contain organic remains. The concretionery mineral frequently occupies only the pore space of the sediment or if may make room for itself by replacement or by physical displacement of the host sediment.

Concretions may result either from the decomposition of organic matter or because of changing solubility of salts in rising ground water. Calcareous concretions form when the carbonate material present in most sediments segregates in an alkaline environment. Calcareous shells act as ideal nuclei. Siderite concretions occur where the O Eh level lies just below the sediment-water interface. Below it, iron dissolves in the pore water only to reprecipitate as it approaches the surface. Phosphatic nodules may either be pure or may grade into calcarious and ferriginous varieties. They are commonly found where phosphate-rich upwellings enter areas of slow sedimentation and are closely associated with organic remains and faecal pellets.

Iron sulphides develop in acid or neutral-reducing environments by the reaction of iron and bacterially-produced H_2S. Both the stable form (pyrite) and the metastable form (marcasite) are found. Siliceous concretions derive their material from the dissolution of siliceous tests, or sponge, spicules, present in the fine fraction of the sediment or from volcanic SiO_2. They often grow at a very early stage of diagenesis in association with organic remains. Mobilisation of silica is possible only in alkaline environments.

Early diagenetic concretions are very important in the preservation of fossils which become encased and protected before they can be destroyed. Fossils within concretions can also undergo reworking with suffering damage.

Deformation

Deformation of fossils refers to shape changes that occur after the fossil in embedded in sediment. Such changes are caused by compaction that results from increasing overburden pressure and dewatering. If the fossil consist of articulated elements, these elements move relative to each other. On the other hand, shells, armour, or bones are rigid units that yield by fracturing. Regular radial, concentric and axial fractures appears along which relative displacements occur. This sections frequently reveal that shells that appear superficially to be intact are internally shattered.

Within skeletal material whose internal structure is so weakened by solution that individual crystals can move relative to each other, deformation can occur without fracturing. In this way curved shells can be squeezed flat and thick bones can be attenuated. Plastic deformation of internal moulds is a widespread phenomenon (pelomorphic deformation). The individual features of the cast are compressed vertically. The ratio V_2/V_1 (V_1 = original volume; V_2 = final volume) is known as the 'compaction ratio'. The ease with which deformation can be recognised depends on the original attitude of the mould.

Tectonic deformation also causes distortion, extension, and disruption of fossils enclosed in sediment but this is not longer to be regarded as diagenesis.

Corrosion

Corrosion, involves partial solutions of shells by water, acids (H_2CO_3, H_2SO_4), or alkalis (NH_4OH). It commonly occurs when buried fossils are re-exposed on the sea floor and frequently causes removal of those parts of the shell that project above the new surface. Such planar corrosion on the sea floor is called 'subsolution'.

Corrosion can also occur within the sediment when organic remains in unconsolidated sediment are attacked by rising pore water. In lithified sediment this can produce partings and stylolites. Weathering also produces corrosion mainly because of CO_2 dissolved in rainwater. The carbon dioxide released from plant roots also dissolves carbonate.

Gaps in the Record

Only a minute fraction (well under 1%) of all organisms are preserved as fossils and different groups fare quite differently. Completely soft-bodied animals and plants survive in only a few lucky cases whereas organisms with skeletal parts stand a better chance of preservation, although even then the record is very fragmentary.

Because of their mode of life, some organism s are hardly ever preserved. For example the plants and animals of the tropical rain forest decompose remarkably quickly after death as do organisms in turbulent seas and flying animals.

Only a very small part of the area over which fossiliferous sediments were once deposited now lies at the Earth's surface. The greater part has either been eroded away or been buried under younger sediments. Moreover the beds at the surface are exposed only in places and only some of the exposures are accessible and have been studied. Many biotopes are also confined to brief time intervals and may not then occur in any of the available outcrops so that organisms confined to these biotopes remain unknown.

Where unsuitable mode of life, lack of outcrops, and apparent absence of a biotope all coincide, whole groups of organisms may leave no record over fairly long periods of time even although they must then have been extant. There are also many gaps in our knowledge where fossils have eluded our search because of the unfavourable size, chemistry or inconspicuous character. Others are so rare that only long and expensive investigations yield any material.

The smallness of the percentage of organisms preserved suggests that we can expect to find only the more common types as fossils. Representatives of rarer groups are found only in exceptional circumstances. Thus we observe only the peaks in the frequency curve of past life and it is astonishing how much valuable material they alone have yielded.

7

European Casts

The thunder of drumbeats can be heard far and near, the smell of roasting meat permeates the air, palm wine and schnaps are passed around. Why has the Chief called to celebrate the yam festival? Nobody really knows. Nevertheless, drumbeats spread the news through out the forest with mounting excitement, the ornamental stools are removed from storage. The people of Debiso in Western Ghana seldom actually see their ancestral inheritance. The chairs are usually kept by the village chief, locked away in his personal storeroom. With the aid of his medicine men, the Chief is capable of speaking with the spirits of the forest, During three days of festivities, ceremonies will be held at specially chosen sites outside the village to appease the invisible beings.

The rainforest rises like a great fortress, behind the small village, in the clearing, simple tin-roofed huts stand unsheltered in these scorching sun. Even the chickens appear to be the smallest of birds against the backdrop of the rainforest. Few villagers have ever seen the end of the forest although the "Trotro" – an old Bedford truck – makes daily trips on the, timber transport route from Kumasi to the border of the Cote d'lvoire. Until a few years ago, the village could only be reached on foot via winding paths through the deep forest.

The Chief wears a heavy gold-plated crown for the ceremony. The same ponderous headdress had been born by his forefathers in leading the Sefwi tribe into battle against the Ashantis in brutal jungle wars of the past. That was before the hunting camp which once stood at this site had evolved into a permanent settlement. Only the name "Debiso" remains to remind one of the place where two forest elephants killed by Sefwi hunters came to fall upon each other. Sacrifices are made to the forest and its spirits every three years at the Yam festival. But the dwarfs must be appeased more often. Many West African tribes believe strongly in the lore of the little people. The dwarfs, invisible creatures who live at secret places in the 'forest, must be given due respect. Otherwise, their playful pranks may easily become evil tricks. They have even' been known to kill on occasion.

Natural clearings are rare in the deep forest, occurring only where the granite base of the African continent rises to the surface: The Sefwis call the clearings "Apaso"– the dwarfs' gardens. The "gardens" are magical places where sacrifices are made with great care according to the wishes of the village chief and his medicine men. It is also they who determine which of the village elders may accompany them along the twisting forest path leading to the secret site.

A stagnant pool of greenish water awaits the visitors. According to tradition they do not speak and have entered from one direction only. The medicine man now summons the dwarfs in a loud voice and invites them to hear his humble words. After a long sermon, the sheep which has been brought from the villages sacrificed - its throat slit allowing its blood to flow upon the colourful gifts of fruits" nuts, eggs and magical objects spread upon the bare rock. Bottles of schnaps are poured onto the ground as an offering to the dwarfs, no less is drunken by the Chief and his companions themselves for they are fearful and alcohol relieves their fear. It is said in Debiso that some men have never returned from the clearing because they were unable to appease the "little people" The dwarfs are the keepers of the forest and of the animals and do not hesitate to punish wrongdoers. The people strongly believe in this tradition and take care not to anger the little sentries. Thus, the responsibility of protecting the forest so vital to the Sefwi tribe is in the hands of these small creatures. But the dwarfs never reckoned with the Europeans.

A Source of Raw Materials for 500 Years

The sea-route from the major harbors at southern England, Germany or Holland to the coast of the Gulf of Guinea extends over 600 kilometers. Neither the rainforests in Latin America nor, the forests of Southeast Asia are, closer. West Africa's proximity to Europe has influenced its trade relations for centuries and left its mark on the rainforest.

Europe's first trade relations with the West African coast date back to the 15th century. The first trade agencies consisted of nothing more than storage houses and fortifications along the coast. A small" strip of land sufficed to develop trade from the coast famed for its rich natural resources. The rest of the African continent was still a mystery to the Europeans. Nevertheless, trade boomed: gold, ivory, cola nuts and slaves were shipped from the coast, of western Ghana in 1700, British, Dutch and Danish trading posts vied for the best ports. Not every trade settlement have been founded at an ideal location and the sailors had to battle against rough waters.

The Europeans did business with Arab slave dealers from the north and certain tribes of rainforest people who delivered goods and slaves from the interior of the continent to the coast. Slavery was not a European custom alone. The Ashanti tribe, for example, kept slaves for their own use. Entire villages in the savanna north of the West African rainforests were taken into bondage if they did not succeed in fleeing their captors. This led to a depopulation of large areas which still remain thinly populated today. The poor uprooted people from the north were considered nothing more than a good to be traded - even by the West Africans. Slaves were a cheap commodity in contrast to salt which was quite expensive. It was imported from the north in caravans and traded like gold. In the interior of Benin, Togo and in the Volta region of Ghana, a handful of salt was known to be worth one to two slaves.

The Danes were more liberal than other Europeans active in West Africa at the time, abolishing slavery in 1802. The British Government followed suit five years later with a ban on slave export which they applied to a number of other trading countries as well. The British confiscated foreign slave ships and freed their unfortunate passengers. The Europeans' new respect for other races however, posed a problem for some forest peoples; the Ashantis were, left with no takers for their wares and opted to settle their captives as planters. The slave trade was replaced by the export of palm oil. Inadvertently, the British were therefore actually responsible for creating export-oriented agriculture in West Africa. In 1850, they began to expand their influence.

Beginning of Commercial Exploitation

In 1879 following 400 years of trade with the West African coast the Europeans only ruled over large segments of the population in French -Senegal and on the British "Gold Coast" (Ghana) Except in Senegal. European administrations never controlled land more than a few miles inland. The British Colonies of Gambia, Sierra, Leone and Lagos were nothing more than small enclaves in a region still largely under African rule. Within but a few years, however, the situation was to change drastically. In 1870, gum copal from trees of the genus *Daniellia* found in closed rainforests was being exported for the manufacture of varnish in increasing quantities. The European demand for rubber tapped from *Funtumia* trees also rose steadily after 1883. Following a number of unsuccessful attempts, the export of agricultural products rapidly increased. The first successful oil palm plantations in southern Ghana exported up to 30000 tons of oil by 1884. In 1985, Ghaba also exported 63 tons of coffee.

During the last quarter of the 19th century, rainforest in Ghana, and Nigera increasingly gave way to cocoa plantations. In contrast, cocoa was not brought to the Cote d'lvoire until 1912 when, it was introduced from Ghana. Lowland rainforests in the southern part of the rainforest belt suffered most from the cocoa boom. Cocoa cultivation was especially successful in the rural areas of Ghana near Accra and in the forest region near Kumasi. In 1911 after only 26 years of profitable production, Ghana rose to become the top exporter worldwide and managed to hold its position for a number of decades while continuingly increasing production.

First attempts at logging in Ghana had been undertaken near Axim in 1800. They were abandoned, however, due to technical difficulties. It was not until 1887 that logging operations were again established on an '"experimental basis". The British, who now ruled the larger, part of southern Ghana, eliminated the political barriers to allow timber to be floated freely to the coast on the major rivers Tano, Ankobra and Pra. The "experiment" was to have serious consequences: Seven years later, the export of tropical timber had reached 450000 cubic feet (approx. 285 m^3) and the colonial government had begun to distribute timber concessions to European investors. By the year 1913, Ghana's exports had gradually risen to three million cubic feel (approx, 85000 m^3), World War I, however, brought difficult times and the export volume fell back to a much lower level. Nigeria was exporting similar quantities of tropical timber and earlicr In the 19th century the country established its own forest service. Only African mahogany (*Khaya ivorensis*) and smaller quantities of sapele (*Entandrophragma cylindricum*) were actually in demand then for European furniture production.

Until World War I, the British largely controlled the export of tropical timber from the Cote d'lvoire as well. Mahogany was felled near the Bia river and the logs were floated to the Abi Lagoon near Assinie. The consequent depletion of the forest along the shore led to trees being felled further inside the forest. The logs were hauled to the nearest rivers along corduroy roads. In time, other African hardwoods such as makore and iroko gained popularity in Europe. Nevertheless, export statistics showed African mahogany to be at the top of the list until 1951. But until after World War II, timber exports, from West Africa showed slow growth. The fluctuating market demand was not the only reason. Transporting logs from distant locations to the coast posed nearly insurmountable technical problems. The interior rainforests were thus spared from large scale exploitation until the 1950s when the mechanization of forest operations changed the situation dramatically. Bulldozers were brought in to push their way through the forest and trucks were able to haul entire trunks from deep within the forest to the coast.

The European demand for tropical timber increased steadily throughout the 1950s. In the Cote d'lvoire, roundwood production sky-rocketed from 400000 cubic meters in 1958 to more than 5000000 cubic meters in the 1970s. Nowhere, along the Gulf of Guinea did the exploitation of the rainforest proceed at such a fast pace as in the Cote d'lvoire. Neighbouring Liberia was not similarly affected until the mid-1960s. While the export of timber from the West African coast drastically increased after World War II, the export of other forest products came to a near standstill. Artificial resin had long since replaced natural gum copal in the production of varnish and the rubber grown in, West Africa today is actually caoutchouc from Brazilian Hevea rubber trees grown on large plantations. Just as for the cultivation of cocoa, wide areas of rainforest were also cleared for Hevea and oil palm Plantations.

Early Attempts at Conservation: Theory

It would be unfair to assume that the Europeans were interested only in exploiting the natural resources of their African colonies with no consideration for their future. Early in the history of commercial timber exploitation, the colonial government in Ghana adopted the Timber Protection Ordinance of 1907 which banned felling commercial species of a lesser diameter.

In Togo, a German protectorate from 1884-1919, forest conservation had been a concern from the start of colonial rule. The country was endowed with a far smaller area of rainforest than other West African nations. In 1907, just as forestry regulations were adopted the Ghana, a conference was held in Berlin on the reforestation of Togo. The Germans decided to invest mainly in teak, an exotic species well adapted to the drier climate In addition, indigenous hardwoods were also chosen: doussie, sasswood, African mahogany and iroko, the kapok-tree (*Ceiba pentandra*) and "chew-stick" (*Anogeissus leiocarpus*). The bark and leaves of "chew-stick" are of medicinal value, only the roots are used as chewing sticks. Thirteen million trees are said to have been planted in Togo early this century.

In Nigeria, the British had begun with reforestation in the Benin District even before the forestry administration was established. Tens of thousands of African mahogany trees were planted It annually. Between 1901 and 1910, trees of several species native to Nigeria were planted: iroko, obeche, limba and mahogany. A few exotic species were also introduced: teak, cedrela and eucalyptus, In 1920, the Senior Conservator of Forests of Nigeria, A.H. Unwin stated, "Despite many failures owing to experiments on bad soil and seasons of extreme drought, the growth of the trees gives the greatest promise of mature trees, or at any rate merchantable trees being grown in a comparatively short period".

Soon after, the British realized that protecting the forest as a whole was even more important than planting individual trees. With the establishment of the forest service, the administration began to demarcate forest reserves, henceforth to provide a source for permanent and controlled timber exploitation. In order to preserve the steady flow of rivers to lower lying regions, watershed areas were especially given reserve status to protect against flooding, erosion and drought. Shelterbelt forest reserves bordering on savannah territory marking the transition to the Sahel were established to keep back the hot desert winds from the north.

In Ghana, the Forest Ordinance of 1911 allowed the colonial governor to give all uninhabited forest territory forest reserve status, a measure which stripped the surrounding inhabitants of their traditional user rights. The people protested strongly since their livelihood depended on just that which was thus prohibited by the ordinance, namely shifting cultivation and gathering forest products such as essences, fibers, fruits and nuts. The Aborigines' Rights Protection Society lodged a complaint with the Governor for disrespecting the traditional land tenure system.

Reality: Forest Inhabitants Resist

West African tradition holds land to be common property belonging to one or more communities. The members of the community are entitled to use as, much land as they need to clear and cultivate. Should land no, longer be needed, it falls back to the community. Sale of property is not allowed since it does not belong solely to the living users but also to their ancestors buried on the land and to future generations. For Akan tribes in the eastern Cote d'lvoire and in Ghana, the earthly resting place of their forefathers souls is symbolized by the "stools". The stools, each cut from a single block of wood and richly embellished with ornamental carvings, represent the common property, also termed "stool land". During rituals held at regular intervals, offerings of food are placed about the stools. Sacred wine and schnaps are used to douse the precious inheritance. The British were well aware of the importance of the stools in regard to property rights. In 1900, upon the conquest of the mightiest Akan group, the Ashanti tribe, British governor Sir Frederic Hodgson demanded possession of the "Golden Stool". In a storm of protest led by the mother of their imprisoned King Kwaku Dua III, the Ashantis attacked the British fortress in Kumasi. The remembrance of that event probably played a role in their reluctance to strictly enforce the new forest regulations. Local chiefs were still quite influential. The legislation was in fact temporarily withdrawn until the newly established forest service published long lists of forest reserves between 1922 and 1926. As one may expect, the forest reservation program was not at all popular with the local inhabitants, who felt they were being denied their traditional rights. By the year 1939, an area of 14800 square kilometers of Ghanaian rainforest had been declared forest reserves, which totalled 19% of the country's rainforest cover.

In 1926, the French colonial power began similar attempt at forest conservation in the Cote d'lvoire. By 1956, 43 000 square, kilometers of forest had been declared protected areas (forests classees). The local reaction was, no more favourable than in Ghana. Many of the Ivorian forest reserves, were so heavily damaged by illegal slash-and-burn clearing that redefinement was necessary in 1966. Unfortunately, the problem was not easy to solve. Illegal farming on reserved land continued largely uncontrolled.

Consequences of Centralization

Throughout the century, West African peoples have increasingly lost authority over their vast areas of forest. Colonial regulations withdrawing their self-responsibility within the newly established forest reserves were not the only reason. Logging activities made the forest more accessible. Transport routes, although crude, not only allowed timber to be carried out of the forest, but also let new settlers come in. Immigrant farmers came from other parts of Africa, especially from the north where the dry climate forced many to look elsewhere for their livelihood. The new settlers, however, were not familiar with the rainforest. In contrast to West African forest peoples who had lived and worked for generations in the forest, they did not know how to deal with the sensitive soil. Traditional West African forest farmers never do long-term damage to the land. The area surrounding a village is cultivated on a rotational basis and only small areas are cleared and planted for a few years before the next ones are cleared. Once a plot of land has been depleted of its nutrients, it is left under fallow for natural regeneration to occur and the rotation continues around the village allowing several years to pass before the same plot is replanted. This form of shifting cultivation is still practiced today in less populated areas in the western Cote d'lvoire in parts of Liberia and in western Cameroon.

The opening up of the forest for logging activities and increased population led to the collapse of traditional land tenure and, land-use systems in most West African forests. Coffee and cocoa plantations contributed in no lesser sense to the disintegration of the traditional systems. Although land is, considered to be community property, many tribes believe that which grows on the land to be the sole property of the planter. Coffee trees, cocoa trees and oil palm trees thus belong to the farmer and not to the community. Should the community wish to reclaim its land, the farmers must be compensated for their fruit trees. Many villages therefore do not allow immigrant farmers to plant fruit trees on the land for which they have received cultivation rights. The result is that the settlers then plant only corn, plantain and cassava, all of which quickly deplete the soil of its nutrients. After only a few years, the planters find themselves forced to move on to a new plot of land. Those who are determined to remain at one location for a longer period usually choose to plant coffee and cocoa. They do their best to gain possession of land against the traditional rules. These circumstances have led to a wider distribution of plantations and villages throughout the forests of West Africa.

West African tribes thus had good reason to resist the centralization of the forest administration at the turn of the century. They tried to defined their culture and their ancestors – without success. But neither did the West African forest services achieve their goals. Forest legislation based on European principles was inadequate in the light of African reality. In trying to encourage the sustainable development of the rainforest, the forest services lost control over the consequences of their ordinances. Even today, West African forestry departments lack the necessary foresight, control and overview to deal effectively with the situation. Law and regulations aimed at protecting the forest often lead to just the opposite.

Large-scale Exploitation and the Consequences

It was not until the 1970s that the inadequate implementation of forest regulations became apparent Following World War II, commercial exploitation had increased to such an extent that no West African forestry department was capable of enforcing the law. In comparison with rainforests in other parts of the world in 1973, Africa showed the greatest area encroached upon by logging activities although African timber production measured only one third that of Asia. This fact signalized very extensive exploitation practices in African rainforests which were continually opening up new areas.

In 1985, the U. N. Food and Agricultural Organization (FAO) estimated 72% of West African rainforests to be fallow land: destroyed forest, plantations or secondary bush. Between 1981 and 1985, the remaining areas of undisturbed forest on the Gulf of Guinea were being opened up at the rate of 1640 square kilometers annually. Many timber companies adopted highly selective logging practices to meet specific demands or merely for reasons of profit. Only the best specimens of mahogany, utile, sapele and makore were extracted from the forest. With few exceptions, however, such selective timber exploitation foresaw a one-time use only and did not aim at long-term sustainable use.

Table 7.1 The status of closed broadleaved forests ("rainforests") in West Africa from Sierra Leone to, Nigeria.

Status of forest	*km²*	%
Undisturbed, productive	21260	4.1
Undisturbed, unproductive	64460	12.5
Logged	45870	8.9
Managed	11670	2.3
Forest fallow	370820	72.1
Total West Africa	514080	100.0

During the past decade, each year has seen an average loss of 7200 square kilometers of West African forest. The major cause is burning and clearing by migrant farmers. But over 90% of the areas destroyed were forests previously opened up by timber companies.

Logging roads 'not only serve to carry timber out of the forest but also pave the way for settlers to come in. What timber companies leave standing is devastated by slash-and-burn farming Unfortunately, the resulting increase in agricultural area is minimal: since the soil quality of the land thus gained affords only a few harvests, the process of burning and clearing continues. The planters follow the timber exploiters further into the rainforest leaving unproductive land behind.

Table 7.2. Annual decrease in area of closed broadleaved forests ("rainforests").

Forest Type	*km²*	*%*
Untouched, productive	210	2.9
Untouched, unproductive	340	4.7
Logged	6650	92.4
Total West Africa	7200	100.0

Today, we look back upon a half a millenium of trade relations with Europe – relations which were rarely in favour of West Africa. European demand dictated West Africa's export of natural resources. During times of war or upon the discovery of an alternative product, Europeans lost interest in trade, Whether dealing in slaves or hardwood, Africa's fate was typical for a deliverer of raw materials. Unlike the eastern and southern parts of the continent at higher altitudes, West Africa did not attract many white settlers, its climate being unpleasantly hot and humid. The whites were interested only in what could be exported. West Africa became Europe's major source of raw materials and remained a loyal trader even in less profitable times. But West Africa was poorly paid for her loyalty. What began as a sustainable use of forest products developed into a timber export economy nothing short of all out exploitation, In the words of tropical forest expert Professor H. Steinlin, "The exploitation can be compared to that of a mine, a natural resource is being excavated without guarantee of a sustainable production". This attitude has had serious consequences for the forest and last but not least for the local people.

8

Finding Fossils

Although an amateur may occasionally discover something of significance by simply looking among the rocks in his own vicinity, the important finds are almost invariably made by the professional palaeontologists, versed in the necessary geology, who lay careful plans for their expeditions. Whether they are searching for particular fossils or intend to investigate the fossil content in rocks of a particular age, they must begin by consulting geological maps to determine where promising rocks of the right age are exposed. Their knowledge of geology goes beyond map reading, in fact, to the understanding of rock formations and recognition of different layers of rock in the field. They must distinguish between igneous rocks, solidified from a molten state, and metamorphic rocks, transformed by tremendous stress (materials that never, or almost never, contain vertebrate (fossils) and sedimentary rocks, which may. The last are formed when particles of ash, silt, clay, or sand settle and are compacted and cemented into a solid mass. The resulting rocks may be tuff, shale, limestone, mudstone, or sandstone, depending upon the nature and size of the particles of which they are composed. Although some sedimentary deposits, like the tuff from volcanic ash, may form on land, most rocks of this type are laid down under water on the bottoms of lakes, streams, and seas and at the mouths of rivers opening onto continental shelves. Very few fossils are recovered while these deposits are submerged. Prospecting takes place where sedimentary rocks have been lifted out of the water. Repeatedly in the history of the earth, areas have risen, spilling off the water which had covered them and exposing the old bottoms to the forces of erosion. These places are the paleontologist's hunting ground.

The best locations for fossil hunting are often not the most comfortable. Greenland and Spitzbergen, where rocks have been exposed by the melting snow in relatively recent times, can be visited only for a few months in summer because of the cold and the violent storms which occur there during the rest of the year. Digging at more temperate latitudes is not necessarily easier. Exposed rocks, of course, support no protective foliage cover. The palaeontologist is usually out in the open under the broiling sun and often in wind as well. In remote areas he must frequently cope with lack of water and an absence of roads. Fossil hunters have exchanged their canteens and horses for more modern equipment where possible, but even today they have to spend a good part of their energy in hiking and hauling supplies. Constructions sites—foundation holes and road cuts—in the midst of settled areas afford access to hitherto buried fossils and, compared to other locations, can be relatively convenient place to work.

When the arrives at his destination, the palaeontologist must rely on his eyes as the tools he uses first. That he looks for fossils may seem too obvious to have to mention, but it should be emphasized that the time spent in looking over the ground can be long indeed. Like a man seeking a four-leaf clover, the palaeontologist walks the area, running his eyes over very detail of the surface under his feet. Even though he has a mental picture of the kind of the telltale signs which indicate the presence of vertebrate remains, his search may net him nothing. Eventually, if he finds no trace of fossils, he must consider the time and supplies that remain available to him and decide whether or not to move to another place. It is difficult to call a halt to an expedition which has been unsuccessful—to pack and leave an area where there may be important fossils lying by chance unobserved. Even a last-minute find by an assistant at that point is sufficient to raise everyone's morale.

Collecting Fossils in the Field

If fossils are found, another phase of the work begins. Collecting involves certain rules that must be followed if the fossils are to be use to the student of evolution. Every curator of a natural history museum has received at some time or other a box with a fossil in it sent by an amateur collector who wants to know if his contribution has any importance. If the fossil is unaccompanied by specific data relating to the location from which it was taken and the rocks in which it as found, the curator must usually explain gently that the fossil can be classified only very generally and might be retained by the collector as an interesting souvenir of an extinct form. Fossils collected on expeditions are labeled systematically to provide information for the use of the worker who will attempt to interpret the nature, age, and habitat of the ancient vertebrate.

Although some fossils are collected by merely being picked up from the weathered or unconsolidated sediments in which they lie, most fossils must be extricated from the rocky matrix that encloses them. If the matrix is extremely hard, trying to free the fossil completely in the field would result in sever damage to the specimen. In such a case, the paleontologist cuts out or detaches a slab of the rock containing the fossil and transports it to the laboratory, where a preparator can remove the surrounding rock under conditions that minimize the danger to the specimen. If the palaeontologists had discovered a large, well-preserved but solidly embedded skull, the work of removing it is bound to time-consuming and backbreaking. The exposed portions of the skull are first protected by strips of burlap soaked in plaster of paris. The water for mixing the plaster often has to be brought from a distance, of course, if the excavation site is in a dry area. The paleontologist and his assistants work with picks, chisels, and hammers to cut and separate the block containing the fossil. When the block is free, it is crated and conveyed to the laboratory. If the digging is being done in his own country, and if modern transport facilities are available, the labors of the palaeontologist may end when he and his co-workers heft the crate into their truck. Sending fossils home from another country is a more complicated affair. The chief of the expedition has then to reach an agreement with the country in which the digging is taking place giving him permission not only to explore but to export what he finds. No palaeontologist wants to risk having his boxes of fossils impounded by a government that suspects him of trying to carry off a national treasure.

Preparing Fossils in the Laboratory

Most fossils that reach the laboratory must undergo some preparatory work before the paleontologist can study them profitably. The techniques used to expose fossils depend upon the nature of the matrix in which the remains are embedded and the condition of the remains themselves. Resistant structures like teeth if buried in clay, may simply have to be washed carefully in water but more often matrices are harder and fossil materials farther removed from their original state. As a rule, animals after death and burial lose not only their soft tissues but also much or all of the organic matter built into their harder parts. Small specimens may be completely carbonized, that is, reduced to a thin, black film of residual carbon through the gradual transformation of their substance into gases that seep away. Bones which retain their shape are nevertheless sometimes rendered friable through the loss of the organic framework that held their mineral material together. Under certain circumstances, skeletal structures may be hardened during their interment. If ground water carries dissolved calcium, silicon, or iron into their interior, crystals are likely to form, filling every interstice. The bony tissue itself is occasionally petrified as the original salts it contained are replaced by new ones. Extensive petrifaction is not common in vertebrate fossils, however. The preparator usually has to work upon thin or relatively delicate remains that are embedded in rock that is more resistant than they. If he has reason to suspect that preserved elements lie in the interior of a slab, he may use x-rays to detect them, but most often he faces the problem of removing from partially visible fossils as much of the covering matrix as possible.

As every detail of a fossil may be important, every square millimeter that can be exposed represents a gain for the investigator. Rarely, the sedimentary rock containing the fossil is so weathered or unconsolidated that it can be washed and brushed off. More usually, especially with older vertebrate materials, the surrounding matrix must be attacked with chisels, small drills, and even needles. Sometimes a small specimen, like the head shield of an ancient jawless fish, is submerged in alcohol or xylol to make fine details visible more clearly. Working on such a fossil, the preparator uses a dissecting scope to lessen further the chance of destroying with his sharp instrument delicate parts of the bony plates or their surface ornamentation. Where chipping or picking proves impractical, careful applications of acid may be effective in weakening the rock. Hydrochloric and hydrofluoric acid have been tried, the former against limy cover and the latter for rock containing a large amount of silica.

Frequently, the preparator knows when he looks at an embedded fossil that removing it from the rock is impossible. This is the case when ancient bone has been infiltrated by mineral material, retaining its distinctive osseous pattern but becoming inseparable from the mass of rock around it. Also, whole skulls are sometimes filled with sediments which, after they have hardened, can never be picked out from the interior. Such fossils were of limited use until W. J. Sollas, a British geologist-palaeontologist, devised a new method for their preparation. He resorted to grinding away the entire mass, fossil and rock, thin layer after thin layer. He studied and drew the pattern of preserved bone that appeared on the ground surface before removing each successive layer. Eventually the entire fossil was reduced to a powder, but Sollas had obtained a series of drawings showing the configuration of the skull bones trapped in the rock. One of these drawings, by itself, was of limited significance, as it represented only a section through the skull at one level. From the whole series, however, Sollas was able to reconstruct an accurate model of the original remains. He did so by reproducing the pattern of bone shown in each drawing in a thin layer of plaster and then assembling the plaster layers—usually several hundred of them—in the proper order. In this way he built from a series of sections, a three-dimensional replica of the fossil skull which has been hidden in the rock. The method used by Sollas was modified by workers who followed him. Although the tedium of making the model could not be avoided, the process was facilitated by using sheets of wax instead of plaster. Now, rather than losing the material of the specimen as it is ground, preparators have found a way to preserve thin sections of it. After grinding away .5 to 2 millimeters of bone and matrix, they etch the surface of the block with hydrochloric acid for less than a minute, wash it, coat it with acetone, and cover it with a sheet of cellulose nitrate. When the preparation is dry, they peel off the sheet of cellulose nitrate. A fine layer of fossil material and sediment adheres to it. Peels, as these sheets are called, can be kept indefinitely and studied and restudied at leisure. Since the particles of bone upon them will accept histological stains, the peels can be coloured and projected or viewed under the microscope. Although many workers prefer to prepare specimens in a recalcitrant matrix by using faster methods employing acid, grinding and making peels has become one of the standard techniques for investigating fossils of small size.

Fossils in Museums

A small number of the fossils undergo further treatment to prepare them for public display. The bones of land vertebrates, freed entirely from the rock in which they were embedded, are assembled in their proper order and fixed in place. In this way the frames of huge dinosaurs emerge from piles of separate bones to tower over visitors in high-ceilinged museum halls, and skeletons of mammal-like reptiles stand again on all fours in the glass display cases. Fossil fishes and flattened amphibian skulls may be exposed in relief against the rocky matrix,

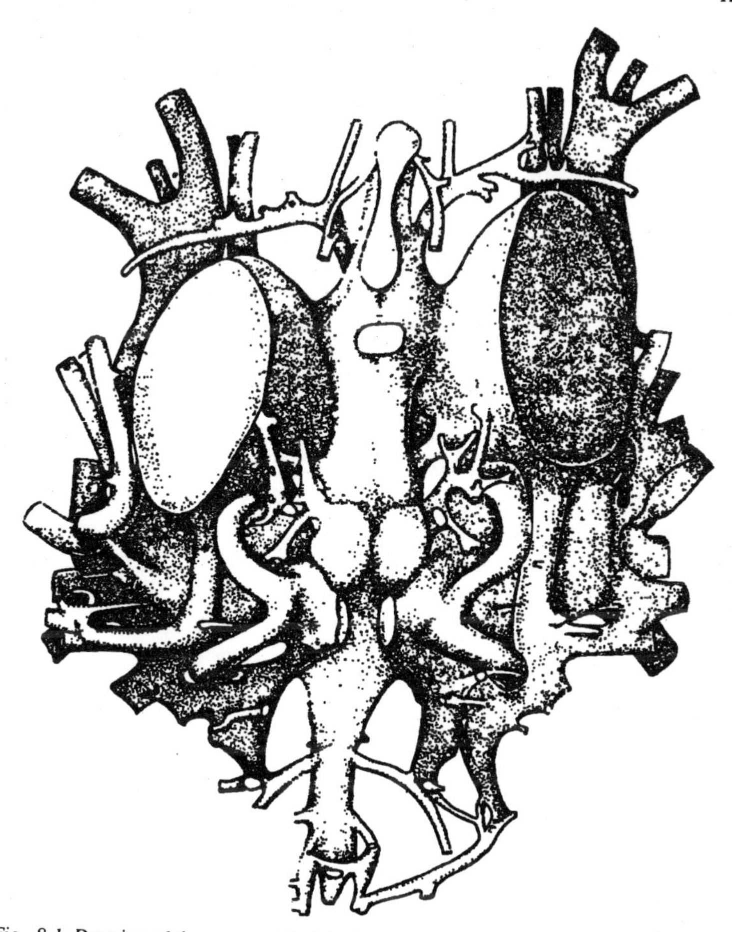

Fig. 8.1. Drawing of the wax model of the interior spaces in the skull of Cephalaspis, an extinct jawless fish (dorsal view).

their bones darkened or outlined to make them more visible to viewer. Whole slabs of rock containing footprints go up on museum walls.

Fossils on display in museums are, not the most part, handsome examples of forms well known to the palaeontologist. Drawers and boxes in storage rooms hold, besides additional examples of recognized genera, specimens only cursorily examined, classified in a general way, and labeled with collecting data. These materials wait until someone has the time to undertake a thorough study of them. Because fossil hunters bring in material faster than it can be described, all but the newest departments of paleontology have a backlog of specimens on which more work must be done. Palaeontologists say, only half in jest, that the key to the solution of several problems in vertebrate history may be lying on the shelf in a laboratory cabinet somewhere.

Studying Fossil Material

An investigator, as he turns his attention to a particular fossil, begins a four-pronged inquiry. Not only does he try to identify and describe the remains he has in hand, but he wants also to determine the conditions under which the animal lived, the period of its existence, if not already known, and the place of the organism in the evolutionary scheme. The results of his study should provide his colleagues with an accurate description of the specimen and an interpretation of its significance.

Identification and description of fossils material require that the paleontologist draw upon his knowledge of comparative anatomy: by recognizing familiar patterns of skeletal design or arrangement, the researcher is able to assign the fossil to one of the classes of vertebrates and to refer it to an order and family. On the basis of its individual peculiarities, he may give it a new generic and specific name. When a fossil consists of a fairly complete skeleton or the greatest part of a skull, the palaeontologist task is almost always easier than when a fossil consists of a fairly complete skeleton or the greatest part of a skull, the paleontologist's task is almost always easier than when a fossil is more fragmentary. Small pieces of dermal bone, for instance, he may decide have probably come from the armor of an early jawed fish, but more precise classification may have to wait until he can match the bits of bone with similar bones of a more complete specimen. If he recognizes isolated spines or teeth as having belonged to some sort of shark, he may establish a new genus under the class Chondrichthyes. However, as such fragmentary remains give no further clue to the identity of the fish that bore them, he must list the new genus, not in a particular order of the class Chondrichthyes, but simply under the heading incertae sedis, of uncertain seat. When new fossils are found with spines or teeth of the same sort attached to other hard parts, it may become possible to reclassify the genus in a known order or to declare that it represents a new one.

The severest test of the paleontologist's ability to recognize anatomical structures comes when a specimen has been crushed or its parts have become dislocated. It is not unusual to find remains of early fishes and tetrapods that have been disarticulated after death by the weight of the sediments in which they were buried or broken by the shifting of rock. A first look at the skull of an ancient lobe-finned fish can be discouraging indeed. An untutored eye may see nothing initially but the black shale in which the fossil is embedded. Little by little the outlines of bones appear, but elements project in every direction and are half hidden by one another. Palaeontologists have no choice, though, but to work with the specimens they have, and their long study of such fossils has enabled them to discover the complicated patterns of bones that distinguish the various types of early vertebrates.

Type Specimens

When a fossil form appears which is different from others that have been described, a new specific name is created for it. The fossil first given the name is known thereafter as the type specimen of that species and serves as a standard of comparison for other fossils thought to be like it. Since the preservation of valuable type specimens is best assured if they remain in one place, palaeontologists find it necessary either to visit laboratories and museums to study collections or to work from plaster or latex models of the originals. The use of photographs and drawings is essential, especially in preliminary studies and in general communication, but for the researcher there is no substitute for examination of the fossils themselves. By looking at specimen after specimen in a particular group, the paleontologist sharpens his ability to identify fragmentary or poorly preserved remains.

Reconstructing Fossil Forms

After describing the fossil that is the subject of his study, a worker tries to obtain some understanding of the appearance of the ancient animal it represents. To this end, he draws a reconstruction—not just a fanciful picture, but one based securely on the anatomical evidence he has. Should the fossil material consist of an entire skeletal, he calculates the bulk of the musculature from the size of the areas of muscle attachment visible on individual bones and draws an outline to encompass both bones and flesh. If the lower jaw is missing, it can be added its size and position estimated from the length of the opposing upper jaw and teeth and from the site of its articulation with the skull. The orbital opening dictates the size and location of the eye. Although the appearance of the surface of the body is often a matter of conjecture, the preservation of scales or imprints of feathers or "mummified" skin enables the wearing away in the second area of layer B before the deposition of the sediments which formed C. Palaeontologists who are trying to establish the relative ages of fossils recovered from widely separated sites must contend with even more complex stratigraphical riddles. No overall solutions are in sight—or possible, for that matter. Palaeontologists continue to regard the correlation of strata in different areas as being among the most difficult problems in the study of vertebrate evolution.

Index Fossils

Palaeontologists and geologists have attempted to decipher the history of rock formations one at a time and then to compare series from different localities. Their studies began, as is always the case, with descriptions of the strata which exist and advanced to the search for similar patterns of deposition over a broad area. Following another of William Smith's principles, workers learned to identify contemporaneous rock layers by certain of the fossils they contain. Index fossils, as these forms are called, are easily recognizable remains of animals that enjoyed a wide distribution for restricted periods of time. They serve as markers in the rock, making it possible to define the temporal relationship of one stratigraphical series to another. Little by little, a general outline has emerged of the succession of sedimentary rocks formed down through the ages and one of the living things that have inhabited the earth in the last 600 million years.

The Geologic Time Scale

As knowledge accumulated, it became possible to draw up a table of geological ages and to associate with each the forms of life which characterized it. The table shows the earth's history divided into five eras. The two oldest, the Archeozoic and Proterozoic, are distinguished entirely by physical criteria and are of interest chiefly to the geologist, since few fossils appear in rocks from those times. Palaeontologists have given more attention to the last three the Paleozoic, Mesozoic, and Cenozoic, because it is in the strata laid down then that almost all fossils are found. The names of several of the periods into which these three eras are subdivided reflect the names of places where palaeontologists as well as geologists worked to define distinctive layers of rock and fossil forms. The smallest units of time shown on the geological table are the epochs into which the Tertiary and Quaternary periods are divided. Individual rock formations, usually named after the locality where they are first described, are assigned to the proper epoch or period of the geological calendar. A paleontologist who wants to collect remains of bony fishes what lived in the Eocene epoch of the Tertiary period may travel to Wyoming to search among the exposed rocks of the Green River Formation, as the strata in that location are known to have formed at that time.

Absolute Age of the Earth

Early Attempts at Absolute Geochronology

After having constructed a geologic time scale on the basis of relative age, it is understandable that geologists would seek some way to assign absolute ages in millions of years to the various periods and epochs. From the time of Hutton, leaders in the scientific community were convinced that the earth was indeed very old, and certainly it was much older than the approximately 6000 years estimated by biblical scholars from calculations involving the ages of post-Adamite generations. But how old was the earth? And how might one quantify the geologic time scale?

To geologists of the 1800's it was apparent that to determine the absolute age of the earth or of particular rock bodies, they would have to concentrate on natural processes that continue in a single recognizable way and that also leave some sort of tangible record in the rocks. Evolution is one such process, and Charles Lyell recognized this. By comparing the amount of evolution exhibited by marine mollusks in the various series of the Tertiary System with the amount that had occurred since the beginning of the Pleistocene ice age, Lyell estimated that 80 million years had elapsed since the beginning of the Cenozoic. He came astonishingly close to the mark. However, for older sequences, estimates based on rates of evolution were difficult, not only because of missing parts in the fossil record, but also because rates of evolution for many taxa were not well understood.

In another attempt, geologists reasoned that if rates of deposition could be determined for sedimentary rocks, then they might be able to estimate the time required for deposition of a given thickness of strata. Similar reasoning suggested that one could estimate total elapsed geologic time by dividing the average thickness of sediment transported annually to the oceans into the total thickness of sedimentary rock that had ever been deposited in the past. Unfortunately, such estimates did not adequately account for past differences in rates of sedimentation or losses to the total stratigraphic section during episodes of erosion. Also, some very ancient sediments were no longer recognizable, having been converted to igneous and metamorphic rocks in the course of mountain building. As a result of these uncertainties, estimates of the earth's total age based on sedimentational rates ranged from as little as a million to over a billion years.

Some types of sediment to permit one to know the rate of deposition quite accurately. For example, shales deposited in lakes often exhibit summer and winter layer that differ in colour and texture and can be counted much like growth rings in trees. The regularly repeated layers are called *varves*. Each summer-winter couplet represents deposition during 1 year, and an entire sequence of varves provides a record of the exact number of years that a particular environment existed. An illustration of the use of varves is provided by a study completed in 1929 by W.H. Bradley. By counting the varves in the shales of Wyoming's Green River Formation, Bradley was able to show that 6½ millions years were required to deposit a thickness of 790 meters of shale.

In yet another geochronologic scheme, investigators attempted to determine the total age of the oceans. They speculated that the oceans' basins had been filled very shortly after the origin of the planet and thus would be only slightly younger in age than the earth itself. The best known of the calculations for the age of the oceans were made by the distinguished Irish geologist John Joly. From information provided by gauges placed at the mouths of streams, Joly was able to estimate the annual increment of salt to the oceans. Then, knowing the salinity of ocean water and the approximate volume of water, he calculated the amount of salt already held in solution in the oceans. An estimate of the age of the oceans was derived from the following formula:

$$\frac{\text{Total salt content in ocean (in grams)}}{\text{Rate of salt added each year (grams per year)}}$$

equals

Age of ocean (in years)

Beginning with essential nonsaline oceans, it would have taken about 90 million years for the oceans to reach their present salinity, according to Joly. The figure, however, was off the mark by a factor of 50, largely because there was no way to account accurately for recycled salt and salt incorporated into clay minerals deposited on the sea floors. Vast quantities of salt once in the sea had become extensive evaporite deposits on land; some of the salt begin carried back to the sea had been dissolved, not from primary rocks, but from eroding marine strata on the continents. Even though in error, Joly's calculations clearly supported those geologists who insisted on an age for the earth far in excess of a few million year. The belief in the earth's immense antiquity was also supported by Darwin, Huxley, and other evolutionary biologists who saw the need for time in the hundreds of millions of years to accomplish the organic evolution apparent in the fossil record.

The opinion of the geologists and biologists that the earth was immensely old was soon to be challenged by the physicists. Spearheading this attack was Lord Kelvin, considered by many the outstanding physicist of the nineteenth century. Kelvin calculated the age of the earth on the assumption that it had cooled form a molten state and that the rate of cooling followed ordinary laws of heat conduction and radiation. Kelvin estimated the number of years it would have taken the earth to cool from a hot mass to its present condition. His assertions regarding the age of the earth varied over 2 decades of debate, but in his later years he confidently believed that 24 to 40 million years was a reasonable age for the earth. The biologists and geologists found Kelvin's estimates difficult to accept. But how could they do battle against his elegant mathematics when they were themselves armed only with inaccurate dating schemes and geologic intuition? For those geologists unwilling to capitulate, however, new discoveries showed their beliefs to be correct and Kelvin's to be unavoidably wrong.

The correct answer to the question "how old is the earth?" was provided only after the discovery of radioactivity, a phenomenon unknown to Kelvin during his active years. With the detection of natural radioactivity by Henri Becquerel in 1896, followed by the isolation of radium by Marie and Pierre Curie 2 years later, the world became aware that the earth had its own built-in source of heat. It was not inexorably cooling at a steady and predictable rate, as Kelvin had suggested.

Radiometric Methods of Dating the Earth

Radioactivity

The radioactivity discovered by Henri Becquerel was a consequence of the fact that some elements, such as uranium and thorium, are unstable. Such elements will decay to form other elements or other isotopes of the same element. To understand what is meant by "decay," let us consider what happens to a radioactive element like uranium 238. Uranium 238 has an atomic weight of 238. The "238" represents the sum of the atom's protons and neutrons (each proton and neutron having a "weight" of 1). Uranium has an atomic number (number of protons) of 92. Such atoms with specific atomic number and weight are sometimes termed *nuclides*. Sooner or later (and entirely spontaneously) the uranium 238 atom will fire off a particle from the nucleus called an *alpha particle*. Alpha particles are positively charged ions of helium. They have an atomic weight of 4 and an atomic number of 2. Thus, when the alpha particle is emitted, the new atom will have an atomic weight of 234 and an atomic number of 90. From the decay of the parent nuclide, uranium 238, the daughter of the nuclide, thorium 234, is obtained. A shorthand equation for this change is written:

$$^{238}_{92}U \rightarrow ^{234}_{90}Th \ ^{4}_{2}He$$

This change is not, however, the end of the process, for the nucleus of thorium 234 is not stable. It eventually emits a beta particle (an electron discharged from the nucleus when a neutron splits into a proton and an electron). There is now an extra proton in the nucleus but no loss of atomic weight because electrons are essentially weightless. Thus, from $^{234}_{90}Th$ the daughter element $^{234}_{91}Pa$ (protactinium) is formed. In this case, the atomic number has been increased by one. In other instances, the beta particle may be captured by the nucleus, where it combines with a proton to form a neutron. The loss of the proton would decrease the atomic number by one.

A third kind of emission in the radioactive decay process is called *gamma radiation*. It consists of a form of invisible electromagnetic waves having even shorter wavelengths than do x-rays.

As alpha and beta particles, as well as gamma radiation, move through the surrounding materials, their energy is transformed into increased activity of the electrons in the atoms of the surrounding medium. The result is heat. This radiogenic source of heat was the unknown entity in Lord Kelvin's calculations of the earth's thermal history.

Clocks in the Rocks (Radiometric Dating)

Nuclear adjustments such as those described previously occur many times before a final, stable daughter element, such as lead, is formed. The rate at which the steps in the process takes place is unaffected by changes in temperature, pressure, or the chemical environment, since these do not involve the nucleus. Indeed, one can confidently assume that the rate of decay of long-lived isotopes has not varied since the earth came into existence. Therefore, once a quantity of radioactive nuclides has been incorporated into a growing mineral crystal, the quantity will begin to decay at a steady rate with a definite percentage of the radiogenic atoms undergoing decay in each increment of time. Each radioactive element has a particular mode of decay and a unique decay rate. As time passes, the quantity of the parent nuclide diminishes and the amount of daughter atoms increases, thereby indicating how much time has elapsed since the clock began its time-keeping. The

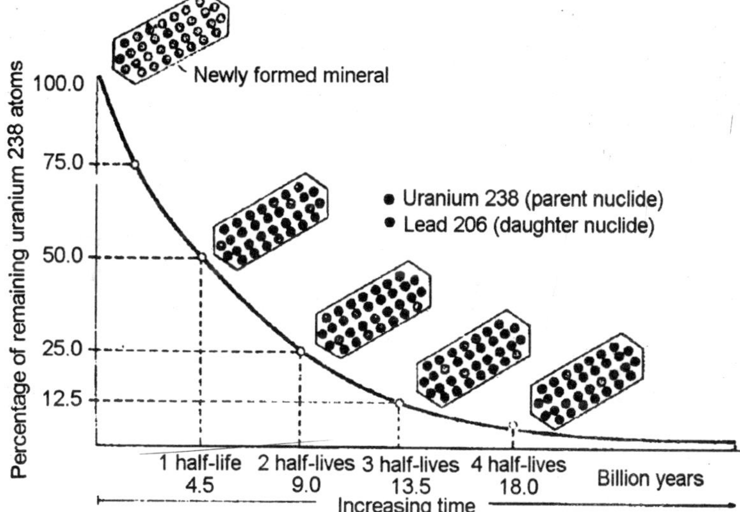

Fig. 8.2. Rate of radioactive decay of uranium 238 to lead 206. During each half-life, one half of the remaining amount of the radioactive element decays to its daughter element.

"beginning", of "time zero," for any mineral containing radioactive nuclides would be the moment when the radioactive parent atoms became part of a mineral from which daughter elements could not escape. The retention of daughter elements is essential, for they must be counted to determine the original quantity of the parent nuclide.

The determination of the ratio of parent to daughter nuclides is usually accomplished with the use of mass spectrometer, an analytical instrument capable of separating and measuring the proportions of minute particles according to their mass differences. In the mass spectrometer, samples of elements are vaporized in an evacuated chamber, where they are bombarded by a stream of electrons. This bombardment knocks electrons off the atoms, leaving them positively charged. A stream of these positively charged ions is deflected as it passes between plates that bear opposite charges of electricity. The degree of deflection is proportional to the masses of atoms.

Not all radioactive decays are measured by means of mass spectrometer. In the case of carbon 14, which decays by beta particle emission, the measurement of nuclides is accomplished in directly by the use of a very sensitive *geiger counter*.

Of the three major families of rocks, the igneous clan lends itself best to radiometric dating. The dates obtained from such rocks indicate the time that a silicate melt containing radioactive elements solidified. In contrast to igneous rocks, sedimentary rocks can only rarely be dated radiometrically. Some dates for sedimentary strata have been obtained from a mineral called *glauconite*, which is believed to form "in place" at the time of deposition. This greenish mineral contains radioactive potassium 40, which decays to argon 40 and can be used in geochronology. Because of possible losses the daughter element argon, care must be taken in interpreting dates, however, in most instances, potassium-argon dates derived from glauconite are considered minimal ages for the enclosing strata. As for classic sedimentary rocks that contain radioactive elements in their detrital mineral grains, the ages obtained refer to the parent rock that was eroded and is older than the sedimentary layer.

Dates obtained from metamorphic rocks may also require special care in interpretation. The age of a particular mineral may record the time the rock first formed or any one of a number of subsequent metamorphic recrystallizations.

There are many other problems that can affect the validity of a radiometric age. If some of the daughter products are removed from the sample by weathering or leaching, its age would be underestimated. If the element being analyzed was a gas, some of that gas (as with argon in glauconite) might have diffused out of the rock. The heat accompanying burial or mountain building might enhance such losses. There is also the possibility that at a later time older rocks may be partially remelted so that the age would be that of the second, rather than the initial, melting event. Clearly, great care must be taken in understanding the field relationships of the rock masses under investigation and in selecting samples.

Once an age has been determined for a particular rock unit, it is sometimes possible to use that date to approximate the age of a adjacent rock masses. A shale lying below a lava flow that is 110 million years old and above another flow dated at 180 million years old must be between 110 and 180 million years of age. Fossils within the shale might permit one to assign it to a particular geologic system or series and, by correlation, to extend the age data around the world.

Half-Life

There is no way that one can predict with certainty the moment of disintegration for any individual radioactive atom in a mineral. We do know that it would take an infinitely long time for all of the atoms in a quantity of radioactive elements to be entirely transformed to stable daughter products. Experimenters have also shown that the decline in the number of atoms is rapid in the early stages but becomes progressively slower in the later stages. One can statistically forecast what percentage of a large population of atoms will decay in a certain amount of time.

Because of these features of radioactive, it is convenient to consider the number of years needed for half of the original quantity of atoms to decay. This span of years is termed the *half-life*. Thus, at the end of the years constituting one half-life, one half of the original quantity of radioactive element still has not undergone decay. After another half-life, one half of what was left is halved, so that one fourth of the original quantity remains. After a third half-life, only one eighth would remain, and so on.

Every radioactive nuclide has its own unique half-life. Uranium 235, for example, has a half-life of 704 million years. Thus, if a sample contains 50 per cent of the original amount of uranium 235 and 50 per cent of its daughter product, lead 207, then that sample is 704 million years old. If the analyses indicate 25 per cent of uranium 235 and 75 per cent of lead 207, two half-lives would have elapsed, and the sample would be 1408 million years old.

The Principal Geologic Timekeepers

At one time, there were many more radioactive nuclides present on earth than there are now. Many of these had short half-lives and have long since decayed to undetectable quantities. Fortunately, for those interested in dating the earth's most ancient rocks, there remain a few long-lived radioactive nuclides. The most useful of these are uranium 238, uranium 235, rubidium 87 and potassium 40. There are also a few short-lived radioactive elements that are used for dating more recent events. Carbon 14 is an example of such a short-lived isotope. There are also short-lived nuclides that represent segments of a uranium or thorium decay series.

Timekeepers That Produce Lead

Dating method involving lead require radioactive nuclides of uranium or thorium that were incorporated into the earth's crust when it congealed. To determine the age of a sample of mineral or rock, one must known the original quantity of parent nuclides as well as the quantity remaining at the present time. The original number of parent atoms should be equal to the sum of the present quantity of parent atoms and daughter atoms. This raises the question of whether or not some of the lead may not have already been in the mineral and, if not detected, cause its radiometric age to exceed its true age. Lead 204, which is never produced by decay, provided a means of detecting original lead. All common lead contains a mixture of four lead isotopes. In most minerals used for dating, the proportions of the lead isotopes are nearly constant, so that lead 204 can be used to calculate the quantities of original lead 206 and lead 207. These quantities can then be subtracted from the total to give the amount due to radioactivity.

As we have seen, different isotopes decay at different rates. Geochronologists take advantage of this fact by simultaneously analyzing two or three isotope pairs as a means to cross-check ages and detect errors. For example, if 234U/207Pb radiometric ages and the 238UU/206 Pb ages agree, then they are said to be *concordant*; there then exists a high probability that the radiometric age is valid. Uranium-lead ages that vary widely are said to be *discordant*. However, even the best of concordant dates do not agree perfectly but are expected to vary within reasonable limits. Unavoidable losses or gains of isotopes by interactions with surrounding solutions or from the heat accompanying geologic processes are the usual causes for variance. Radiometric ages should be considered reasonable approximations of true age.

Radiometric that depend upon uranium/lead ratios may also be checked against ages derived from lead 207 to lead 206. Because the half-life of uranium 235 is much less that the half-life of uranium 238, the ratio of lead 207 (produced by the decay of uranium 235) to lead 206 will change regularly with age and can be used as a radioactive timekeeper.

The Uranium Method

The solution of the problem of determining geological age finally came from a most unexpected source—the radioactive disintegration of the atom. Radioactive disintegration of the element uranium has given the most reliable measurement of geological time that we now possess.

To say that uranium is radioactive is to say that uranium is unstable. In the course of time it loses electrons and particles from the nucleus. It should be recalled that the chemical identity of an element depends upon the numbers and type of particles of which it is composed. When uranium loses these particles it is not longer uranium, but has become something else. Most natural uranium has an atomic weight of 238; it disintegrates, through a number of intermediate products, to finally become lead with an atomic weight of 206. (Lead 206 is a rare form; most lead has an atomic weight of 207).

The time that any particular uranium atom will disintegrate is uncontrolled and unpredictable, but when large numbers of atoms are studied, it is possible to give a highly accurate figure for the average life expectancy, of 'uranium atom'. The figure is a fantastic one: There is a 50-50 chance that given uranium atom will have disintegrated in 7,600,000,000 years. This means that in a sample of many uranium atoms, half of them would disintegrate in this length of time. This value is called the half life, in this case the half-life of uranium.

We could express this phenomenon in still another way. If it were possible for us to place a 10-grm sample of uranium in a container, leave it for 7,600,000,000 years, and then measure what was left, we would find that 5 grams of uranium remained and that 5 grams had been converted into lead 206. During the next 7,600,000,000 years half of the remaining 5 grams of uranium would have disintegrated into lead. This process would continue, and with the passage of each half-life period of 7,600,000,000 years, half of the uranium present at the beginning of the period would have disintegrated. An original 10 grams would be reduced to 5, then to 2.5 and to 1.25 grams in three half-life periods. The rate of spontaneous radioactive disintegration is not affected by any conditions that are encountered in nature. Therefore, radioactive disintegration occurs at a rate so constant that it can be used as a measure of time. If, for example, 10 grams of uranium had been sealed in some sedimentary rocks formed 7,600,000,000 years ago, an analysis today would show 5 grams of uranium and 5 grams of lead 206. The ratio of uranium to lead 206 can be used to determine the age of rocks containing uranium. The formula is :

$$\text{Age in years} = \frac{\text{amount of lead 206}}{\text{amount of uranium}} \times 7,600,000,000$$

In 'a' a is region where sedimentary rocks are forming is shown. This could be at the delta of a river or the bottom of a lake or ocean. In 'b' a volcanic eruption has occurred and lava has flowed out over the area where the sedimentary rocks are forming. This lava would contain many chemical substance, and for the sake of our discussion let us assume that in this case it includes uranium. It would be unlikely to contain lead would be lead 207. In 'a' the lava has been covered with more sedimentary rocks. If a geologist were to discover rocks like those shown in 'c', a determination of the ratio of lead 206 to uranium in the lava would reveal when these rocks were formed.

This is a splendid method, but unfortunately there are only a few cases where all the geological requirements for accurate measurement

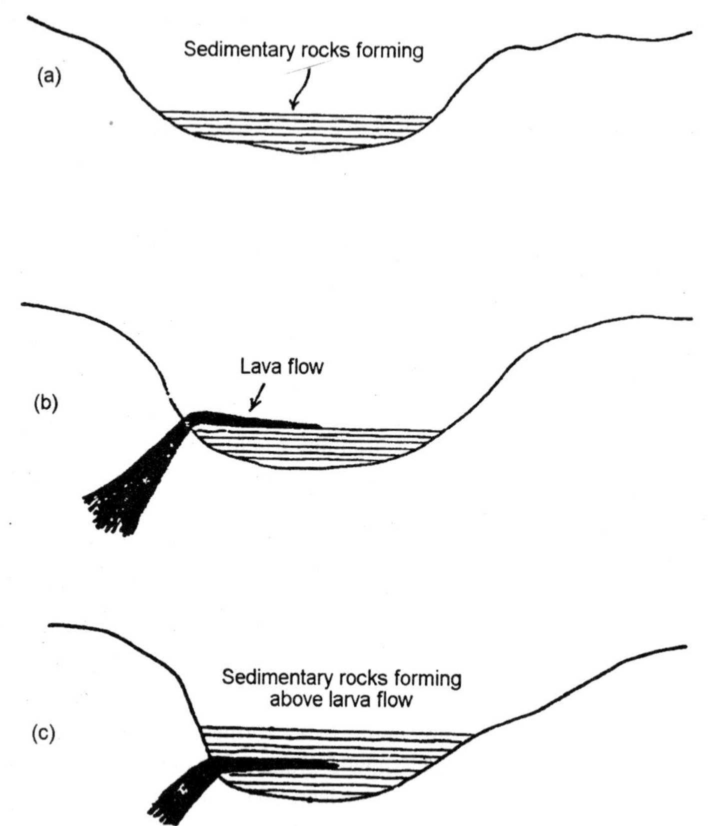

Fig. 8.3. The uranium method for determining the age of sedimentary rocks.

have been met. The time interval of greatest interest for those concerned with evolution is that covered since the beginning of the Cambrian. There are only five reliable age determinations for this interval. These are as follows:

Late Cambrian	440 million years ago
End of Ordovician	350 million years ago
End of Devonian	255 million years ago
Late Carboniferous	215 million years ago
Early Tertiary (Paleocene)	58 million years ago

The oldest rocks so far discovered are about two billion years old. These belong to the Proterozoic era and were found in Canada. The age of the earth is thought to be much greater. Several independent methods suggest a figure of approximately 4,500,000,000 years.

It is possible to use these data, together with the estimates of thickness of sedimentary rocks, to obtain tentative values for all the geological periods. For example, there is a 95-million-year interval between the end of the Ordovician and the end of the Devonian. This interval includes both the Silurian and the Devonian periods. How long was each? It has been found that the amount of sedimentary rock formed in the Devonian period is greater than in the Silurian. The Devonian accounts for 61 per cent and the Silurian for the remaining 39 per cent of the total formed in both periods. If we assume the rate of rock formations to be the same in the two periods, then 39 per cent of the 95-million year interval, or 37 million years, would be the length of the Silurian period, and the remaining 58 million years would be the length of the Devonian period. These are tentative values that can be used until more accurate ones are available from the uranium method.

By similar methods, the beginning of the Cambrian, which is the earliest time from which well-preserved and abundant fossils are found, is placed at approximately 520 million year ago.

In the years since the second, World War, many new methods for determination of geological age have been introduced and are being tested. Many additional measurements of the age of geological formations can be expected in the near future.

The Potassium-Argon Method

Potassium and argon are another radioactive pair widely used for dating rocks by means of electron capture (causing a proton to be transformed into a neutron), about 11 per cent of the potassium 40 in a mineral decays to argon 40, which may then be retained within the parent mineral. The remaining potassium 40 decays to calcium 40 (by emission of a beta particle from a neutron, thereby transforming it into a proton). The decay of potassium 40 to calcium 40 is not useful for obtaining radiometric ages, because radiogenic calcium cannot be distinguished from original calcium in a rock. Thus, geochronologists concentrate their efforts on the 11 per cent of potassium 40 atoms that decay to argon. One advantage of using argon is that it is inert—that is, does not combine chemically with other elements. Argon 40 found in a mineral is very likely to have originated there following the decay of adjacent potassium 40 atoms in the mineral.

Another advantage to the potassium-argon scheme for dating rocks is that potassium 40 is an abundant constituent of many common minerals, including mica, feldspars, and hornblendes. However; like

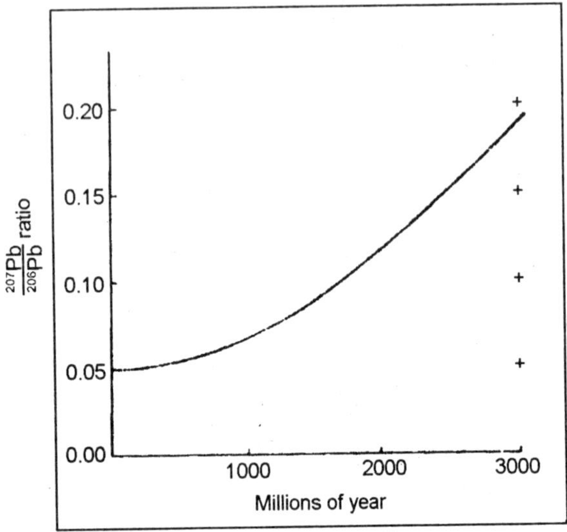

Fig. 8.4. Graph showing how the ratio of lead 207 to lead 206 can be used as a measure of age.

all radiometric methods potassium-argon dating is not without its limitations. A sample will yield a valid age only if none of the argon has leaked out of the mineral being analyzed. Leakage may indeed occur if the rock has experienced temperatures above about 125°C. In specific localities, the ages of rocks dated by this method reflect the last episode of heating rather than the time of origin of the rock itself. A less serious problem is mechanical entrapment of atmospheric argon in flowing lavas.

The half-life of potassium 40 is 1251 million years. If the ratio of potassium 40 to argon 40 is found to be 1 to 1, then the age of the sample is 1251 million years, if the ratio is 3 to 1, then yet another half-life has elapsed, and the rock would have a radiogenic age of two half-lives, or 2502 million years.

Potassium-argon is widely used in deciphering various types of geologic problems. Geologists are now using the method of studies relating to sea floor movements. For many years, scientists have been curious about the alignment of the major Hawaiian Islands and the adjacent seamounts. With the advent of the theory of sea floor spreading, scientists developed the concept that these volcanic islands were built over a relatively fixed "hot spot" deep in the upper mantle. Conduits from the "hot spot" brought lavas up to the sea floor, where eruptions periodically occurred. Volcanoes that developed over the "hot spot" were then conveyed along by sea floor movement, and new volcanoes were produced over the vacated position. Geologists reasoned that if this process had taken place in Hawaiian Islands, then potassium-argon radiometric ages should change in sequence along the island-seamount chain. The dates do indeed support the theory, and they even suggest that the direction of movement changed from a more northerly trend to northwesterly trend about 40 million years ago.

The Rubidium-Strontium Method

The dating method based on the disintegration by beta decay of rubidium 87 to strontium 87 can sometimes be used as a check on potassium-argon dates, because rubidium and potassium are often found in the same minerals. The rubidium-strontium scheme has further advantage in that the strontium daughter nuclide is not diffused by relatively mild heating events, as is the case with argon.

In the rubidium-strontium method, a number of samples are collected free from the rock body to be dated. With the aid of the mass spectrometer, the amounts of radioactive rubidium 87, its daughter product strontium 87, and strontium 86 are calculated for each sample. Strontium is an isotope not derived from radioactive decay. A graph is then prepared in which 87Rb/86Sr ratio in each sample is plotted against the 87Sr/86 Sr ratio. From the points on the graph, a straight line is constructed that is termed as *isochron*. The slope of the isochron results from the fact that, with the passage of time, there is continuous decay of rubidium 87, which causes the rubidium 87/strontium 86 ratio to decrease. Conversely, the strontium87/strontium 86 ratio increases as strontium 87 is produced by the decay of rubidium 87. The older the rocks being investigated, the more the original isotope ratios will have been changed, and the greater will be the inclination of the isochron. The slope of the isochron permits a computation of the age of the rock.

The rubidium-strontium and potassium-argon methods need not always depend upon the collection of discrete mineral grains containing the required isotopes. Sometimes the rock under investigation is so finely crystalline and the critical minerals so tiny and dispersed that it is difficult or impossible to obtain a suitable collection of minerals. In such instances, large samples of the entire rock may be used forage determination. This method is called *whole-rock analysis*. It is useful not only for fine-grained rocks but also for rocks in which the yield of useful isotopes from mineral separates is too low for analysis. Whole-rock analysis has also been useful in determining the age of rocks that

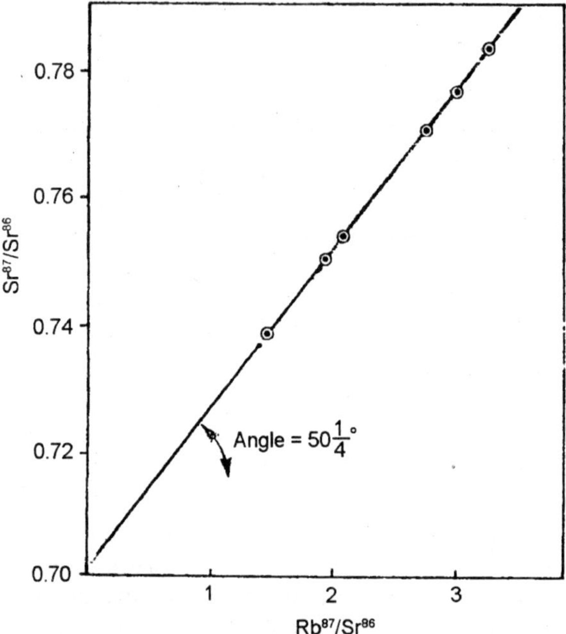

Fig. 8.5. Whole-rock rubidium-strontium isochron for a set of samples of a Precambrian granite body.

have been so severely metamorphosed that the potassium-argon or rubidium-strontium radiometric clocks of individual minerals have been reset. In such cases, the age obtained from the minerals would be that of the episode of metamorphism not the total age of the rock itself. The required isotopes and their products, however, may have merely moved to nearby locations of large chunks of the whole rock may provide valid radiometric age determinations.

The Carbon 14 Method

Techniques for age determination based on content of radiocarbon were first devised by W. F. Libby and his associates at the University of Chicago in 1947. It has become an indispensable aid to archaeologic research and frequently is useful in deciphering the very recent events in geologic history. Because of the short half-life of carbon 14—a mere 5730 years—organic substances older than about 40,000 years no longer contain carbon 14 in measurable amounts.

Unlike uranium 238 and rubidium 87, carbon 14 is created continuously in the earth's upper atmosphere. The story of its origin begins with cosmic rays, which are extremely high-energy particles (mostly protons) that bombard the earth continuously. Such particles strike atoms in the upper atmosphere and split their nuclei into smaller particles, among which are neutrons. Carbon 14 is formed when a neutron strikes an atom of nitrogen 14. As a result of the collision, the nitrogen atom emits a proton and becomes carbon 14. Radioactive carbon is being created by this process at the rate of about two atoms per second for every square centimeter of the earth's surface. The newly created carbon 14 combines quickly with oxygen to form CO_2 which is then distributed by wind and water currents around the globe, it soon finds its way into photosynthetic plants, because they utilize carbon dioxide from the atmosphere to build tissues. Plants containing carbon 14 are ingested by animals, and the isotope becomes a part of their tissues as well.

Eventually, carbon 14 decays back to nitrogen 14 by emission of a beta particle. A plant removing CO_2 from the atmosphere should receive a share of carbon 14 proportional to that in the atmosphere. A state of equilibrium is reached in which the gain in newly produced carbon 14 is balanced by the decays loss. The rate of production or carbon 14 has varied somewhat over the past several thousand years. As a result, corrections in age calculations must be made. Such correction are derived from analyses of standards such as wood samples, whose exact age is known.

The age of some ancient bit of organic material is not determined from the ratio of parent to daughter nuclides, as is done with previously discussed dating schemes. Rather, the age is estimated from the ratio of carbon 14 to all other carbon 14 already present in the once living organism begins to diminish in accordance with the rate of carbon 14 decay. Thus, if the carbon 14 fraction of the total carbon in a piece of pine tree buried in volcanic ash were wound to be about 25 per cent of the quantity in living pines, the age of the wood (and the volcanic activity) would be two half-lives, or 11,460 years. To allow for unavoidable error, the age of the wood might be expressed as 11,460 ± 250 years.

The carbon 14 technique has considerable value to geologists studying the most recent events of the Pleistocene ice age. Prior to the development of the method, the age of sediments deposited by the last advance of continental glaciers was surmised to be about 25,000 years. Radiocarbon dates of a layer of peat beneath the glacial sediments provided an age of only 11,400 years. The method has also been found useful in dating the geologically recent uppermost layer of sediment on the sea floors. For the deeper and older marine sediments, however, the thorium method described further on in employed.

Methods Involving Thorium 230

The past 2 decades of intensive exploration of the sea floor has prompted the development of yet other dating methods are the especially useful for oceanic sediments too old to be dated with carbon 14. These new techniques utilize isotopes that are produced in the intermediary stages of the uranium decay series. Scientists who developed these methods recognized that most of the uranium brought to the oceans by streams remains in solution. While in solution it decays, eventually producing thorium 230. The thorium isotope is precipitated and becomes a component of ocean floor sediments. Thorium 230 itself decays with a half-life of 75,000 years. Because lower levels of the sediment have been undergoing decay longer, geochronologists are able to detect the expected decrease in quantity of thorium 230 at greater depths in a cored column of sediment. The thorium 230 concentration of each measured interval is compared to the quantity in the surface layer, and a time scale relating age to quantity of remaining parent nuclides is then formulated.

Deep sea sediments can also be dated by means of a procedure based upon thorium and protactinium. Thorium 230 has a half-life of 75,000 years and is in the decay series from uranium 238. Protactinium 231 has a half-life of only 34,300 years and is in the line of descent from uranium 235. Both parent nuclides are precipitated in the same proportions, but because of their different rates of decay, the ratio of the two changes regularly with time. Thus the greater the differences in the quantity of undecayed parent isotopes, the older the sediment.

Nuclear Fission Track Timekeepers

Nuclear particle fission tracks were discovered about 20 years ago when scientists using the electron microscope were able to examine the areas around presumed locations of radioactive particles that were embedded in mica. Closer examination showed that the tracks were really small tunnels—like bullet holes—that were produced when high-energy particles of the nucleus of uranium were fired off in the course of spontaneous fission (spontaneous fragmentation of an atom into two or more lighter atoms and nuclear particles). The particles speed through the orderly rows of atoms in the crystal, tearing away electrons from atoms located along the path of trajectory and rendering them positively charged. Their mutual repulsion produces the track. The tracks are only a few atoms in width and are impossible to see without an electron microcopy. Therefore, the sample is immersed for a short period of time in a suitable solution (acid or alkali), which rushes up into the tubes, enlarging the track tunnel so that it can be seen with an ordinary microscope.

The natural rate of track production by uranium atoms is very slow and occurs at a constant rate. For this reason, the tracks can be used to determine the number of years that have elapsed since the uranium-bearing mineral solidified. One first determines the number of uranium atoms that have already disintegrated. This number is obtained with the aid of microscope by counting the etched tracks. Next, one must find the original number of uranium atoms. This quantity can be determined by bombarding the sample with neutrons in a reactor and thereby causing the remaining uranium to undergo fission. A second count of tracks reveals the original quantity of uranium. Finally, one must know the spontaneous fission decay rate for uranium 238. This information is determined by counting the tracks in a piece of uranium-bearing synthetic glass of known date of manufacture.

Fission track dating is of particular interest to geochronologists because it can potentially be used to date specimens only a few centuries old as well as to date rocks billions of years in age. The method helps to date the period between about 40,000 and 1 million years ago—a period for which neither carbon 14 nor potassium-argon methods are suitable. As with all radiometric techniques, however, there can be problems. If rocks have been subjected to high temperatures, tracks may heal an fade away.

The Age of the Earth

Anyone interested in the total age of the earth must decide what event constitutes it "birth". Most geologists assume "year 1" commenced as soon as the earth had collected most of its present mass and had developed a solid crust. Unfortunately, rocks that date from those earliest years have not been found on earth. They have long since been altered and converted to other rocks by various geologic processes. The oldest rocks on earth that have been dated thus far include 3.4 billion-year-old granites from the Barberton Mountain Land of South Africa, 2.7-billion year old granites of southwestern Greenland, and Minnesota. These dates permit us to say the planet is at least about 3.7 billion years.

Meteorites, which many consider to be remnants of a disrupted planet that originally formed at about the same time as the earth, have provided uranium-lead and rubidium-strontium ages of about 4.6 billion years. From such data, and from estimates of how long it would take to produce the quantities of various lead isotopes now found on the earth, geochronologists feel that the 4.6-billion-year age for the earth can be accepted with confidence. Substantiating evidence for this conclusion comes from returned moon rocks. The ages of these rocks range from 3.3 to about 4.6 billion years. The older age determinations are derived from rocks collected on the lunar highland which may represent the original lunar crust. Certainly, the moons and planets of our solar system originated as a result of the same cosmic processes and at about the same time.

9

MICROFOSSILS

Microfossils are the remains of organisms, whole or fragmentary the study of which requires the use of a microscope. Most are unicellular, including members of the Kingdom Monera (i.e. the procaryotic cyanobacteria and bacteria) and of the Kingdom Protoctista (i.e. eucaryotes comprising the highly diverse phytoplankton and zooplankton). Calcareous algae are also included here following the present practice of treating algae as multicellular protoctists rather than as plants. There are also the minute remains of small multicelled animals, the ostracods and conodonts. A further category includes microscopic fossils of uncertain affinity, the acritarchs and chitinozoans. Pollen grains and spores, the study of which is known as palynology, are dealt with in the chapter on plants.

KINGDOM MONERA-PROCARYOTES

Procaryotes are microscopic, one-celled, simple organisms distinguished by their lack of an organised nucleus and organelles. They comprise cyanobacteria or cyanophytes or blue—green algae and bacteria. They have no skeleton and their cells are rarely preserved. Yet they are of major geological significance, especially in studies of the Precambrian and the evolution of early life.

Cyanobacteria

Cells of cyanobacteria are rounded (coccoid) to cylindrical in shape and range up to about 25μm diameter. They occur singly or as groups of cells held within a mucilage-covered cellulose sheath; often cells are arranged in series to form single or clustered filaments. They contain chlorophyll and make food by using light for photosynthesis

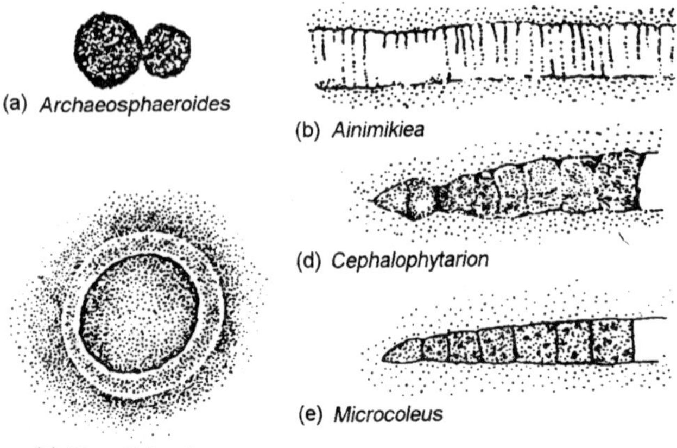

Fig. 9.1. Cyanobacteria, single cells and filaments, greatly magnified. (a), (b), from the Gunflint Chert, Precambrian. (c), (d), from the Bitter Springs Chert, Precambrian. (e), recent form to compare with (d).

releasing oxygen from water in the process. Some fix atmospheric nitrogen Reproduction is by simple fission.

Cyanobacteria are tolerant of a wide range of conditions in water. They are resistant to extremes of temperature ranging from hot springs to pools in ice; some lives in soils; some associate with other organisms, as for instance, in symbiosis with fungi in lichens; some are common in reefs. Their resting spores are resistant to desiccation. They tolerate low oxygen levels and ultraviolet radiation. They contain a blue pigment, phycocyanin, which enables them to photosynthesis at very low light levels; they have been found at depths of 1000 m in clear tropical waters of the Indian Ocean.

Cyanobacteria are among the most ancient known organisms dating back some 2800 my, possibly 3500 my. Over this period of time they show little change and some Precambrian fossils are remarkably like extant forms. They are involved today in the formation of various, aqueous sedimentary deposits, especially in the build-up of *Stromatolites*. These structures are of particular geological interest since they also occur in sedimentary rocks ranging in age back to early Precambrian times. Interpretation of these fossil stromatolites is based on our knowledge of the modern examples and the role of cyanobacteria in their formation.

Stromatolites

A small example of a stromatolite is a bun-shaped structure, a few centimetres across, built of numerous concentrically arranged laminae about 1 mm thick and usually consisting of calcium carbonate. Each lamina is formed by a sticky surface layer of filaments and cells of cyanobacteria with other bacteria and green algae, forming an algal mat which traps sediment washed over its surface. It may also precipitate calcium carbonate as the result of photosynthesis and the consequent abstraction of carbon dioxide from the water. As sediment accumulates, the organic layer grows through and over it to maintain its position at the surface, thus building the laminated structure. Because of the role of organisms in their formation, stromatolites are termed biogenic structures. The process of their formation can be seen today at, for instance, Shark Bay in Western Australia, an area of salty, warm intertidal shallow water. Stromatolites may also occurs in brackish or fresh water, in alkaline lakes and in pools at hot springs.

Fossils stromatolites show considerable range in form and size: they may be small and bun-shaped, or be extensive thick sheets of complex structure limestones or dolomites and sometimes in cherts. Modern examples are restricted, presumably by competition , to harsh environments hostile to most other organisms but in which the algal mat communities can flourish. In the Precambrian, however, they seem to have been more widely distributed, implying an absence of competition.

Laminated structures can, of course, be formed by purely physical processes and caution is needed in identifying them as true stromatolites in the absence of microfossils.

Fossils are extremely rare in Archaean rocks but stromatolites with microfossils are found in the Warrawoona Group in Western Australia. They are about 3500 my in age, one of the oldest examples so far discovered. The rock in which they occur is dark, carbonaceous chert which contains dark-walled coccoids and a variety of filamentous structures, some septate, which resemble bacteria and possibly cyanobacteria. Similar fossils are also found in cherty parts of stromatolitic limestones in the Fortescue Group, 2700 my old, in the same area.

In Proterozoic rocks, the lower limit of which is placed at 2500 my stromatolites are more frequent occurrence and authentic microfossils relatively common and well preserved. Examples occur in the Gunflint carbonaceous chert in Canada, about 2000 my old, which contains a diverse range of single cells and filaments. Comparable with living cyanobacteria and bacteria. Still younger stromatolites in the Bitter Springs black chert in Australia, about 900 my old, contain a much greater diversity of microfossils, about 30 species, which have been identified as cyanobacteria, bacteria, green algae and other forms unascribed.

These examples cover a span of about 2600 my. The record of genuine fossils in this period is now such as to demonstrate a gradual increase in diversity of organism associated with stromatolites. This is consistent with an increase in diversity of the stromatolites themselves, as shown by increasing complexity of the laminated structures, reaching a peak in the late Proterozoic. This development must reflect differences in types of algal mats, each reacting in characteristic way with its environment.

It is generally held that the early atmosphere was anoxic and the earliest organisms were anaerobes. With the arrival of oxygen-releasing cyanobacteria came the start of a notable change in the atmosphere which led to a gradual build-up in the level of oxygen which it contained. Oxygen is, of course, essential for the life processes metazoa which, appearing and diversifying in the late Proterozoic, competed with and disturbed the algal mats to an increasing degree. Invertebrates doubtless grazed the mats or burrowed through them. In the lower Cambrian, archaeocyathans built extensive reefs in shallow waters, thus usurping the procaryote niche. Later, in the Ordovician, other reef-builders, mainly sponges and calcareous algae, appeared and the stromatolites declined. They did not again become prominent in the marine environment.

Calcareous cyanobacteria

Calcium carbonate is deposited around the filaments of some cyanobacteria during photosynthesis, thus forming calcified tubes. One such is *Girvanella* (Cambrian-Recent) which occurs as small nodular masses. A thin-section shows a tangled mass of flexuous tubes 8-30 μm in diameter. *Girvanella* occurs in shallow-water marine limestones and reefs.

Bacteria

Bacteria are diverse procaryote cells occurring singly or as aggregates. They are rod-like, spherical or spiral in shape, measuring about 0.25-2 μm in diameter. Many are motile, moving by one or more lashing threads, flagella. They multiply by simple fissions and some, in unfavourable conditions, from spores which can withstand desiccation and, for a time, high temperatures. This is known as a resting stage and will be followed by regeneration of the motile stage if conditions become favourable again.

Most bacteria are *Heterotrophs*, i.e. they live on organic matter, dead. Some are *Autotrophs*, i.e. they synthesis their organic constituents from simple inorganic materials.

Bacteria are abundant and ubiquitous on land, fresh water and sea. They are mainly aerobic but some are anaerobic. Most do not contain chlorophyll and do not photosynthesis; they do not tolerate bright light or ultraviolet radiation. They are however, tolerant with regard to salinity, temperature and pressure, and are found living, for example, at the high-temperature hydrothermal vents of the East Pacific oceanic ridge where the water pressure is about 250 atmosphere.

Bacteria form an important part of the food chain either as a direct food source for animals such as protozoans, worms, sponges, filter-feeders and mud-eaters, or indirectly, by breaking down organic matter into substances which can then be used by higher organisms. Some can fix nitrogen from the atmosphere thus making it available to plants.

Of great geological interest is bacterial activity in sediments which results in chemical and mineralogical changes. Organic matter in muds is partly decomposed by methanogens belonging to the primitive archaebacteria, producing methane; ferric hydroxide is reduced to the ferrous form, and sulphate is reduced to hydrogen sulphide, these two product reacting to form ferrous sulphide and, eventually, pyrites. In different circumstances bacterial activity may result in the formation of calcium carbonate from calcium sulphate or of sulphur from calcium sulphate; or the direct precipitation of ferric hydroxide by 'iron' bacteria. Bog iron ore is formed in this way. In all these cases the presence of bacteria is recorded not so much by their presence as fossils, but by their products.

Fossil bacteria will obviously be extremely rare but forms similar to extant bacteria have been recorded from rocks, mainly cherts, ranging in age back to the Archaean. One example is the Gunflint Chert, 2000 my old; another is carbonaceous chert in the 3500-my old Warrawoona Group which contains a variety of bacteria-like fossils. Other instances have been recorded form limestones, iron, and manganese ores, oil shales, coals and in plant and animal tissues.

Kingdom Protoctista

Protoctists include a diversity of aquatic, mainly marine and aerobic eucaryotes. It is convenient to distinguish two groups: plant-like *Autotrophs*, 'algae', with chloroplasts, able to photosynthesis their food, and animal-like *Heterotrophs*, 'protozoans' which feed on other organisms. Only those which secrete a skeleton are considered here. The skeleton may be organic, calcareous or siliceous.

The algae range from the one-celled, largely planktonic forms which make up the phytoplankton to the many-celled benthic seaweeds. They are confined to the photic zone and are very important in marine ecology; they provide food, directly or indirectly, for all aquatic creatures and, by their photosynthesis, oxygenate their environment.

The protozoans are one-celled and include both planktonic forms, the zooplankton, and benthic forms.

The primarily unicellular forms are sometimes segregated in a more restricted Kingdom, Protista, which excludes the seaweeds.

Phytoplankton

Phylum Dinophyta (Silurian-Recent)

Dinophytes or more usually, dinoflagellates, are plant-like by virtue of having cellulose in the cell wall and chlorophyll. They also have reddish, light-absorbing pigment from which they take in alternative name: pyrrophyte (fire plant). The shape is varied: it may be spherical, ovoid or polygonal and there may be horn-like projections. They measure about 20-200μm. Some are fresh-water forms but typically they are marine phytoplankton. They are mostly motile, moving by means of two flagella which lie in furrows and work respectively transversely and lengthwise. Non-motile forms include *Zooxanthellae* which live in symbiosis with for instance, corals. The forms which occur as fossils are those which have a resting, encysted phase as *Cysts*. The cyst, formed inside the cell, has a two-layered wall of tough organic material resistant to decay. Its surface may be smooth or sculptured with granules, spines, ridges or processes. Each has an 'escape' pore (archaeopyle) through which regeneration can take place.

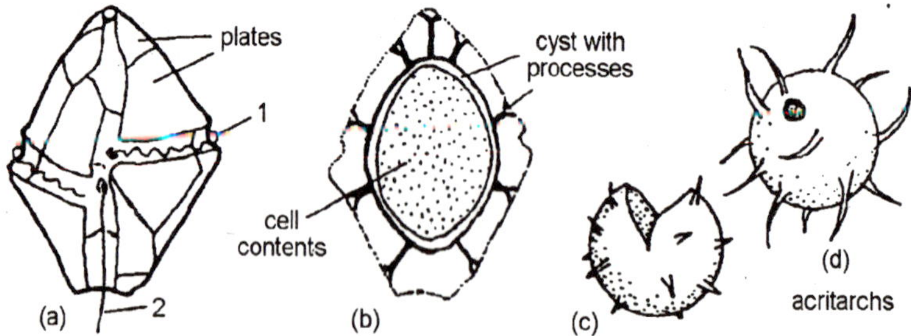

Fig. 9.2. Dinocysts and acritarchs. (a) Dinoflagellate, ventral view of motile cell showing 1, transverse and 2, longitudinal flagella. (b) Cyst formed inside cell wall which then decays. (c), (d)–Acritarchs.

Dinoflagellates live in the upper 50 m of the photic zone and are one of the major primary producers at the base of the food chain. They are commonest in warm, open ocean areas, preferring temperatures between 18 and 25°C, though they tolerate a wider range. They have a roughly latitudinal distribution, being most diverse in low latitudes. Some genera can have a devastating effect on other organisms in shallow water when they 'bloom' in such vast numbers as to form a 'red tide' and their toxic wastes can kill fish and benthos such as mussels.

Fossils cysts occur mainly in clays and shales. Possible examples are recorded from the Silurian but their main history starts in the Permian. From late Trias times they were common and their diversity increased during the Mesozoic reaching an acme in the late Cretaceous. At the end of this period they declined sharply but were again important in the Eocene. Their numbers were low in the Pliocene but, today, many species are known, though few of these encyst.

Group Acritarcha (U Precambrian–Recent)

Acritarchs are minute, cyst-like fossils of uncertain origin. They are single-celled and hollow, have a decay-resistant organic wall of one or more layers, which contain a substance similar to sporopollenin found in spores of vascular plants and measure about 10-150 μm in diameter. The variable shape, reminiscent of cysts, spores or eggs, may be spherical, ovoid, triangular or polygonal. The surface may be smooth or sculptured with granules or spiny projections which may branch. There is an 'escape' hole in some; others must have opened by splitting. They possibly represent cysts of various sorts of planktonic algae such as dinoflagellates.

Acritarchs usually occur in fine-grained rocks such as shales and also in limestones, often in association with marine fossils. A few have been recorded from fresh-water Pleistocene peats. In marine rocks they are geographically widespread and seem to have been deposited away from shorclines, suggesting a planktonic mode of life. If they do represent phytoplankton they must have been an important food source.

Acritarchs are found in the late Precambrian and are stratigraphically useful there. They are mainly spherical forms. Spiny form appeared in the early Cambrian and the group as a whole reached its acme in the Ordovician. They remained important and world-wide in the Silurian, but in late Devonian times a decline set in and they are generally rare in the latest Palaeozoic. They are insignificant in the Mesozoic and Cainozoic.

Phylum Chrysophyta—silicoflagellates (L Cretaceous—Recent)

These are marine, flagellate, photosynthetic protists, about 20-50 μm in diameter. They secrete, within the cell, a delicate framework of opaline silica tubes with radiating spines. This framework may be ring-like, hexagonal or form a hemispherical lattice.

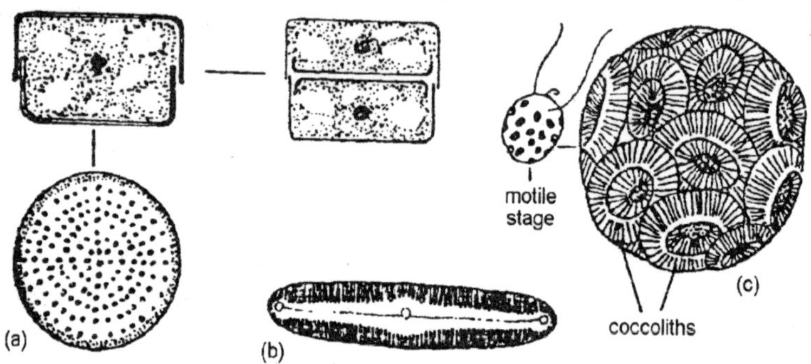

Fig. 9.3. Phytoplankton.

They are minor constituents of the phytoplankton of upwelling silica-rich waters in high latitudes and other areas, their remains being found with those of the diatoms. Some species are temperature sensitive and their fossils are useful as indicators of temperature fluctuating.

They first appeared in the Lower Cretaceous and were most abundant in the Miocene.

Phylum Bacillariophyta—diatoms (L. Cretaceous—Recent)

Diatoms are minute, non-flagellate, photosynthetic forms, the cell wall of which is impregnated by opaline silica to form a delicately sculptured, reticulate skeleton, the *frustrule*. They consist, pillbox like, of two partly overlapping valves punctured by fine pores. They range in size between about 20 and 200 μm. They may be single or may join in mucusbound filaments or chain. Two groups are distinguished by shape *centric* diatoms, circulation in plan with radial symmetry; and *pennate* diatoms, elliptical in plan with bilateral symmetry. They occupy many habitats in sea and fresh water. Most centric forms are marine and planktonic, living most densely in the upper 30 m of the photo zone. Most pennates are benthic and live in fresh water (and soil and also in shallow seas.)

Growth of marine, planktonic diatoms is seasonal with 'blooms' in spring and late summer, and they are of major importance as primary producers near the base of food chain. They are enormously abundant in the seas of subpolar and certain other regions where upwelling deep waters provide the necessary nutrients and dissolved silica. On death the majority of the skeletons quickly dissolve but a small proportion accumulate on the ocean floor as diatomaceous ooze subsequently lithified as chert.

Deposits also accumulate in fresh-water lakes in high latitudes, diatomaceous earths of Pleistocene age being found, for example, in northern Britain. Such deposits are of economic value in various ways, for example as fine abrasives. Deposits in dried-up lakes may become wind-borne as fine dust and widely redistributed on land and sea.

Centric diatoms first appeared in the Cretaceous and underwent a major radiation in the Palaeocene reaching a peak in the Miocene, after which they declined in numbers. Pennate forms appeared in the Palaeocene and increased gradually through the Tertiary, being numerous today mostly as delicate fresh-water species.

Many living species have a fossil record and their changing geographical distribution in the rocks of the Cainozoic can be used to infer changes in oceanic conditions.

Phylum Haptophyta—Coccolithophorida (Trias—Recent)

Coccoliths are minute calcite scales embedded in the cell walls of coccolithophores, marine, photosynthetic forms, spherical or ovoid in shape and about 5-20 μm in diameter. They are motile with two flagella and a filiform appendage which may be coiled, the *Haptonema*. The coccoliths, about 1-15 μm in diameter, are round, oval or stellate discs which show elaborate patterns of radially arranged calcite plates and rods. These are shed from, and replaced by, the parent cell as it grows.

Coccolithophores are found throughout the photic zone but are at their maximum numbers from a depth of a few meters to about 50 meters. They are most abundant and diverse in the warm, open oceans between latitudes 45°N and S, but some species live in colder, temperate and subpolar waters. Most have a fairly narrow temperature range and some found as fossils in the young rocks, can be used as indicator of temperature fluctuation in the past. Like diatoms, they are major primary food source.

Many warm-water species have a restricted geological range but are also widespread and hence of use in the biostratigraphy of the Mesozoic and Cainozoic rocks.

Coccoliths were rare in the late Trias. Their diversity and numbers increased though the Jurassic to reach a sharp high peak in the late Cretaceous, the Chalk owing much of its character to an exceptional abundance of coccoliths and coccolith debris. They were, however, almost wiped out in the end-Cretaceous extinctions and only a few species survived. New forms appeared in the Palaeocene followed by a rapid diversification and increase in numbers in the Eocene. In the later Eocene numbers again dropped and continued to decline apart from a temporary surge in the Miocene. Today, they are moderately abundant and their remains contribute to deposits of calcareous ooze on the ocean floors.

Seaweeds and Other Algae

Many-celled seaweeds are included by some taxonomists in the plant kingdom, but in the Protoctista by others. They are separated into phyla on the basic of their mode of reproduction and cell chemistry including the type of photosynthetic pigments they contain. The forms considered here occur as fossils, often in fragmented state; they are identified by their microstructure.

Seaweeds have a simple body, the *thallus*, which in many-celled forms may be filamentous, branching or a flat membrane. They occur in fresh, brackish and sea water, and are benthic, anchored by 'holdfast', or encrusting a hard surface. They live in the photic zone and are most abundant down to about 50 m. A number of calcareous algae secrete calcium carbonate and are potential fossils in which shape, tissue structure and sometimes reproductive structures may be presented. The carbonate may encrust the thallus as the result of carbon dioxide being extracted from the water by photosynthesis; it may also form as an intercellular deposit by metabolic processes.

Phylum Rhodophyta—red algae

These are mainly marine, living in both warm and cold waters. They are coloured by reddish pigments which mask their green chlorophyll and, by absorbing blue light, enable them to live at depth to which longer wavelengths do not penetrate. They are most abundant just below low-tide level but have been found on sea mouths at a depth of 268 m where only 0.0005 % of surface light remains. They use light much more efficiently than the shallow-water forms.

'Solenopora' (Cambrian—Cretaceous) belongs to a group of calcite secreting algae which were common in the Palaeozoic and Mesozoic. It forms irregular, encrusting nodular masses of close-packed radiating filaments, more or less polygonal in cross-section, and with cell structure. Some species referred to '*Solenopora*' have been identified as unrelated organisms, e.g. cyanobacteria.

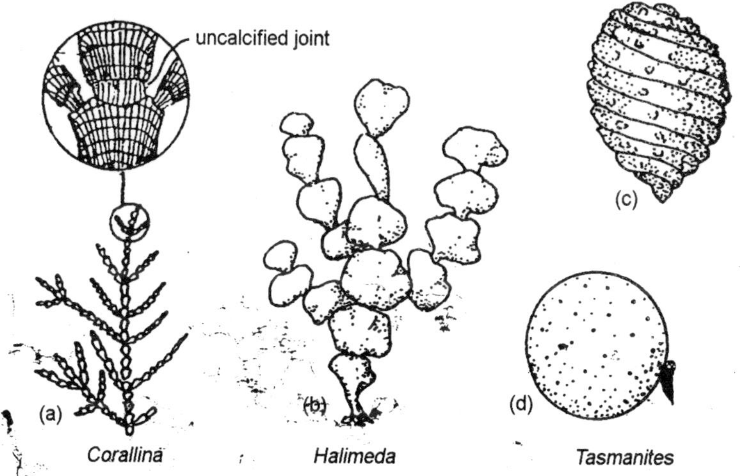

Fig. 9.4. Algae. (a) Coralline red alga with enlargement to show uncalcified joint. (b) Calcareous green alga. (c) Calcified reproductive body of an Eocene charophyte. (d) A cyst, with perforate organic wall, found in some oil-shales.

The other important group, the coralline algae, are found first in the Carboniferous and are of increasing importance in the Mesozoic and cainozoic because of their association with coral reefs. *Porolithon* and *Lithothamnion* are important components of modern reefs. They encrust, helping to bind and cement their coral substrate, forming a purplish-red algal ridge on the exposed, turbulent edge of the reef. *Lithothamnion* (Cretaceous-Recent) is encrusting and may have irregular branches a few centimeters high. It also grows in cold waters. Others, like *Corallina* (Cretaceous-Recent) live in quieter waters such as back-reef areas. Here, the thallus is erect, much branched, and consists of a series of calcified segments with intervening non-calcified 'joints'. It fragments readily, adding to the sediment of the reef area.

Phylum Chlorophyta—green algae

These are very diverse and widespread. They are mainly fresh water (including damp places) organisms but do occur in blackish and sea water. Marine forms live in shallow water (to about 100 m) and are especially abundant intertidally. The thallus is made of an interwoven tissue of non-septate filaments, often freely branching. A number secrete aragonite which is a factor in their preservation as fossils.

Green algae occur as fossils from Cambrian times onwards, often in association with reefs. *Halimeda* (Cainozoic) is commonly found today in sheltered lagoons in coral reefs down to depths of about 60

m. It forms erect, branching growths a few centimetres high, consisting of a series of articulated, calcified discs. It fragments readily and acids greatly to the sediment flooring the lagoon.

Fresh-water green algae include the stoneworts (charophytes). An example is *Chara* (Eocene–Recent) with a calcite-encrusted thallus, forming a slender stalk with whorls of branches bearing male and female sex organs. The female organ, the oogonium, contains the egg cell; it is surrounded by spirally coiled filaments which may become calcified. These are common fossils in fresh-water deposits in the Mesozoic and Cainozoic. Primitive forms are found in the Devonian.

Non-calcareous Chlorophyta

Botryococcus is a one-celled alga which secretes globules of reddish oil in the cell. It lives in fresh or brackish water and divides to form colonies about 10-100 μm across. The algae 'bloom' in season, quickly multiplying to form a dense cover over the surface of the water which becomes oxygen deficient and consequently lethal to fish. The algal remains may accumulate on the lake floor to a thickness of 5 or 10 m and such deposits in time are converted to a dark, compact, hydrocarbon-rich rock, thin-sections of which show amber-coloured masses of similar algae. Such rocks are known in general as oil-shales but, in examples where fossil algae are conspicuous and abundant, special names have been given. One such is 'torbanite', examples of which are found in the Carboniferous of Scotland and the Permian of Australia.

A recent deposit of similar origin is known as 'coorongite'. It occurs in Australia where salt lakes and lagoons shrink or dry up in periods of drought. It is soft, brown, rubbery substance formed by the dried-up accumulations of Botryococcus colonies in the resting stage, in which condition the algae contain up to 76% of hydrocarbons.

Other one-celled algae may be associated with coal and oil-shales *Pachysphaera*, a living marine planktonic alga, forms a resting-stage cyst which is spherical and has a two layered organic wall about 8–10 μm thick. It is similar to and may be related to *Tasmanites* Cambrian–Recent, cysts of which occur in large numbers in oil-shales e.g., in Alaska. Jurassic and Cretaceous) and is prominent in Permian coal in Tasmania. The cysts are about 100-600% μm in diameter with a two-layered wall perforated by pores.

Zooplankton–Phylum Sarcodina

This group comprises animal-like protoctists alternatively classed as 'Protozoa'. Members of two groups, the radiolarians and foraminiferans, secrete skeletons and are of major importance as fossils because of their immense abundance, diversity and wide distribution in marine rocks.

Radiolarians—Subphylum Actinopoda (M Cambrian–Recent)

These are one-celled, free-floating, marine protoctists which secrete a delicate openwork endoskeleton composed, typically, of opaline silica; in one group, however, this is largely organic. They occur singly as a rule but may associate in colonies.

The term 'Radiolaria' includes members of two classes; the Polycistina with silica only in the skeleton; and the Phaeodorea with a mainly organic skeleton containing a small amount of silica. A third class, the Acantharea is sometimes associated with the radiolaria and is worth mentioning because, unusually, its skeleton is made of celestine (strontium sulphate). The Polycystina have a substantial fossil record; the other two groups are insignificant as fossils, are confined to the Cainozoic, and are not considered further.

Polycystina

The protoplasm of the cell includes an inner endoplasm enclosed in a porous organic membrane, the central capsule, and an outer ectoplasm. Long straight, thread-like pseudopodia, some with a silica stiffening, radiate from the capsule to the outside. They are sticky and serve to trap passing food such as bacteria and diatoms. The food is passed in a stream of protoplasm into the capsule for digestion. The ectoplasm is frothy with vacuoles. It ;secretes the skeleton and, in forms living in the photic zone, contains symbiotic algae, zooxanthellae.

The skeleton is symmetrical either spherical or discoidal. Order Spumelarida, or bell-shaped (Order Nassellarida). It is built of spines, spicules and rods of silica which unite as a meshwork, from the surface of which more spines radiate. The spumellarid skeleton has one or more concentric, reticulate shells, frequently with spines projecting from the surface of the outermost one. The nassellarid skeleton is like a spinose tripod or a conical bell-shaped lattice, open at the base.

As planktonic animals, radiolarians are at the mercy of currents and rough weather. It is essential for them to maintain buoyancy and the body is designed to this end. Features which aid floating include the tiny size; the spherical or bell-shape; the open-work skeleton, light but strong; the long pseudopodia which increase resistance to sinking; secretion of oil globules; and the frothy ectoplasm with vacuoles which can be collapsed or expanded to vary their volume.

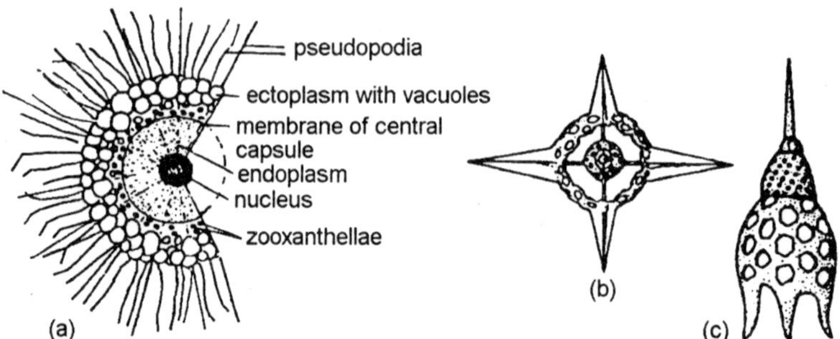

Fig. 9.5. Radiolaria. (a) Cross-section to show cell structure (skeleton omitted). (b) Spumellarid. (c) Nasselarid.

Radiolarians are mainly oceanic and are widespread from subpolar regions to low latitudes. They are most diverse in warmer seas and most profuse in nutrient-rich where their food supply is ample and the necessary silica is available. They tend to 'bloom' seasonally as their silica and food supply waxes and wanes. Different species inhabit distinct depth zones in the ocean, most living in the photic zone but occurring down to about 4000 m depth. They also show temperature preferences and, as with foraminifera, these are useful in elucidating past movements of water masses.

Radiolarians range from the Middle Cambrian when spumellarids appeared. They showed a gradual increase through the Palaeozoic becoming more diverse in the Carboniferous. Their numbers dropped sharply in the Permian and remained low in the Triassic. The Jurassic saw a radiation of new forms, including nassellarids, which led to a peak of numbers in the Cretaceous. Numbers slumped at the end of this period and they were generally less abundant in the Cainozoic. Today, however, they are very prolific and include many forms with skeletons so delicate that they are unlikely to be preserved in the fossil record, but dissolve immediately on death. Only small proportion survive as fossils, either as a minor constituent of ocean sediments or, where the population was extremely large and other sediments absent, as radiolarian ooze later lithified to chert.

Foraminifera (Subphylum Rhizopoda)

These are aquatic, mainly marine protoctists which secrete a shell, the *test*, which consists typically of calcium carbonate but may be organic or built of agglutinated detrital particles. Most foraminifera are tiny, less than 1 m diameter; a few are large and may reach several centimetres.

The soft tissue, *protoplasm*, lies mainly within the test but some extends through one or more openings, *apertures*, to enswathe it and form sticky thread, *pseudopodia*. These are used for movement and to trap and engulf food particles such as bacteria, diatoms, small metazoa and organic debris. Certain large, reef-dwelling and planktonic foraminifera enjoy a symbiotic association with photosynthetic algae.

The test consists either of a single *chamber* or, more usually, of several chambers (*multilocular*) which are separated by *septa*. Each chamber communicates with the next one via an opening, the *foramen* in the septum between them. The opening to the outside the *aperture*, is in the wall of the final chamber. The chambers are arranged in varied ways; they may lie in a single or double row or, commonly, are coiled in a plane or helical spiral.

In many species two forms are found which are known, from living foraminifera, to represent stages in a complex life history involving alternation of sexual and asexual generations. In the sexual generation the first chamber is relatively large (*megalosphere*) yet the whole test may be smaller, with fewer chambers, than that of the sexual generation which starts with a small chamber (*microsphere*).

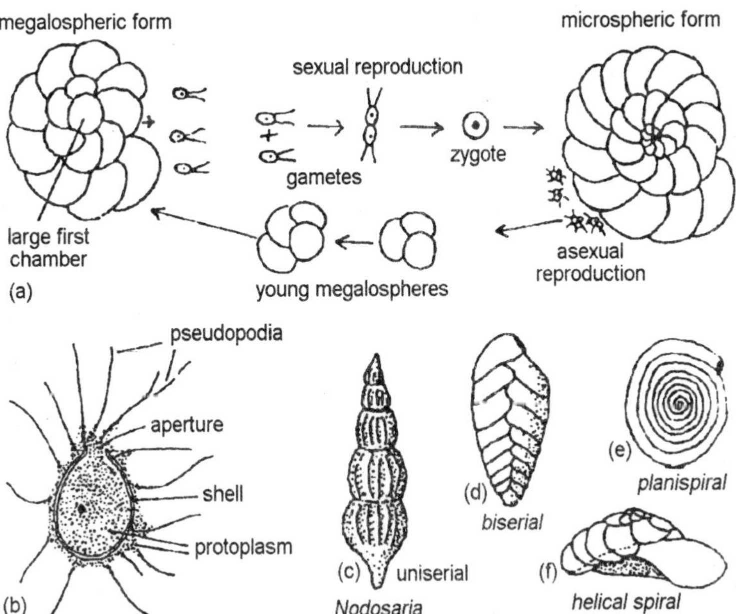

Fig. 9.6. Foraminifera. (a) Life-cycle of a foraminiferid. (b) Relationship of protoplasm and shell in a form with one chamber, (c)-(f) Varied arrangement of chambers.

Test wall

The composition and structure of the test wall is the basis for separating foraminifera into five suborders. Thus, the test may be entirely organic, consisting of a horny, proteinaceous substance (tectin); such forms (Allogromiina) are rare as fossils (range late Cambrian–Recent). In those with an agglutinated test (Textulariina), detrital grain such as sand or shell fragments are held in an organic cement (range Cambrian–Recent). Calcified tests, the majority, also have an organic constituent. There are three distinct wall structures; (i) microgranular Fusulinina) with more or less randomly arranged, equidimensional grains of calcite giving a sugary appearance (Range Ordovician–Trias); (ii) porcelaneous (Miliolina) with tiny calcite needles, most lying at random in a organic matrix but those at the outer surface lying parallel to give a smooth, shiny, milk-white surface (range Carboniferous–Recent); (iii) hyaline (Rotaliina with crystals, usually of calcite, but aragonite in some, arranged in one or more layers, giving a glassy appearance, the walls may be *perforate*, pierced by fine pores through which pseudopodia emerge (range Permian–Recent).

Foraminifera are mainly marine; a few live in brackish water (estuaries) where the salinity may fall to between 15% and 5% and a number tolerate hypersaline water. They are enormously abundant in the sea at all depths from the shoreline to the abyssal zone, in temperatures from 1 to 50°C, and in high and low latitudes. They are prey to many creatures, yet in parts of the abyssal regions where little other detrital sediment accumulates, their remains may build up extensive deposits of calcareous ooze. Fossil forms constitute the bulk of some limestones, as in the nummulitic limestone from with the Egyptian Pyramids were, in part, built.

Mode of life

Most foraminifera are benthic but a number, like Globigenna, are planktonic. *Benthic* foraminifera may be sessile, or mobile and use their pseudopodia to crawl sluggishly. Diversity of species generally increases offshore, is high over the continental shelf but decreases in deeper water and lower temperatures. While some forms are cosmopolitan, many species are restricted to a narrow range of physical conditions. They may be sensitive to salinity, temperature, light, depth or to the nature of the substrate, e.g. whether hard, sandy or muddy. Organic-rich muds may be densely populated by foraminifera feeding on bacteria. Foraminifera with agglutinated tests are characteristic of near-shore and brackish-water especially in colder waters, and they are the predominant forms below about 5000 m. Those with calcareous tests are commonest in warmer regions and shallow waters. Species with large tests are mainly restricted to well-lit shallow, warm waters (Above 20°C) including reefal habitats, probably because of their association with minute photosynthetic algae, zooxanthellae, which live embedded in the animal's protoplasm. This is a symbiotic situation similar to that of reef-corals and zooxanthellae. The algae use waste products as nutrients and, in their uptake of carbon dioxide, may facilitate calcite secretion in the large foraminiferal test. The host may in part be nourished by soluble byproducts of photosynthesis and may also digest some of the algae.

Planktonic foraminifera are widespread in the open ocean away from coastlines. There are relatively few species but these occur in enormous numbers. There may be a high seasonal peak in numbers as in the Arctic Ocean in the late summer. Most live in the food-rich surface waters between 10 and 15m an numbers decrease with increasing depth; few live below 1000 m. Different species live at different depths and, for each, the density of the test must be compatible with the preferred depth. While, in general, the test is relatively small and thus slow to sink, buoyancy may be improved by high porosity of the test wall or by a growth of long spines which increase resistance to sinking. In general, smaller forms live near the surface and larger forms in deeper water. Some species may migrate downwards as they mature; juveniles with small thin tests live near the surface and mature forms with larger, thicker tests live near and reproduce in deeper water. Some forms show a diurnal rhythm of vertical movement.

Diversity is highest in tropical waters and falls polewards. The distribution of many species is controlled by temperature and so there are 'warm'-water and 'cold'-water species which characterise faunal provinces. However, species characteristic of one province may also be found, in less abundance, in adjacent provinces. The faunal provinces are roughly parallel to latitude but surface currents and deeper water counter-currents, which influence temperature, complicate the pattern. In addition, cold-water species found at shallow depths in high latitudes also occur in deeper water of the same temperature in low latitudes.

In a few species the test is coiled, in some of the left and in others to the right. An example is *Neogloboquadrina pachyderma* which lives polewards of 25° N and 25°S. Over 90% of specimens in polar waters are coiled to the left while, in more temperate waters, most are coiled to the right. The temperature at which coiling direction is reversed varies from about 7°C in some regions to 10°C elsewhere. It is not understood why these changes occur.

Some Examples of Common Benthic Foraminifera

Saccammina. Test agglutinated; simple; globular shape with aperture opening at the end of short, projecting neck. *Rhabdammina* (Ordovician–Recent), a related form, has tubular branches, each with a terminal aperture, radiating from a central area (Silurian–Recent).

Textularia. Test agglutinated; multilocular and biserial with chambers in two alternating series; slit-like aperture is terminal (U-Carboniferous–Recent).

Fusulina. Test calcareous, microgranular, perforate; spindle-shaped tapering at each end, with many whorls of chambers coiled planispirally so that each completely envelops the preceding one; wall layered and cut by small pores; septa fluted transverse to the axis of coiling, in such a way that forward folds of one septum may meet backward folds of the next to form 'chamberlets' aperture slit-like (U-Carboniferous).

Quinqueloculina. Test calcareous, porcelaneous, imperforate; multilocular, coiled with sausage-shaped chambers added at angles of 144° so that five can be seen; aperture simple with a tooth-like projection (Jurassic-Recent).

Lagena. Test calcareous, hyaline, perforate; consist of one flask-shaped chamber with a long neck carrying the aperture. Nodosaria (Permian-Recent), a related form, has several chambers arranged in a series.

Nummulites. Test calcareous, hyaline, perforate; biconvex lens-shape with many simple chambers coiled planispirally, each whorl completely enveloping the previous one; the walls are thus layered; they carry a system of canals which open at the surface through pores; aperture slit-like and curved.

Heterostegina (Palaeocene–Recent) is a reef-dwelling nummulite with the chambers divided into chamberlets. Study of this form has explained the rolė of the elaborate canal and pore system. It is a means of communication between the chamberlets and the sea water by which the varied biological processes, including locomotion, reproduction and protection are carried on. *Heterostegina* lives in shallow nutrient-poor but well-lit tropical waters. It contains symbiotic algae on which it depends for much of its nourishment; and its test, flattened and translucent is well designed to allow easy penetration of light for the algae within to photosynthesis.

Some Examples of Common Planktonic Foraminifera

Globigerina. Test calcareous, hyaline, coarsely perforate; consists of several almost globular chambers coiled in a helical spiral (trochospiral) in life bears many very long fine spines; large basal aperture (Palaeocene–Recent).

Orbulina (Miocene–Recent) is a related form in which the final chamber is spherical and encloses the Globigerina-like earlier chambers. The spherical shape is an efficient design to promote buoyancy.

Globorotalia (Palaeocene–Recent) is similar to Globigerina but chambers are smooth-surfaced, without spines, less inflated, and are "pinched"-out on the periphery into prominent keel. Species of all three have limited stratigraphic ranges and are, therefore, valuable for correlation purposes.

Geological History

Cambrian–Recent—Foraminifera appeared in the Cambrian. Early forms were benthic and typically agglutinated. By Silurian times they had become diverse and accompanied by forms with microgranular tests and these became the dominant forms in the Upper Palaeozoic, the fusulines being typical. Forms with porcelaneous tests appeared in the Carboniferous; and with hyaline tests in the Permian.

The fusulines include relatively simple lens-shaped forms, the endothyrids, which were important in the early Carboniferous; followed by giant fusulines, spindle-shaped forms with highly complex internal structure in the late Carboniferous and Permian, characteristic of warm, shallow seas and reefs. These are of great stratigraphic value and have been widely used in North America. Russia and the Far East. They died out in the massive extinctions of organisms in the late Permian.

In the Mesozoic there was renewed diversification, mainly of forms with hyaline tests, but agglutinated and porcelaneous tests were also common. In the later Jurassic a new niche was occupied when planktonic forms appeared. A spectacular radiation occurred in the Cretaceous, notably of large porcelaneous, benthic forms living in shallow tropical seas; and of planktonic forms in the widespread Chalk seas. The Cretaceous ended with a large-scale disappearance of both benthic and planktonic forms.

A resurgence of numbers in the Palaeocene included survivors from the Cretaceous and newcomers like the planktonic globigerinids (still prominent today) and the large, benthic nummulites. The latter contributed massively eastwards from the Mediterranean area to the

East Indies. Several species of Nummulites occur in the lower Tertiary rocks of the Hampshire and Paris Basins. Foraminifera are less diverse today though they are very abundant in warm seas.

Palaeoecology

The ecology of modern foraminifera is a direct guide to the living conditions of their fossil ancestor in the newer rocks; and an indirect guide to the palaeoecology of similar forms in older rocks. Data concerning temperature and depth of water, for example, can be used to establish fluctuating conditions in Quaternary sequences.

More general guides may also be used; today, large, calcareous foraminifera with symbiotic algae live in shallow, tropical waters and are likely to have done so in the past; open ocean waters have a high ratio of planktonic to benthic forms; extremes of salinity are indicated by low diversity; in brackish water mainly agglutinated forms with only an occasional hyaline form, such as *Ammonia*, are found; hypersaline waters deter most foraminifera except porcelaneous forms such as the miliolines.

Uses of Foraminifera

Foraminifera, like other fossils which show evolutionary changes with time and are geographically widespread, are of great value in establishing stratigraphical ages of rocks and making correlations on regional or continental scales.

In the oil industry, exploration drilling and oil-field development involve constant monitoring of the stratigraphic age of the rocks being penetrated; foraminifera are valuable for this purpose. As with other microfossils, they are particularly useful simply because of their small size, many specimens being obtainable from a small chip of rock. Such small samples are often the norm in drilling operations which comminute the rocks and flush the bits to the surface. Larger fossils are fragmented by this process but microfossils, if present can be sampled adequately. Even when rock cores are obtained these are one only of very small diameter and the same considerations apply.

Exploration of the ocean floors by deep-sea drilling has involved the stratigraphic use of foraminifera, together with coccoliths (nannofossils) and radiolaria, on an extensive scale. These are the common fossils in widespread oceanic sedimentary deposits, normally up to a few hundred metres thick and ranging in age back to the upper Jurassic.

As one example we may take the testing of the sea-floor spreading theory. This postulates formation of new basaltic crust at mid-ocean ridges and its subsequent lateral movement away from the ridge as yet more new crust is formed. If the theory is correct, the age of the basaltic crust must increase away from the ridge. In practice, the most effective way to determine the age of the basalt is to determine, from the fossils, the stratigraphic age of the sedimentary rocks resting immediately on it. Then, by referring to the stratigraphic radiometric time scale, one can convert the stratigraphic age into an age expressed in years. To test the theory holes were drilled at increasing distance from the mid-Atlantic ride in the South Atlantic, and the basal stratigraphic ages, hole by hole, over a total distance of 1300 km, provided to be Upper Miocene (11 my); Lower Miocene (24 my); Upper Eocene (40 my); Middle Eocene (489 my); Upper Cretaceous (67 my). The results strongly support the theory and also allow an average spreading rate to be calculated 2 cm per year.

As a second example were take the interpretation of a sedimentary sequence found at a hole drilled in the Pacific floor, about 41 °N and about 900 km east of Japan in a water depth of 5600 m. The hole was 300 m deep and penetrated the top of the basaltic crust. The stratigraphic age of the basal sediment is early Cretaceous (120 my) which dates the underlying basalt. Palaeomagnetic determinations give the latter a palaeolatitude of 6°S at the time of its formation. The crust, therefore, has moved, like a conveyor belt, at a rate of a few centimetres a year, from that latitude of its present position at 41°N in 120 m. This is consistent with the general motion of the plate in a northwesterly direction towards the trench in which it is being consumed. In the following account the different members of the sedimentary sequence are mentioned in turn, starting at the base.

The sediments immediately above the basalt are calcareous and siliceous oozes of planktonic foraminifera, coccoliths (nannofossils), and radiolaria, formed above the CCD form surface waters rich in nutrients, in the equatorial region. At this time the newly formed crust was located on the flanks of the ridge at a comparatively shallow depth. As it moved further away from the ridge it cooled and subsided below the CCD with formation of siliceous, radiolarian ooze only. Continued motion northwards then took it away from the nutrient-rich equatorial waters, formation of ooze ceased, and only thin clays, sparsely fossiliferous, accumulated, below the CCD. Finally, nutrient-rich waters in the north-west Pacific were reached with the formation of another siliceous ooze of radiolaria and diatoms, below the CCD, and extending to the present ocean floor.

This combination of palaeontology and palaeomagnetics has proved a very effective technique in unravelling the history of the ocean floors.

PROBLEMATICA

Chitinozoa (Precambrian, Ordovician–Permian)

Chitinozoans are microscopic fossils of unknown affinities. They are hollow, flask or bottle-shaped vesicles, open at one end and with a wall consisting of a chitinous organic substance, pseudochitin. The wall, the only part preserved, is two-layered, dark coloured and highly resistant to both chemical and physical changes. Chitinozoans have, for instance, been recognized intact in deformed rocks like slates. They are generally between 50 and 300 μm in size but may be longer.

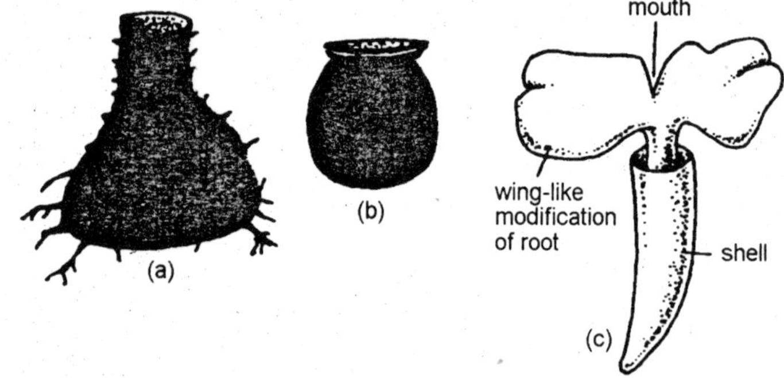

Fig. 9.7. Chitinozoans and a pteropod. (a), (b)–Chitinozoa. (c)–A recent pteropod.

The vesicle varies in shape and structure but in cross-section shows radial symmetry. Usually one end is drawn out into a neck with an aperture which is closed by a lid, the *operculum*; this may be recessed within the neck or by terminal. The surface may be smooth, striate, tuberculate or bear hollow spines. Sometimes the vesicles occur as clusters.

Chitinozoans are generally regarded as animal in origin because of the chitinous substance of the wall. Suggestions as to their nature include invertebrate egg cases and protective shells for animal protoctists.

Chitinozoans are confined to and widely distributed within marine deposits of Palaeozoic age. They occur commonly in shallow-water silts and shales and are also found in limestones, graptolitic shales and cherts, and in slates.

Possible chitinozoans have been reported from the late Precambrian but none are known from the Cambrian. They became common after a major radiation in the Ordovician. In the middle and late Devonian they suffered extinctions and were rate in the Carboniferous. Only a few species have been reported from the Permian.

KINGDOM ANIMALIA ('ANIMAL MICROFOSSILS')

Gastropoda–Group Pteropoda (Cretaceous–Recent)

Pteropods are very small, marine, free-swimming gastropods some of which secrete an aragonite shell. They swim using the foot which is modified to form two wing-like fins at the anterior end. The surface of the fin is ciliated and so directs currents of water with food, e.g. diatoms to the mouth.

The pteropods shell is thin, translucent and either conical or plan spiral. It measures about 2.5 to 10 mm or more.

Pteropods are most abundant and diverse in warm tropical oceans and their numbers fall polewards; only one species is found in polar waters. Most occur in the upper 500 m and they show diurnal rhythm, rising at dusk and falling at dawn. In places they occur in enormous shoals and their remains contribute to calcareous oozes in depths down to about 2500 m, below which aragonite, being more soluble than calcite, is unlikely to the preserved. An example of pteropod ooze is found in the mid-area of the South Atlantic.

Crustacea–Class Ostracoda

Ostracods are tiny, aquatic crustaceans which first appeared in the Cambrian, are common as fossils, and are still widely distributed today, mainly in the sea but also in brackish and fresh water. They are typically ovate or kidney shaped in outline, and between 0.5 and 5 mm in size though some are much larger up to a few centimetres.

Ostracods have a laterally compressed body which is indistinctly segmented; the head and thorax are large, the abdomen rudimentary. The body is enclosed by a two-valved *carapace* which is periodically

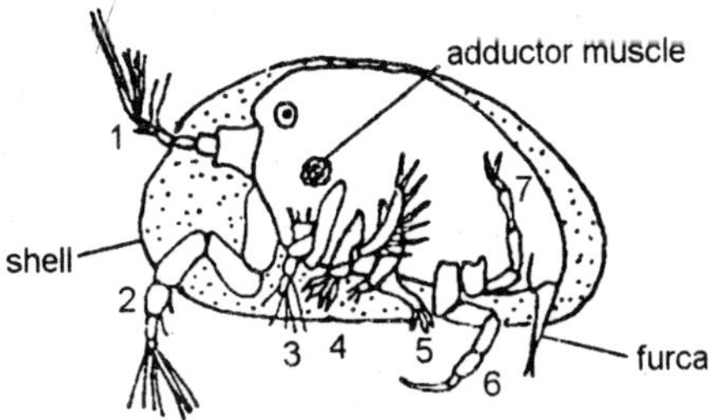

Fig. 9.8. Ostracods. A recent ostracod showing appendages. 1-4, head appendages. 5-7, thoracic appendages.

shed and regrown. This develops as chitinous outgrowths extending as *valves* on either side of the body. The carapace usually becomes calcified except along the dorsal margin where the chitin acts as an elastic ligament serving to open the valves. A hinge structure of teeth and sockets, or ridges and grooves, controls articulation. The valves are unequal in size and one overlaps, the other. They are closed by adductor muscles which run transversely from one valve to the other muscle scars on the inside of each valve mark their position. The carapace is pierced by canals which appear on the surface as pores through which tactile bristles, setae, pass to the outside. In addition, many, usually shallow-water forms, have eye spots or eye tubercles or ridges, and often it shows dimorphism, the size and shape differing in the male and female. When the carapace opens the limbs extend. There are five to seven pairs of joined and bristled limbs, either one-branched (uniramous) or two-branched (biramous). These are modified according to function. The head carries four pairs: two anterior uniramous pairs used in swimming or walking; and two biramous pairs beside and behind the mouth, concerned with feeding. The thoracic limbs serve for walking or digging.

Reproduction is sexual. The female may brood her eggs in brood pouches within the carapace, or may lay them separately in the water. The eggs of fresh-water forms are often resistant to desiccation and thus survive periods of drought.

Mode of life

Most ostracods live in normal sea water salinity (about 35%) and may be found in temperatures ranging from 0 to 50°C. They are most diverse in the neritic zone and in tropical waters; and are mostly benthic in habit though some are pelagic.

Benthic forms generally crawl or may burrow in sediment. They feed by filtering organic detritus or organisms like diatoms, bacteria and protozoans; some scavenge, e.g. on dead fish. The nature of the carapace may reflect the animal's habitat. For instance, in turbulent near-shore areas it is relatively stout, with eye spots and marked sculpture; and is streamlined in those forms which burrow in fine silts or muds. The small number which live in deep water, where it is uniformly cold 4–6°C and dark, have relatively large, thin-walled, sculptured carapace. These species occupy conditions which are widespread in the oceans, and tend to be cosmopolitan.

Pelagic forms swim using their long, bristled antennae. They filter fine food particles and some capture small animals like copepods or tiny fish. Their carapace is weakly calcified and larger than average, up to 3 cm. They are most numerous in nutrient-rich currents, mostly in the upper waters, numbers and diversity falling with increasing depth. They show rhythm, rising at night and falling by day.

Some ostracods are tolerant of a range in salinity and a few such species may colonise brackish water (up to 30%, or hypersaline water (over about 40%). The number of individuals tends to be very large in each case. Living forms with a long fossil record are thus useful as indicator of palaeosalinity.

Fresh-water ostracods live in lakes, rivers and ponds including some which may be ephemeral. Most belong to one family (Cyprididae and tend to have smooth, thin-walled carapace. Some are scavengers of plant or animal matter.

Some common examples

Leperditia. Carapace stout and well-calcified. Purse-shaped with long, straight hinge margin; right valve overlap the left; surface smooth; distinct eye tubercles; adductor muscle scars large. Widespread in shallow-water, marine deposits; benthic (Silurian-Devonian)

Beyrichia. Carapace bean-shaped with long, straight hinge line and convex ventral margin; surface granular; three distinct lobes, the central one being the smallest; anterior lobe is enlarged in the female shell to form a brood-pouch. Widespread in shallow-water, marine deposits; benthic (Silurian–Devonian).

Bairdia. Long-ranging, marine, benthic, found today living in coarse-sands in shallow waters. Thick, smooth carapace, strongly convex on

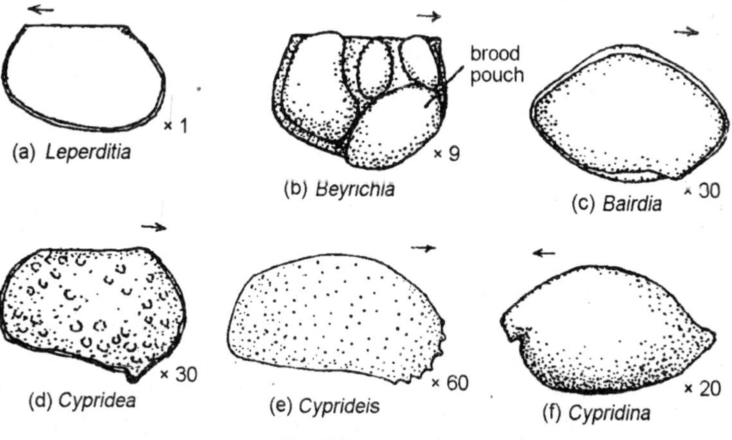

Fig. 9.9. Ostracods.

the dorsal margin and with a pointed posterior end: left valve larger and partly overlaps the right (Ordovician–Recent).

Cypridea. Lives in fresh or slightly brackish water. Carapace thin and chitinous or weakly calcified, kidney-shaped to ovate with almost straight dorsal and ventral margins; there is a 'beak' at the antero-ventral angle where the lower antennae emerged; surface pitted and usually tuberculate (U Jurassic–L Cretaceous).

Cyprides. Is tolerant of salinity and may be found commonly in both brackish and hypersaline waters. Smooth, subovate carapace (Miocene–Recent).

Cypridiza. Marine, pelagic form, Carapace smooth, thin-walled and ovate in shape with convex dorsal and ventral margins; prominent anterior process, the rostrum, accommodates the antennae which are used actively in swimming (U. Cretaceous–Recent).

Geological history (Cambrian–Recent)

The ostracods appeared early in the Cambrian. The first forms had thin, chitinous valves but in the early Ordovician a major radiation brought a diversity of new forms with stouter, well-calcified valves. These have been separated-into four orders, of which two were confined to the Palaeozoic and two were more long-ranging. One order, of which *Leperditia* is typical, died out at the end of the Devonian. Beyrichia belonged to a much larger, more important order which continued beyond the Devonian, though in lessening numbers, until it finally disappeared in the Permian extinctions. Of the remaining two orders, only a few representatives survived into the Mesozoic and their numbers did not increase until the Jurassic. The majority were benthic and these show great diversity in the Cretaceous. Numbers were low at the start of the Cainozoic but then increased. Ostracods are very abundant at the present day. The group to which most pelagic forms belong are not generally common as fossils perhaps because the rather thin carapace is not easily preserved.

Fresh-water ostracods appeared in the Devonian and were common in non-marine episodes from then on.

Phylum Chordata–Conodonts

Conodont elements are microscopic, tooth-like, phosphatic fossils, brownish in colour, glassy in appearance and mostly measuring between 0.2 and 3 mm in length. They occur in marine rocks either singly or, rarely, in assemblages each of which represent the mineralised apparatus of an individual animal. Fossils of complete conodont animals have been found with conodont apparatuses preserved in situ in the mouth region where they were presumably part of a food-processing device. These animals are now referred to the phylum Chordata.

Conodont elements are translucent structures made up of calcium phosphate together with organic proteinaceous matter. They resemble teeth and are described using terms appropriate to teeth. The upper, oral, surface bears one or more cusps or denticles; the lower, aboral, surface is followed out to form a basal cavity. Sometimes the basal cavity contains an inner cone, the basal body.

Conodont elements are found in three basic shapes:

1. Simple cones, or *coniform* elements, each shaped like a curved rose them with a sharp pointed cusp on the oral surface and an expanded base aborally. The convex edge is anterior and the concave edge posterior.
2. Compound, *ramiform*, elements have the base extended to form bar-like processes bearing smaller, conical denticles in front of, or behind the main cusp, or on one or both sides of it. The processes may be directed downwards from the main cusp so that the element, as a whole, is arched. The basal cavity is deepest under the main cusp.
3. In compound *pectiniform* elements, the processes form a laterally flattened blade, or may be extended at the base to form a broad shell or platform on one or both sides. The processes, with small denticles, form a central ridge, the keel-like carina; commonly the anterior process remains blade-like. Other, more complex shapes are produced by the development of lateral processes with platforms. The surface of the platforms may be sculpted by ridges or tubercles. The basal cavity may be larger or very small; the basal body, if preserved, may be quite large.

While some compound conodont elements are bilaterally symmetrical in plant, most have a curved outline with a convex outer and concave inner side. They may be found in pairs which are mirror images and are regarded as right- and left-had elements of a bilaterally symmetrical arrangement.

Mode of Growth

Studies of the ultrastructure of conodonts show that they grew by accretion lamella laid down in one of three growth patterns which define three conodont types. It is not known how these are related, if at all. Two types, protoconodonts and paraconodonts, were short-lived and numerically insignificant. The third, the euconodonts (true conodonts) was a major and long-lasting group.

Protoconodonts (U Precambrian–L. Ordovician) are simple cones translucent phosphate with much organic matter and with a deep basal cavity. The lamellae show a cone-in-cone arrangement with the newest layer at the base. They were obviously superficial structures and show close similarities in morphology to the grasping spines of chaetognaths (arrow worms, common marine pelagic animals). *Paraconodonts* are similar but the cone-in-cone structure is modified by a slight overlapping of the newer lamellae up the sides of the cone, indicating that they were partly embedded in soft tissue.

Euconodonts (U Cambrian - U Trias) include coniform, ramiform and pectiniform elements, and comprise conodont and basal body. The lamellae, closely spaced, were added over the entire oral and aboral surface of each part, producing a concentric growth pattern which indicates that the entire element was enclosed in soft tissue. While generally translucent, most form contain some 'white matter', vesicular patches within older layers in the denticles. Only euconodonts are considered further.

The Conodont Animal

Fossils of conodont-bearing animals have been found in the lower Silurian of Wisconsin, U.S.A, and in the lower Carboniferous of east Scotland. The latter were discovered in a thin limestone, associated with varied fossils such as shrimps, ostracods, nautiloids and fish. Three specimens contain conodont elements in situ. The fossils are of a soft-bodied, elongated and very slender animal rather like a miniature eel. The best preserved is about 4 cm long by 1.8 mm wide. The head end is slightly wider (1.95 mm) and is bilobed. The assemblage of conodonts present in the mouth region includes ramiform elements in front and pectiniform elements at the back. Towards the posterior end there are faint traces of segmentation with segments which 'Vee' forwards as in amphioxus (a cephalochordate). Traces of fin-rays at the tail end suggest the presence of caudal fins. The picture thus presented is of a tiny, elongate animal, bilaterally symmetrical, perhaps laterally flattened, and designed for swimming.

Affinities of Conodont Animals

The shape and segmentation of conodont animals, together with the phosphatic composition of the elements, suggests that their affinities may lie with primitive chordates, possibly with the jawless agnathans. Conodont elements themselves have no counterpart in other chordates,

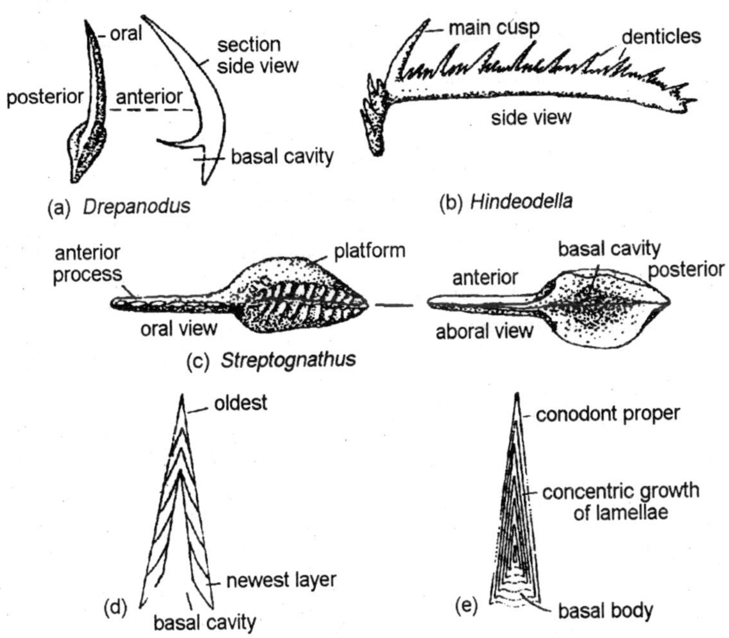

Fig. 9.10. Types of conodont elements: (a) conform; (b) ramiform; (c) pectiniform; (d) protoconodont; (e) euconodont.

but their nature and location suggests that they had a grasping, cutting and grinding function concerned with feeding.

Living agnathans are eel-like forms. They include lampreys (with a record going back to the upper Carboniferous,) and hagfish (a fossil of which has recently been found in the Mazon Creek fauna (U Carboniferous) of Illinois, USA. Hagfish differ from lampreys in many ways. They have a feeding mechanism which suggest a model for understanding how the conodont teeth may have operated.

The hagfish, up to 40 cm. long, is marine, living in a burrow from which it can emerge and swim. It feeds on animals such as polychaete worms and fish, alive or dead. Its sucker-like mouth has an extensible tongue, supported by cartilage and bearing a many-toothed horny plate along either side. The hagfish attaches to its prey and extends its tongue with the plates spread out against the soft flesh. As the tongue is retracted, it folds so that the 'teeth' interdigitate and tear off flesh. The hagfish is not, of course, an exact model. Its 'teeth' are horny and lie on the tongue whereas the phosphatic conodont elements were embedded in tissue which carried out growth and repair but which was not sufficiently robust for preservation.

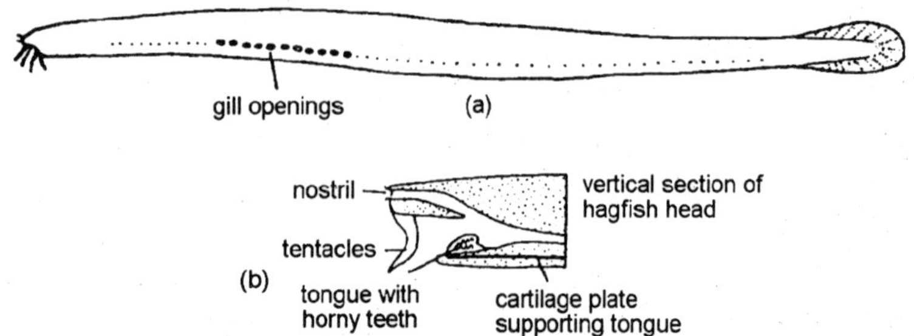

Fig. 9.11. Hagfish and conodont animal.

Geological Occurrence

Conodonts are widespread in marine rocks including limestones from which they can easily be extracted by chemical means since they are resistant to dilute organic acids. They are stratigraphically valuable since they show rapid evolutionary change. Some were cosmopolitan and may occur in a wide range of facies, including relatively deep-water shales. Perhaps with limestones and cherts. Here they are associated with pelagic fossils: e.g. graptolites or goniatites according to the geological age of the rocks concerned but, usually with, no benthic fossils. Such conodont animals were probably also pelagic, perhaps, nektonic, a life-style borne out by the shape of the body and the presence of fin-rays in the fossils. Other conodonts were relatively restricted in distribution. They may occur in near-shore deposits associated with benthic fossils of various sorts. Often such elements were larger and more robust than those occurring in deeper water facies.

Geological History

Euconodonts (U Cambrian–Trias) had a long and varied history. They underwent a major radiation in the early Ordovician, reaching a peak of diversity and widespread distribution by the mid-Ordovician. At this time the elements were predominantly of ramiform and coniform types. In the late Ordovician numbers fell and were at a reduced level until late Devonian when there was a moderate resurgence of ramiform and pectiniform elements which continued into the Carboniferous. Coniform elements are not found after the Devonian. In the late Carboniferous a decline set in, and elements are generally smaller and less common in the Permian and Trias. They survived longer in the Tethys region than elsewhere, finally disappearing in the late Trias.

10

TRACE FOSSILS

Trace fossils are signs of animal activity preserved in rocks: a footprint is the obvious example; the bones of the foot would form a body fossil. They are often the only evidence for the presence of soft-bodied animals not otherwise preserved. Their study, *Ichnology*, involves identifying the activity and making deductions about the animal responsible for it.

Trace fossils are made by both skeletal and non-skeletal organisms. They include signs of locomotion (footprints and crawling trails); resting and dwelling (surface depressions, burrows and borings); feeding processes (tooth marks, grazing furrows, 'mining' burrows, stomach contents, gastroliths), faeces (coprolites, faecal pellets); reproduction (nests); and use of tools (flint implements). They are found in marine and non-marine rocks, and are best preserved in sandy sediments, preferably calcareous, and mudstones or shales. In some rocks they may be the only evidence of life at the time, and have then a special value.

A trace, such as footprints, made on soft but firm substrate, is preserved when covered, soon after, by further sediment preferably different in texture and colour. The eventual lithified fossil may be exposed both on the original substrate as in impression, and as a natural cast on the underside of the covering rock.

Rarely, the animal responsible for a trace found as a body fossil, e.g. a crawling invertebrate at the end of its trail. Usually, however, the trace-maker cannot be directly identified and circumstantial evidence is used. This is based largely on studies of modern organisms.

Marine Trace Fossils

In the modern sea the various animals in an area of sea-floor of constitute a community, and the component members change with depth of water and type of substrate. The sea-floor can, therefore, be divided into broad zones, each characterised by particular types of potential trace fossils. In shallow water, the traces are often those of infaunal filter-feeders, whereas in deep water they are made by deposit feeders, mainly surface dwellers. Generally, conditions are well oxygenated but there are some organisms which can tolerate relatively low levels of oxygen.

Marine environments fall broadly into four regions: the intertidal, the sublittoral (extending over the continental shelf), the bathyal and

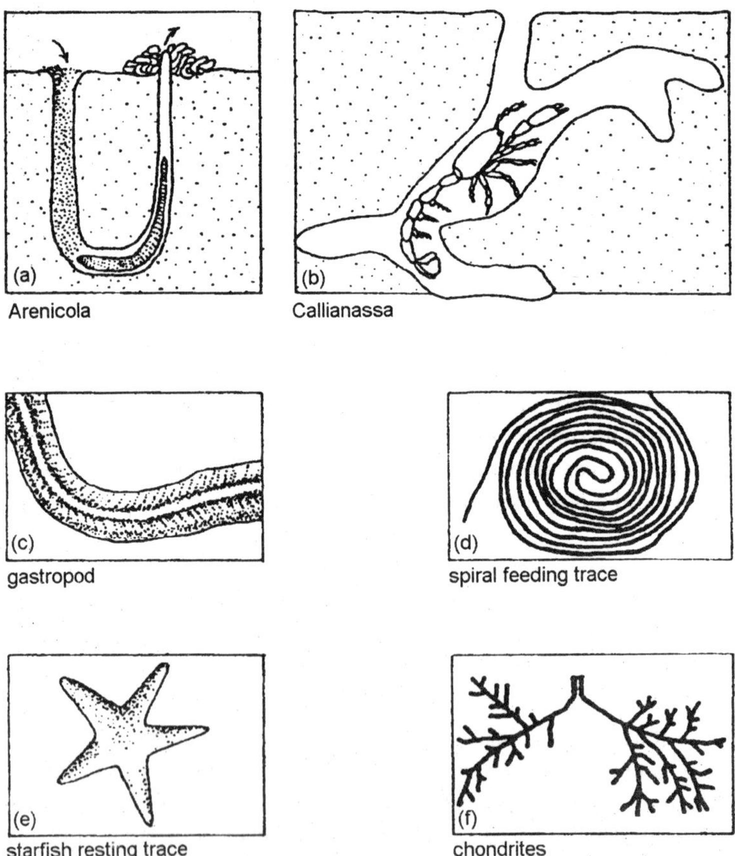

Fig. 10.1. Invertebrate animal traces, recent and fossil.

the abyssal. Each grades into the next without any sharp demarcation; lithology rather than actual depth is the controlling factor. Similarly, there is overlap in the distribution of organisms and the traces they make. The intertidal area is a high-energy zone with a variety of substrates ranging from rocky foreshores and sheets of well-sorted, shifting sands to the more sheltered bays and inlets where tidal flats of muddy sands and silts are found. It is now well-represented in the fossil record. Organisms leaving traces are not particularly diverse though individuals may be numerous.

On hard substrates some organisms, such as bivalves and echinoderms, may carve out cavities, sometimes with a constricted entrance, e.g. *Pholas*. In some cases the boring organisms make their dwellings in shells, e.g. the polychaete *Polydora*, which lives in tubes, about 1-2.5 mm in diameter, bored in oyster shells. The common boring sponge, *Clionia*, makes a network of tunnels with openings about 2 mm in diameter in shells and soft rocks. Feeding traces abound: shells which have been bored by predatory gastropods such as *Natica* and *Mura* (neogastropod), or gastropod shells with the aperture damaged by predating crabs, are common.

On sandy substrates traces are most likely to be of infaunal filter-feeders living in burrows, vertical or U-shaped. A well-known fossil example is found in the Cambrian 'piperock' of north-west Scotland which takes its name from the abundant vertical burrows known as Skolithus. A variety of living 'worms' occupy vertical burrows; some line it with a stabilising mucus, others bind detritus to form a tube or, like *Sabella*, secrete a membranous tube.

On tidal flats some of the sediments have a high content of organic matter, a source of food for deposit feeders which may be present in large numbers. An example is the polychaete worm, *Arenicola*, which constructs a U-shaped burrow. About two-thirds of this is lined with mucus, and the remaining third, lying above the worm's head, is filled by sand and organic matter being drawn into the burrow and ingested by the animal. The unwanted sand is defaecated at the entrance to the tail shaft as a spiral faecal casting. Periodically the worm pumps water into the burrow via the tail shaft, and this preforms several functions; it provides oxygen, introduces an additional food source in the form of suspended micro-organisms, and is finally used to mobilise the sand in the head shaft. These burrows are marked at the surface by the spiral casting and also by a funnel-shaped depression above the head shaft.

The sublittoral regions one of moderate to low energy. It passes, as depth increases, to a region of quiet water unaffected even by storms, and finally merges into the bathyal zone. Similarly, changes in traces fossils are not sharply marked. Deposits are variable but generally finer than in the intertidal area; fine sands, silts and muds, usually rich in organic matter which has settled from higher levels in the water. The fauna is diverse and abundant, including surface and infaunal suspension and deposit feeders. Where sedimentation rates are very low the surface layers may be worked and reworked by burrowers over a long period of time producing a chaotic pattern of structures. This process is known as *Bioturbation*. Conversely, rapid sedimentation may result in better preservation of traces.

Traces include crawling trails, resting races and vertical and oblique feeding burrows. Vertical pipes and U-shaped burrows are abundant and made by many different organisms including worms and arthropodous. They retain their form readily if they have been plastered with mucus or if a tubular covering has been secreted by the organisms. Sometimes a funnel-like depression is present or faecal deposits such as worm castings are found. Modern examples include *Chaetopterus*, a polychaete worm which lives in a parchment-like U-shaped tube in mud or sand and potentially might be preserved as an entire U-shaped trace. U-shaped burrows, complete or otherwise, are found throughout the Phanerozoic. Some burrows are inclined or more-or-less horizontal. *Callianassa*, a type of crab living and feeding in muds and sands at the present day, constructs a complex system of cross-connected branching tunnels, the walls of which are supported by mucus. It keeps open a pipe to the surface through with it circulates water with oxygen. Similar burrows in Mesozoic rocks are called *Thalassinoides*. Deposit feeders include certain echinoids, e.g. *Echinocardium*, which excavate horizontal tunnels, backfilling as they advance.

Surface markings are usually preserved under quiet water conditions. They are essentially grooves which are semicircular in cross-section and which were impressed in soft but firm sediment. Some are food-searching trails made by free-living, mobile organisms such as snails, trilobites and worms, and may be random, winding, sinuous or a regular meandering pattern. Some snails have a groove along the length of the foot and, as they crawl., leave a double trail. Signs of their feeding are not usually obvious. *Cruziana* is the name given to another type of double trail common in Lower Palaeozoic rocks. This consists of two parallel grooves, separated by a median ridge, and showing numerous

(a) The footprints were made by a large bipedal dinosaur, probably *Iguanodon* Diagram (a) shows how closely the right foot skeleton fits into one of the footprints

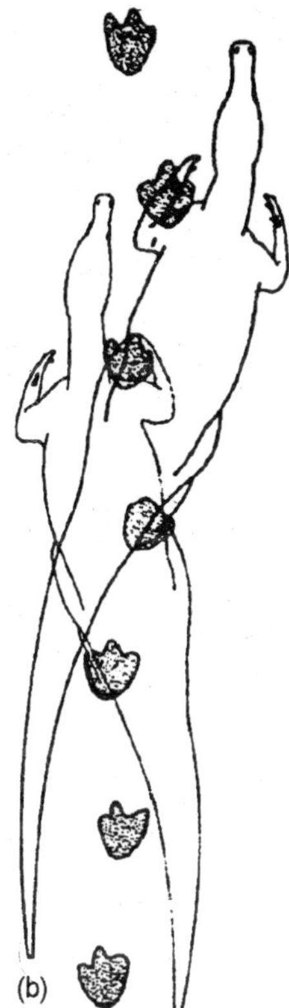

(b) The disposition of the footprints sheds some light on the way *Iguanodon* walked. The impressions though clearly alternate, are almost in line. This suggests that at each stride the body swayed from side to side, the head and tail region remaining parallel to the direction of movement. This kind of movement involves some rotation of the foot relative to the body which would normally take place at the ankle. The absence of a tail groove indicates that the tail was held clear of the ground to counter-balance the body.

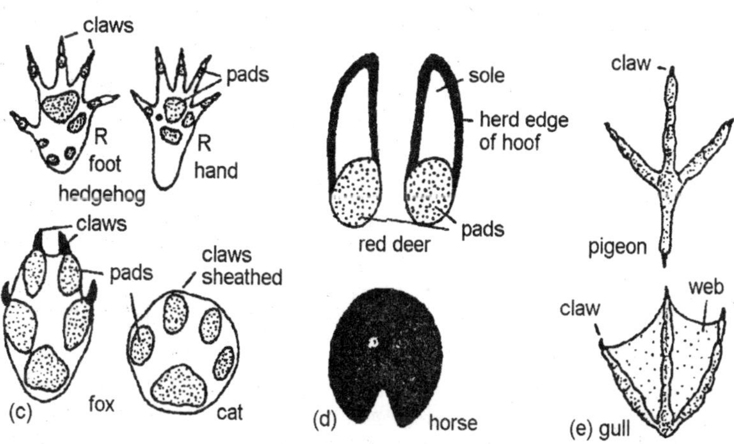

Fig. 10.2. Vertebrate animal tracks.

scratch marks made by the ventral appendages of an unknown animal (probably an arthropod) as it ploughed into a firm substrate of sand overlying mud. Resting traces, called *Rusophycus*, are thought to have been excavated by trilobites; they are more or less bilobed depressions with scratch marks.

The sublittoral region merges into the bathyal region, an area of low energy where muddy fine sands, silts and muds rich in organic matter are laid down. In places the oxygen level may be low. The diversity of organism is low though individuals are abundant. Most are deposit feeder which graze the surface or mine along shallow tunnels parallel to or inclined to the substrate as if the animal was following an organic-rich seam within the sediment. Feeding traces may be quite simple or moderately complex, ribbon-like or spiral patterns.

The bathyal region merges into the abyssal region where the deposits are very fine muds. Conditions are quiet though at times disturbed by turbidity currents flowing down the continental slope bringing sediment with organic matter which augments that which continually rains down from the upper waters. It is dark, pressure in excess of 400 atmospheres, and it is uniformly cold, about 4-5°C. There is adequate oxygen. A wide range of different, sometimes unusual, organisms lives here and may be abundant, but they live at a slow tempo. They are mainly deposit feeders or scavengers. Traces include crawling, grazing and shallow feeding -cum-dwelling traces.

Burrows are often elaborate and complex in design and sometimes three-dimensional, indicating a systematic searching for food. Oxygen supply is ensured by short breathing tubes reaching to the surface. Patterns include ribbon-like meanderings, intricate spirals and honeycomb structures, and dendritic systems. *Chondrites* is a dendritic system of descending and radiating tunnels common in the Cretaceous and Tertiary flysch in Europe. It was made, perhaps, by a thin worm-like animal which made an initial tunnel from which it excavated side branches alternately on either side starting at the far end and backtracking towards the entrance. The tunnels are back-filled with excreted sediment.

Trace fossils connected with feeding processes include *Coprolites*. These represent the fossilised faeces of fish and other aquatic vertebrates and are generally phosphatised. Faecal pellets are smaller and produced by invertebrates of many types. Coprolites and faecal pellets are abundant and widespread. They may contain remains of food which has resisted digestive juices. Examples are belemnite hooks in coprolites; and foraminiferal tests or coccolith plates in faecal pellets made by copepods feeding in the plankton. In the latter case the calcareous material is partly protected from dissolution in the deeper waters to which it falls.

Continental Trace Fossils

On the continents there are few environments in which trace fossils are likely to be preserved. Low-lying areas where sediments can accumulate in river flood-plains, deltas and lakes are the most favourable. Animal populations are likely to be high in such areas, especially near communal watering places. Volcanic areas are also a possibility: imprints in damp volcanic ash, quickly covered by another ash-fall, are now well-known.

The traces are most commonly of locomotion but traces of feeding and dwellings are sometimes preserved. The main animal track-makers include invertebrates such as snails, insects and other arthropods: and vertebrates (fish, amphibia, reptiles, birds and mammals). Plant traces are usually of root systems.

Animal tracks are instructive. Over a period of geological time they record the anatomy of the foot and the increased efficiency of locomotion as indicated by chances in foot and limb structure. In the case of vertebrates, the records is of a change from sprawling, undulating movement on four legs sometimes with tail trails, to the development of speed such as is shown by the horse or the cheetah.

Early vertebrate signs include markings apparently made by the lobe-finned fish in shallow-water rocks and found, for instance, in the middle Devonian of Orkney. Amphibian tracks appeared in the upper Devonian, for instance in Victoria, Australia, where tracks, possibly of an ichthyostegid show left and right 'hand' and 'foot' marks, set wide apart, each with splayed-out short digits. Some also show an undulating tail mark between the left- and right-hand tracks. Amphibian tracks are common in the Carboniferous but after the Permian are rarely found.

Reptile tracks are found in the upper Carboniferous and, while rare at first, become increasingly common and widespread in the Mesozoic. The early tracks are of a sprawling, clumsy gait. Later forms indicate more efficient striding with quickening movement: the trackway becomes narrower and the individual prints more widely separated. To a varying extent the characteristics of the track-makers can be deduced from the trail and the individual imprints: quadrupedal or bipedal; walking plantigrade (full foot) or digitigrade; shape of food and number of digits; whether claws or hoofs were present; general indication of size. The tracks show, too, whether the animal was moving slowly or running, alone or in herds. The size of footprints is a rough guide to the size of the animal which made them: and the distance between successive prints made by the same foot, the *Stride*, may give some idea of the speed at which the animal was moving. Estimates are based on the relationship between length of stride and length of limb to the hip. For a given size, the longer the stride the faster the movement. For instance, a large sauropod which made huge, deeply impressed footprints arranged in a narrow trackway with a stride of 2 m is judged to have been walking at about 3.6 kph.

11

APPEARANCE OF CHORDATES

Chordates originated from non-chordates. Members of Hemichordata (*Balanoglossus*) have common characters of both the groups Non-chordata and Chordata. Chordates have following characters:

Notochord

Vertebrates, or to use the more comprehensive term, chordates, have several diagnostic characters which are absolutely distinctive, separating them sharply from all other forms of life. This is an internal axial stiffening running lengthwise of the trunk and serving to resist the bodily shortening which the contraction of the muscles would otherwise cause. In its most primitive form the notochord is membranous, composed of cellular connective tissue, the cavities of which are so demanded with fluid as to render the whole structure turgid, resistant to pressure, but highly elastic. Later the notochord becomes cartilaginous, to be replaced in higher forms by the bony vertebral column consisting of a number of short but often complex vertebrae separated, for mobility, by cartilaginous intervertebral discs.

Perforated Pharynx

The development of apertures known as gill-slits though the walls of the pharynx or throat cavity is the second chordate character. These vary in number from a pair to more than a hundred (Amphioxus) and are always present, but by no means always retain their ancient respiratory function, for with ourselves and other mammals, the slits, of which there are several pairs in the embryo, are reduced in number until but one pair is left and these form the eustachian tubes which serve to equalize the air pressure on either side of the ear drum by connecting the middle ear with the cavity of the throat.

Neurocoel

The third diagnostic character is a hollow nerve cord, the so-called spinal cord, which lies immediately above the notochord or the vertebral column. This may be a very simple structure, or again its anterior portion may increase and develop until in its highest expression, the human brain, it has formed what is probably the most intricate thing in nature. In every case, however simple or complex it may be the internal canal or neurocoel persists, though exceptions may be said to exist in the tunicates or sea-squirts, which are striking examples of degeneracy resulting from sedentary life. In them the active larva has a nervous system which conforms to our definition, but in the adult it is reduced to a single ganglion, or mass of nerve matter, with no trace of the neurocoel.

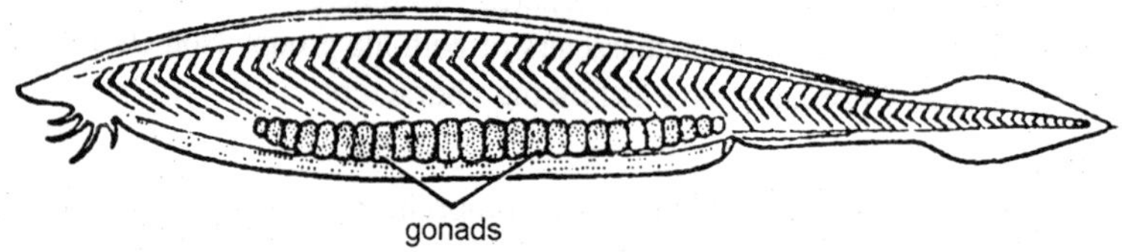

Fig. 11.1. Amphioxus. External features.

Other Characters

Other distinctive features are usually shown by chordates, They are generally segmented, the segmentation showing in the nervous system, gill-slits, vertebral column, ribs and breast-bone, and in the muscles of the trunk. When paired limbs are present, their number never exceeds four, while in the invertebrate there is no such limitation. Finally, there is apparently a reversal of surfaces, for whereas in the invertebrate the bulk of the nervous system lies below the gut and the blood system above, in the chordate the reverse is true. In the vertebrate there is a remarkable thickening or concentration of the muscles along the dorsal side within which lie the notochord and spinal cord.

It is, therefore, this group of forms, comprising the degenerate tunicates, the unprogressive *Amphioxus*, the fishes, amphibians, reptiles, birds, mammals, the last of course including man, that we wish to consider, and our immediate problem is to learn, if possible, the time, place, and source of vertebrate beginnings.

Time of Origin

The chordates are a very ancient race, dating back probably to the beginning of Paleozoic time, although the tangible record of their existence commences with the fragmentary remains of armored "fishes" (ostracoderms) found in Middle Ordovician rocks near Canyon City, Colorado, in the Big Horn Mountains of Wyoming, and in the Black Hills of South Dakota. But these relies are those of creatures which had already traveled far along the evolutionary road and, according to most authorities, do not represent the most primitive members of the chordate stem. Hence we may safely say that the time of origin was not later than the beginning of the Ordovician, and probably long before. There is however, little chance of finding the geologic record of the ancestral forms, if, as we may suppose, they were soft-bodied, delicate organisms without hard parts for fossilization. An apparent exception lies in the little "fish plate," *Eoichthys howelli*, discovered in the Cambrian shales of St. Albans, Vermont, in 1926. Later study shows this fragment to have been misinterpreted, so the Ordovician record still stands.

Place of Origin

The main contrast between invertebrate and vertebrate animals seems to be that, as a whole, the former are static organisms with little or no power of locomotion, while the latter are essentially dynamic. There are of course exceptions in each group, for the cephalopod molluscs, squid, etc., are splendid locomotor types, and some chordates have become sluggish or even fixed as an outcome of sedentary habits. Whether or not this contrast of type is a result of the physical environment, invertebrates to static marine waters and the vertebrates to dynamic fresh-water streams, as Chamberlin held, is not at all clear. The most primitive chordates existing to-day—tunicates, *Amphioxus*, etc.—are marine, inhabiting for the most part the flatsea, where they lead a wholly or partially sedentary life.

That this is therefore the ancestral habitat seems at first sight plausible, yet within this area there seems to be lacking the necessary physical or external stimulus to impel the evolution of the chordate characteristics, especially segmented body muscles and the notochord. Perhaps the strongest stimuli would be escape and pursuit and the need to remain in the environment. The cephalopod locomotion, acquired in the sea, is largely for the former purposes and is based on an entirely different principle from that of the chordate. There are, however, among the segmented worms, marine forms which swim by a wriggling or undulatory movement comparable to that of the vertebrates, except that it is up and down or dorsoventral, instead of from side to side or lateral. There is no reason, however, why the sea might not have produced the one as well as the other. Briefly, Chamberlin invoked the dynamic rivers which would give the undulatory movement, extrinsically, to a passive elongated animal temporarily anchored by the mouth. He imagined that the creature might learn to produce the same movements actively, in order to avoid begin swept out of the environment into the sea, and thus develop motor organs accordingly, an idea which has had very little general acceptance. Frankly, we do not known the place of origin, nor is there any direct evidence which can be brought to bear upon the problem.

Theories of Origin of the Chordates

At the present time there are a number of theories for the origin of the vertebrates. These are as follows :

1. The Coelenterate ancestry
2. The Nemertean (flat worm) ancestry
3. The Annelid ancestry
4. The Arachnid or Ammocoetes-Limulus ancestry
5. The Echinoderm ancestry
6. Amphioxus ancestry.

The Coelenterate Ancestry

According to this theory Amphioxus, supposedly the primitive pre-vertebrate, was derived from a coelenterate, such as Hydra or Jelly-fish, by the addition of a third layer of cells. This third layer of cells is believed to have developed because it came into contact with the substratum and then to creep forward This theory suggests that the vertebrates became separated from invertebrate stock at a very early stage, i.e., before the nerve cord and heart developed in both or either. But there is no positive evidence that they have given rise to any of the more advanced Phyla including the chordate.

The Nemertean (flatworm) Ancestry

Kofoid suggested that the chordates have been derived from the flatworms (nemertean) ancestors because of the arrangement of the

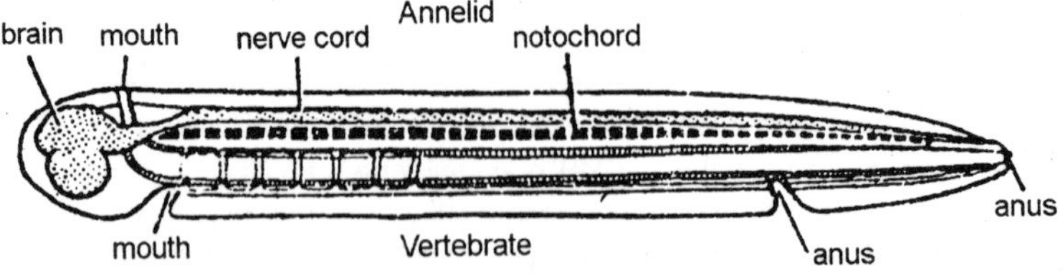

Fig. 10.2. Diagram showing the supposed transformation of an annelid worm into a vertebrate.

nemertean nervous system in eight longitudinal cords. He proposed that the development of the two dorsal cords at the expanse of the ventral and lateral cords. It has been suggested that the long proboscis and sheath of the flatworm developed into the vertebrate notochord. The horizontal lateral blood vessels are shift to form the main dorsal and ventral vessels. However, this theory is not supported by any evidence and intermediate stages are unknown. Hence it is discarded.

Annelid Ancestry

The hypothesis of annelid for the vertebrates derives the primitive chordate from the phylum Annelida, typified by the earth- and marine-worms. In many of the principal organs there is a marked correspondence, with the exception of the general reversal of the relations of the various parts to one another. For, as we have seen, the relative position of blood and nervous systems is diametrically opposite in the vertebrate and invertebrate groups. But by postulating a physiological reversal of the animal––and we know that in the flounder and squid such a change from the morphologically normal posture can take place—the various organs of the worm may be brought into almost complete harmony with those of the vertebrate. Perhaps the greatest difficulty lies in the development of the notochord, but even this seems to have its annelid prototype in "the 'Faserstrang,' a bundle of fibers running along the nerve chain and serving as a support. This and the notochord lie in a precisely similar position in relation to the other organs, and in both cases they are enclosed with the nerve cord in a common sheath of connective tissue".

The reversible diagram shows quite clearly this correspondence of parts. In the annelid position we see the mouth at *m*, from which the oesophagus arises, passing through the nervous system and connecting with the long straight gut, *HH*, which terminates at the posterior end of the body at the anus, *a*. The nervous system consists of the large supraoesophageal ganglion or brain from which nerve connectives run, one on either side of the oesophagus, to the ventral nerve chain. The main blood-vessel lies dorsal to the gut and another lies beneath it. These are connected in the anterior region by semi-circular pulsating vessels or "hearts" which cause the blood to flow forward in the dorsal vessel and aft in the ventral one. Reverse our diagram and the form becomes a vertebrate, the blood now flowing forward in the pulsating ventral aorta which serves as a heart, the ancient semicircular "hearts" having relinquished their primal function for that of respiration, since the gill-slits arising between them make them the branchial vessels. As in the annelids, the mouth is again on the ventral side and this can only be brought about through the abandonment of the old and the formation of the new one by an inpushing of the body-wall at *st* until communication with the gut is effected. This stomodaeum is balanced at the hinder end of the trunk by the new hind gut, the proctodaeum, *pr*, the ancient intestine in the tail region being aborted.

The brain and nerve cord are the homologues of the supraoesophageal ganglion and ventral nervous chain of the annelid. Indications of the ancient mouth are seen in several structures such as the neuropore in the embryo of Amphioxus, which forms in this place a direct communication between the cavity of the nerve cord (neurocoel) and the exterior and is otherwise unaccounted for. Other indications of the early mouth and its oesophagus are the fourth ventricle of the brain, a cavity which lies exactly in the place where in the diagram the annelid oesophagus pierces the nervous system, and also the hypophysis, a structure attached to the lower side of the mid-brain, part of which is pushed up from the alimentary canal, and for which there is as yet no satisfactory explanation.

Add to all this the remarkable correspondence of the kidney tubes or nephridia of the annelids and vertebrates, and the evidence is presented. Wilder says in summation: "Convincing as these comparisons

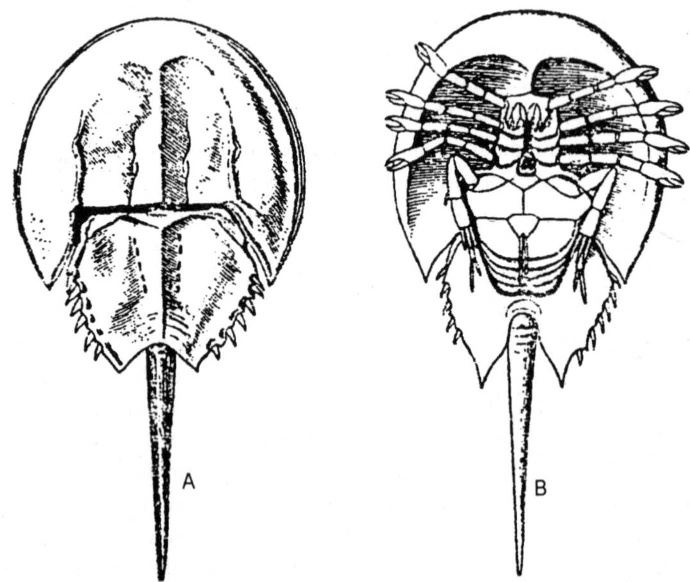

Fig. 10.3. Horseshoe crab, Limulus. A–dorsal, and B–ventral aspect.

seem when taken by themselves, the influence of later investigation has tended rather away from the annelid hypothesis, and at present, although there are many investigators who seek the ancestor of vertebrates in some worm-like form, there are few who wish to definitely assert that this ancestor was an annelid."

Arthropod Ancestry

In addition to the annelid theory, some authorities have tried to prove vertebrate descent from/Arthropoda, especially from the more primitive arachnoids such as are now represented by the scorpion and the horseshoe crab (*Limulus,*) and formerly by the extinct Merostomata. By this hypothesis we must set aside as primitive such forms as Amphioxus and the cyclostomes and start with the highly specialized ostracoderms which lived in Ordovician and Devonian times and thus were contemporaneous with and in general appearance and probable habits quite similar to the Merostomata. The soft parts of the Merostomata are of course unknown, but is reasonable to suppose that they were not unlike those of the related scorpion and Limulus, and, as Patten has shown, especially in the brain and cranial nerves of

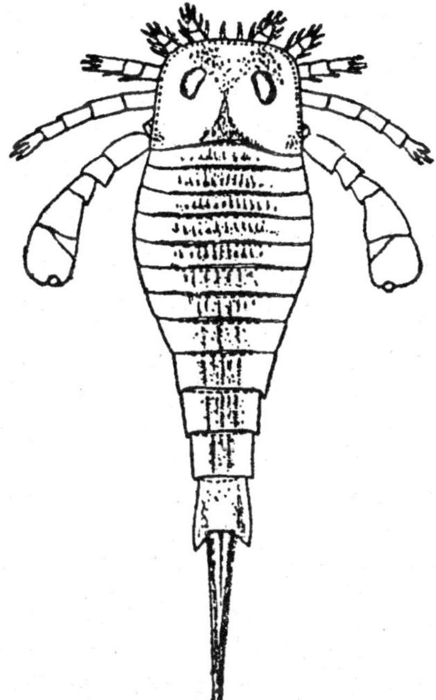

Fig. 10.4. Merostome, Eurypterus fischeri, Silurian.

vertebrates and the fused cephalothoracic ganglionic mass found in such arachnoids, there are many points of resemblance. Then, too, the sense organs, especially the eyes, are more or less comparable, and there is in Limulus an internal skeletal piece known as the "endocranium" or sternum which serves to protect the central nerve complex, and which in general form and in its relation to other parts resembles the primordial vertebrate skull. Similarities also exist between the heart and arterial systems of each group, and the appendages may e compared. There are, again, the very arthropod-like jaws which Patten has demonstrated in the ostracoderm *Bothriolepis*, a type which, on the other hand, shows many vertebrate-like characteristics; and the general arrangement of the plates by which the cephalothorax is covered is also very similar in the ostracoderms and contemporary arachnoids, but unfortunately for the argument Bothriolepis is a highly specialized end-form from the Upper Devonian. Nevertheless, while the arachnoid theory has been set forth by Gaskell (*The Origin of Vertebrates, 1908*) and by Patten (*The Evolution of the Vertebrates and Their Kin,* 1912), the main thesis has received thus far but little recognition, although the evidence, especially in Professor Patten's book, is based upon an admirably executed piece of research.

Echinoderm Ancestry

Some zoologists believe that the chordates have originated from echinoderms. But no typical chordate characters are found among these radially symmetrical animals and they have no internal skeleton. Echinoderms have too many peculiar and complicated organs of their own. Inspite of these dissimilarities between the echinoderms and the chordates, there are some embryological similarities between the two:

1. Embryonic development is similar in echinoderms and primitive chordates.
2. In both coelom originates from enteron, enterocoelic.
3. Bipinnaria larva of echinoderms (starfish) is bilaterally symmetrical, minute and transparent, resembling the tornaria larva of Balanoglossus which is a chordate. Muller (1850) regarded tornaria larva as the larva of echinoderm due to this close similarity.
4. Ciliated bands and coelomic cavities in both are similar.
5. Skeletons in both groups originate mesodermally.
6. Serological tests show that blood proteins are similar.
7. The same kind of phosphorous compound plays an essential part in the energy cycle of muscular action and differ from that in other animals.

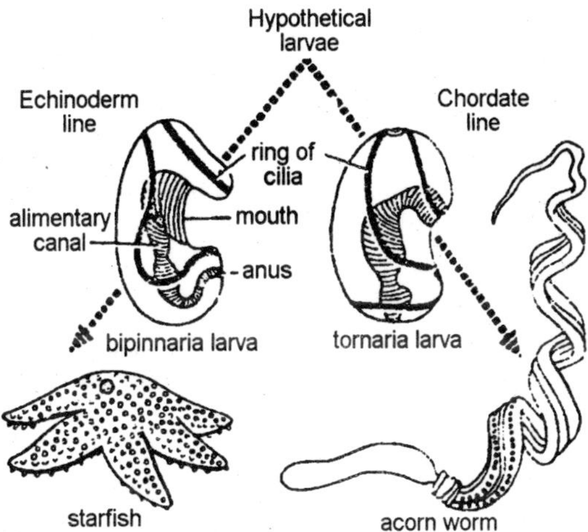

Fig. 10.5. Diagram of the larvae. A–Tornaria larva of Balanoglossus. B–Bipinnaria larva of starfish

Cambrian and Ordovician Carpoidea (primitive echinoderm) have some similarities to some ostracoderms but their possible significant as indication of close connection is exceedingly remote.

Willy assumed that echinoderms have descended from bilaterally symmetrical pelagic ancestors which represented in their development such larval forms as the bipinnaria and auricularia. Auricularia of echinoderm resembles with the tornaria larva of Balanoglossus. But Balanoglossus is a protochordate having gill slits and dorsal tubular nervous system. It requires only the notochord and segmental myotomes to approximate the amphioxid type. The evidence from these larvae strongly suggests that in the beginning of world, there existed some type of small, bilaterally symmetrical animals of very simple structure having many features of the larval echinoderm or acorn worm (*Balanoglossus*), but lacked the specialization of either vertebrate group or echinoderm. These forms by assuming radial symmetry and sessile mode, gave rise to echinoderms, and on the other hand these forms retained their original bilateral symmetry, acquired specialized breathing organs (gill slits), powerful musculature for motility, a notochord for its support and dorsal nerve cord for the better control of body activities. These chordates finally gave rise to vertebrates

Auricularia larva → Tornaria larva →
tunicate larva → Amphioxus →
Ostracoderms (earlier fishes).

Hypothetical chordate ancestry.

It is assumed that ascidian tadpole larva does not metamorphose and undergo further evolution to give rise to vertebrates.

Amphioxus Ancestry

The theory of Amphioxus ancestry places especial emphasis upon the notochord, the gill-slits, and the dorsal position of the central nervous system, and by means of these has traced the line of vertebrate ancestry though a series of transitional forms, externally very unlike one another and each somewhat isolated in its systematic position. Of these, *Amphioxus*, the lancelet, stands nearest the true vertebrates, in fact it is nearest the diagrammatic vertebrate of any living type, although, owing to certain specializations, it can hardly be considered a true stem-form. The lancelet was first described in 1778 as a shell-less snail or slug, and was named *Limax lanceolatus*. It is an inhabitant of the shallow sea, being found off the coasts of all parts of the world. There are sixteen known species most of which are recorded from tropical and *sub-tropical* shores, mainly between latitudes 40° N. and 40° S. In habits they are very sedentary, living for the most part partly buried in the sand or mud in a nearly erect posture, with the anterior end protruding. Aside from their primitive character, their world-wide distribution, coupled with sedentary habits, points out a very great antiquity. As fossils, however, they are thus far entirely unknown and probably always will be.

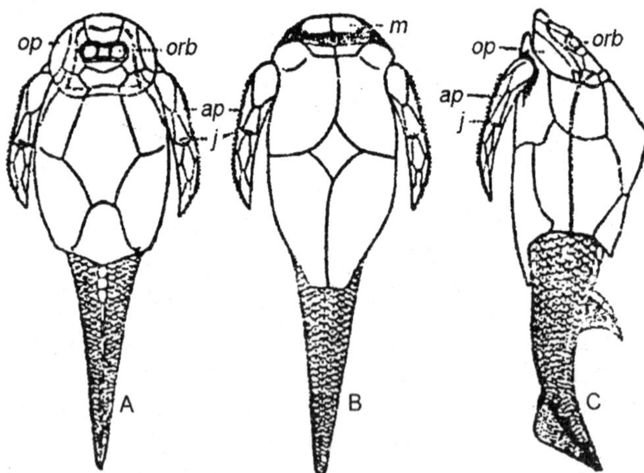

Fig. 10.6. Ostracoderm, Pterichthys milleri, Lower Old Red Sandstone, Scotland. A-dorsal, B-ventral, and C-lateral aspects. ap, pair of lateral appendages; j, joint in appendage; m, supposed upper jaw, with notches for narial openings; op, operculum; orb, orbits.

It will thus be seen that *Amphioxus* is a very simple "vertebrate," specialized a little along certain lines, but with several structures of such fundamental importance that they must be borne in mind in our search for yet more primitive forms. These structures are the notochord, the dorsally situated nerve cord, and the pharynx perforated by gill-slits and provided with an endostyle.

The only other living creatures (except Balanoglossus,) which possess these structures during any part of their career are the *tunicates*, some of which are planktonic, others meroplanktonic, in that while they have active larvae they soon settle down and become wholly sedentary in their habits. These sedentary forms, curiously enough, have retained more of their primitive characters than those which are planktonic, as their larvae are comparatively undifferentiated. The free-swimming forms on the other hand, are often modified in a remarkable way and may have so complex a life-history that the old-time chordate characteristics have almost entirely disappeared. They always, however, possess the gill-slits except in certain locomotive individuals among the colonial types, which have lost all organs except those of propulsion, that is, the muscles, nerves, and sense organs. The name tunicate comes from the test or tunic which surrounds the entire animal and is

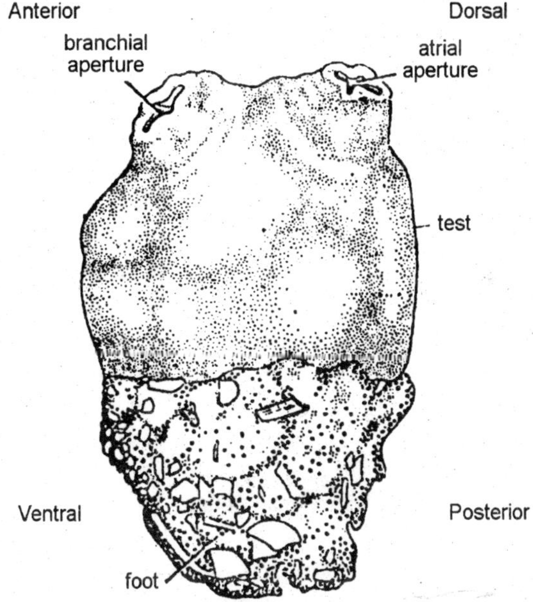

Fig. 10.7. Herdmania pallida. External features.

comparable to the shell of a mollusc in that it is formed by the body-wall or mantle. This test is unique in being made up of a substance closely comparable to the cellulose of plants.

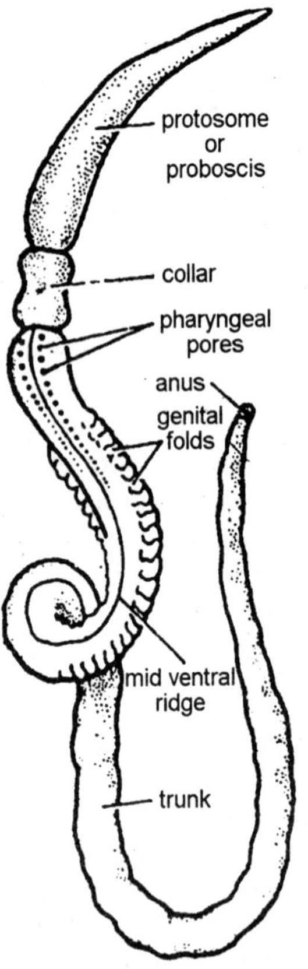

Fig. 10.8. Balanoglossus.

While the adult tunicate shows certain Amphioxus-like characteristics, it is the larva in which these are particularly emphasized, for in this stage the creature, which is tadpole-like, possesses a well developed notochord, segmented muscles, and a prolonged nerve tube with a brain-like vesicle forward which contains a pigment spot and another organ, possibly for balancing. The gill-slits are much fewer in number than in the adult and the endostyle lies in its normal position. The heart also lies ventrally and just behind the oesophagus.

But this comparatively high organization is retained for a very brief time, a few hours only. Then the creature settles down on a pair of adhesive papillae and undergoes a marked retrogressive metamorphosis during which it loses tail, notochord, and segmental muscles; the nerve tube is reduced to a single ganglion, the sense organs disappear, the gill-slits increase in number, and the animal, after relinquishing practically all of the organs that serve to link it with the vertebrates, degenerates into what is virtually an invertebrate form. It is however, evident that the tunicates represent a group more or less closely allied to Amphioxus, and hence to the other vertebrates, but that since the time of the common ancestor they have taken a divergent road, resulting in a type of degenerate adult whose real affinities are masked by its specializations. Thus, as Wilder says, "The ancestor that we here seek is better seen in the larva than in the adult, and we may believe that there once existed an adult animal with attributes like that of the tunicate larva of the present day, and that this animal was the direct ancestor of that group of which Amphioxus is now the only living representative."

Back of the tunicate ancestor there is but one known form which may or may not be near the main ancestral line. This creature is *Balanoglossus*, a marine worm that lives between high and low water marks in fragile tubes of cemented sand.

There is but a single rather slender clue to the ancestry of Balanoglossus, and that is again afforded by its embryology, for here there is a peculiar ciliated larva, the so-called *Tornaria*, which shows a very marked resemblance to the larvae of the echinoderms, and the universal occurrence of this larva within the latter group shows that, whereas they are all to-day, with rare exceptions, either sedentary or vagrant benthos, as their radial symmetry implies, they are descended from a pelagic bilaterally symmetrical ancestry. Thus, according to this belief we may "accept as a very ancient common ancestor of both echinoderms and vertebrates the form which all these larvae may be said to copy; a form having the characteristics common to all, including bilaterality, minute size, transparency, locomotion by bands of cilia, and pelagic life. The lineal descendants of this hypothetical ancestor chose two paths, the one leading to the *Echinodermata*, the other to *Balanoglossus*, the Tunicata, *Amphioxus*, and eventually the Vertebrata"

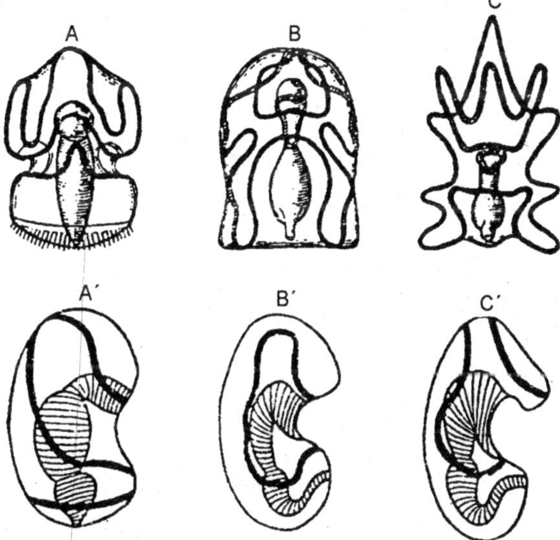

Fig. 10.9. Comparison of Tornaria larva with larval echinoderms. Main ciliated bands in black, lesses systems cross-linked. Ventral aspect: A–Tornaria; B–Auricularia (sea-cucumber); C–Bipinnaria (starfish). Lateral view: A′–Tornaria; B′–Auricularia; C′–Bipinnaria.

While this theory is incomplete in many details, it has strength where the other hypotheses are weakest in that it is based not alone upon adult structure but upon ontogeny as well. The weakest link in the chain of evidence is that which binds *Balanoglossus* to the echinoderms—the *Tornaria* larva—because the adult structures are so remote and echinoderms give not the slightest, clue in their bodily make-up to chordate affinities. As Matthew says: The origin of chordates "is still an unsolved problem. We cannot yet point to our ultimate ancestor in the Cambrian fauna. But whether known or unknown, and the latter is by far the more likely hypothesis, we can be pretty sure that we had an ancestor at that time, and that this evolutionary status was not above that of the trilobite, and may have been no better than a worm."

12

FIRST FISHES

Vertebrates certainly arose from some sort of invertebrate ancestor, but the exact ancestral group is uncertain. The fossil record provides no clues because the earliest fossil vertebrates, bone fragments of primitive fish found in Middle Ordovician rocks, are already fully differentiated from their invertebrate ancestors. Our knowledge of the origin of water-dwelling vertebrates must therefore depend entirely on indirect evidence provided by some present day soft-bodied marine animals that appear to be related to the ancestral vertebrates. These animals have a backbone like structure at some stage of their life history, but, unlike that of the vertebrates, it is not divided into separate segments or "vertebrae". Instead it is a solid rod, called a "notochord", made up of stiff, gelatinous organic matter. A similar solid notochord is found in the early embryological development of the vertebrates, but it is replaced by separate vertebrae in later stages. These notochord bearing animals have almost no fossil record, although three different kinds are found in modern seas. The three notochord bearing groups (subphyla) of Chordata can be thought of as "invertebrate chordates", since they have a backbone-like structure but lack separate vertebrate.

INVERTEBRATE CHORDATES

The most revealing of the invertebrate chordates are the *lancelets* that make up the subphylum Cephalochordata. Lancelets are small animals (three or four inches long) that live partially buried in the sand along the shores of warm seas. In addition to a well known notochord, they have many gill slits behind the mouth; along with their elongate shape and ability to swim when not buried, these slits make them very fish-like in appearance. In most aspects of their anatomy, they are much simpler than even the simplest vertebrate. Lacking jaws and teeth, they feed by extracting fine particles from the water as it passes through the gills. They have simple nervous and digestive systems and lack paired fins, eyes and a differentiated head and brain, all of which are characteristic of fish. Lancelets probably are some what specialized descendants of the early Paleozoic ancestors of the vertebrates.

The other two subphyla of invertebrate chordates are also probably descended from the ancestral vertebrates of Early Paleozoic time, but they have become more specialized than the lancelets and lack their suggestive fish-like characteristics. One group, the subphylum Urochordata, is made up of attached or floating marine suspension-feeders that superficially resemble sponges. A notochord is present in the larvae, but not in the adult. The other group, the subphylum Hemichordata, or "acorn worms", are worm-like animals that live in tubes that they construct in mud and sand on the sea floor. Like the lancelets, they have many gill slits that serve to filter small particles from the sea water. The adults appear to have a rudimentary notochord, but it is the early larval stages that are of greatest interest, for they give a clue to a still earlier step in vertebrate history-the relationship of these invertebrate chordates to the other invertebrates.

Evolutionary History of Fishes

The earliest vertebrates, from which all others arose, were primitive fishes. The first fragments of fossil fish bone are found in Middle Ordovician rocks, remains of fishes are rare and poorly preserved until the end of the Silurian period, when fishes began an explosive evolutionary radiation that lasted throughout the succeeding Devonian period. The fish evolution was dominated by the primitive and now mostly extinct classes Agnatha and Placodermi. From these early groups arose the two dominant classes of modern fishes, the chondrichthyes (sharks) and osteichthyes (bony fishes). Most of the evolutionary expansion and replacement of the four fish classes took place during the Devonian period.

The Old Red Sandstone in England and Wales

In the southwest part of England and in neighboring Wales, there lie exposed layers of rock that geologists agree were formed some 400 million years ago in Devonian times. Palaeontologists known these deposits as the Old Red Sandstone and value them as one of the few formations in which the passage from late Silurian to early and middle

Devonian sediments is recorded without major gaps occasioned by erosion. The lower layers were formed as silt and sand settled on the sea bottom. The invertebrate animals preserved in the oldest rocks are recognisable as relatives of modern general which are still marine in habit. The sediments must have accumulated in the shallows of a subsiding sea, however, because the invertebrate fauna at the next higher level are a group characteristic of brackish, estuarine waters. Finally, the uppermost rocks of the Old Red Sandstone are composed of materials that were deposited in fresh water, dropped perhaps as the current slowed in rivers about to empty into the sea. It was in the rocks of this formation that palaeontologists found, only a little over a hundred years ago, fossils belonging to the earliest known group of vertebrate animals.

Early Vertebrate Fossils in the Old Red Sandstone

The remains of these animals consisted of bony plates that could be retrieved from every level of the Old Red Sandstone. Despite the differences in the microscopic structure of the plates, all the hard material was recognizable as vertebrate bony tissue and quite distinct from the shells and exoskeletons of the invertebrates found in the same deposits. The first fossils found were fragmentary and gave few dues to the general form of the animals. As palaeontologists looked elsewhere—in Scotland, Norway, and Spitzbergen and later in other places—they discovered remains that were less broken and scattered. It became obvious that these early vertebrates had been fish-like in shape. Some had been streamlined like free-swimming fishes of today; others were flattened and presumably bottom-dwellers. They varied in size from a few inches to more than a foot in length. All were encased in a bony armor which was so designed as to leave no doubt in palaeontologists' minds that the animals lacked movable jaws and paired fins like those of later backboned forms. Because they shared these characteristics, the ancient vertebrates, or jawless fishes, were grouped together as ostracoderms, animals with a shell-like skin.

Varieties of Ostracoderms

Continued study of fossil material made apparent the existence of many different types of ostracoderms. E. R. Lankester, who wrote a description of the forms dug from the Old Red Sandstone before 1868, divided the group into the Osteostraci and the Heterostraci. Thirty years later, R. H. Traquair reported the discovery of two additional kinds of ostracoderms, the Anaspida and the Coelolepida. Modern workers in the field continue to find this classification valid and are still investigating the structure of the animals in each of the four subdivisions.

The Osteostraci

That the Osteostraci are the type about whose anatomy most is known is due largely to the research of E. A. Stensio. Since the early 1920, he and his technical assistants in laboratories at the University of Stockholm have picked the rock away from the fossilized armor and scale of numerous specimens and ground hundred of sections to determine the nature of the internal structures of the head in these forms. In all the Osteostraci the head is covered dorsally with a shield of bone ornamented over its surface by denticles. The shield shows a characteristic pattern of openings and depressed areas: close to the midline in the center of the head or somewhat forward there lie a

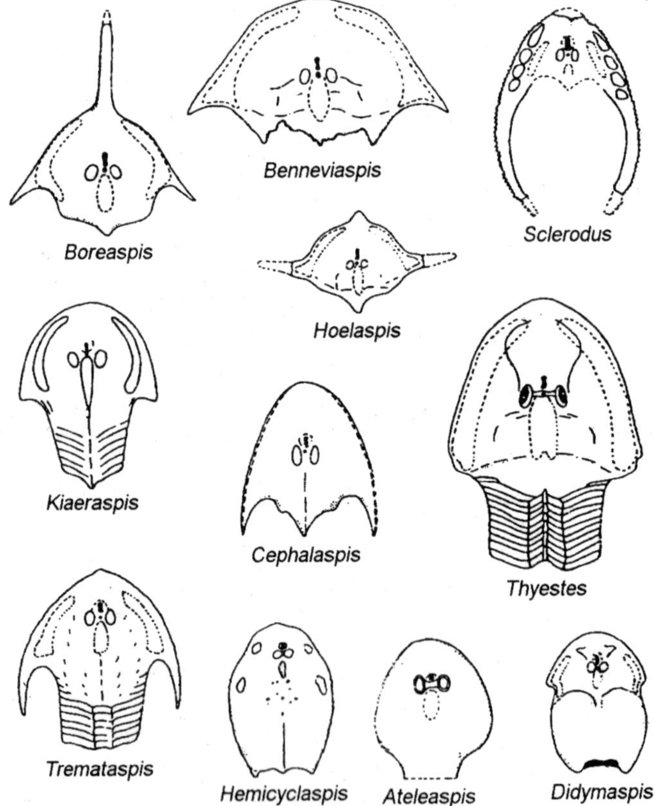

Fig. 12.1. Head shields of various osterostracan ostracoderms in dorsal view.

pair of circular holes through which, in life, the eyes must have stared directly upward. Between these eyes a small pineal opening is invariably present, and in front of it a single aperture for the naso-hypophyseal canal. Although the purpose of the four openings in the head shield is obvious and proved without a doubt by the anatomy of the underlying internal structures, the significance of the two elongated depressed areas on either side of the shield and of the median one behind the pineal opening is still not certain. There is a pavement of small tile-like scales in the floor of each depression and, beneath, a number of bone-lined channels which lead inward to the ear region of the brain. Stensio concluded that the depressed areas held electric organs, but other workers have guessed instead that they housed sense-receptor cells which detected disturbances in the water. An aggregation of sensory structures on the dorsal side of the head suggests that the osteostracans may have swam on or near the bottom, and the fact that the underside of the head was flattened reinforces that possibility. The head shield lapped over to protect the edges of the ventral surface, the remainder of the area being covered by little scales fitted tightly together. Between the ventral tip of the head shield and the front of the scaly region lay a very small mouth, surely useless for ingesting any but the most minute particles. On each side of the mouth a curved row of external gill-openings scratched to the posterior margin of the shield. In most of the Osteostraci, the shield was prolonged at its lateral or posterior edges into a pair of prongs, which may have helped to stabilize the animal as it swam. The stabilizing effect was enhanced, it seems by paddle-shaped flaps, which projected behind the rigid prongs in several forms. Posterior to the head shield, the body was covered more flexibly with rows of elongated scales. It is certain that these ancient vertebrates swam as modern fishes do by side-to-side movement of the tail for not only was the body built to bend, but the caudal fin showed the same heterocercal design that still characterizes a number of fishes today.

Internal Anatomy of the Head Region

Although palaeontologists had early made the assumption that the fossil ostracoderms, so fish-like in form, were true vertebrates, Stensio's revelations concerning the internal anatomy of the head region of the Osteostraci provided unequivocal evidence that the conclusion was correct. When the Swedish scientist made ground sections of the head shield, he found that the outlines of what was probably a cartilaginous endoskeleton were preserved by a very thin layer of bone. It was

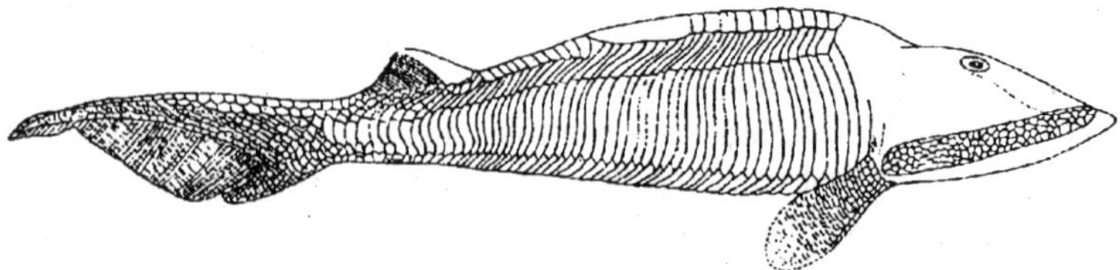

Fig. 12.2. Reconstriction of Hemicyclaspis, a member of the Osteostraci.

possible to discover the configuration of the spaces within the head skeleton and thus to know the shape of the soft organs which had filled them. The brain housed by the cranium was similar to the brain of modern jawless, or agnathous, vertebrates: it was middorsal in position and subdivided into a forebrain, midbrain, and hindbrain. Connections between the forebrain and the eyes, nostril, and hypophysis were evident as well as a cerebellum atop the hindbrain and an internal ear with two semicircular canals on each side. Cranial nerves could be followed to the region of the gill pouches, as they can in any piscine vertebrate. The most peculiar structures were the large tubular passages extending from the cavity of the internal ear outward and upward to the depressed areas in the dorsal surface of the head shield. The endoskeleton that protected the brain was continuous with that which supported the gill pouches. The large size of the pouches, especially the anterior ones, seemed to confirm the idea that the animals must have been filter feeder, sucking water through the small mouth, straining it of organic matter, and expelling it over the gills and through the openings on the underside of the head. Despite the rigidity of the head shield, the ventral patch of fitted scales could have been flexible enough to allow the muscular contractions necessary to maintain such a flow of water. The thin layer of bone which made it possible to learn something about the structure of the gill pouches and the brain did not continue, apparently, posterior to the head shield. As a result, neither Stensio nor any other investigator has been able to describe the internal skeleton that supported the trunk and tail of Osteostraci. Doubtless, these animals, like all primitive vertebrates, had an unmineralized but turgid rod-like structure, called a notochord, beneath the nerve cord, but whether they possessed any skeletal elements in the vicinity of the notochord and the nerve cord is a question that cannot be answered at the present time.

Habitat and Fate of the Osteostraci

Once the vertebrate nature of the Osteostraci had been determined, palaeontologists sought to abstract from the fossil evidence some idea of the habitat and history of these primitive fishes. Even as they drew their conclusions, they remained aware of the small amount of material on which they had to base their thinking. Although distribution of these forms may have been worldwide, collector have dug fossils only from northern Europe and widely separated sites in North America; and, of course, they can know nothing of the large number of osteostracans that were not preserved or whose fossilized remains eroded away. All the described forms but one seem to have lived in fresh or brackish water. The single marine form, *Sclerodus*, found in the Old Red Sandstone of England, may represent an aberrant subgroup or the only evidence remaining of a marine branch of the Osteostraci whose members swam over the sea bottom in other parts of the globe. The gradual extinction of a osteostracans is proved by the decreasing number of fossils found in middle and late Devonian rocks. They disappeared from the earth's waters at the end of the Devonian, undoubtedly because they were displaced by the more progressive fishes—forms with jaws and paired fins that were more efficient than the osteostracans and the other ostracoderms in swimming and food getting. The Osteostraci, in fact, seem to have been suffering in competition with the burgeoning jawed, or gnathostome, fishes throughout Devonian time. Gradually, they were restricted to relatively undisturbed areas at the bottom of streams and no longer frequented more open waters near the floor of lagoons and lakes. Although later osteostracans show some advance in the development of flaps and spines that would have increased their stability in swimming, there occurred a steady decrease in the length of the head shield that was ominous. A loss in dermal bone has been a sure sign of decline in several vertebrate groups.

Origin of the Osteostraci

The origin of the Osteostraci is much more mysterious than their disappearance. An investigator who seeks to know the steps by which these forms came into existence finds himself confronting one of the most perplexing problems in the study of vertebrate evolution. The earliest known osteostracans, those found in Silurian deposits, show all the typical characteristics of the group. No one has found any older fossils whose structure is clearly antecedent to the osteostracan pattern. When a vertebrate group seems to burst upon the scene in this fashion, palaeontologists can say only that they may possess fossils of the ancestral forms which they have not properly identified or that the progenitors of the group were too few in number or too fragile to leave remains that were likely to be found. The latter explanation is probably the true one for the Osteostraci. Early forms of the head shield may have been thinner than the known ones—so thin that they crumbled away after the death of the animals that produced them. It is also possible that the evolution of the Osteostraci may have taken place relatively rapidly, the oldest recognisable forms having developed from a small ancestral stock which inhabited a restricted area. Although palaeontologists may find many specimens of the successfully established form which gained a wider geographical distribution, there is a strong probability that the resting place of the rarer ancestral types will never be discovered.

The Anaspida

There is only one known group of ostracoderms, the Anaspida, that seems related to the Osteostraci. The forms it includes were contemporaries of the osteostracans and thus are surely cousins rather than parents of the latter. Researchers trying to asses the degree of relationship between the anaspids and osteostracans must base their conclusions upon a comparison of the external anatomy of the two types of jawless fishes, as the internal structure of the anaspids is unknown. The chief reason for postulating kinship between the two groups lies in the anaspids' possession of the dorsally located and characteristically arranged single nostril, pineal opening and pair of eyes. Although the eyes are somewhat farther apart in the anaspids than they are in the osteostracans, the similarity of the pattern is unmistakable.

Mode of Life of the Anaspids

The remaining structural characteristics of the anaspids suggest that although they too seem generally to have inhabited brackish or fresh water, their mode of life in that environment was quite different from that of the osteostracans. The anaspids were not flattened bottom-dwellers. They possessed a stream lined body that achieved some stability through a projecting ridge of dorsal spines and in some cases a ventrolaterally placed pair of elongated folds. The caudal fin was peculiarly designed to give an upward thrust to the head as the anaspid swam: the lower of its two lobes was the longer and was probably stiffened by the extension into it of the axial supporting element of the body; the upper lobe, shorter and more flexible, would have driven the tail downward when the fish moved forward. Such a caudal fin may have substituted effectively for a pair of pectoral fins in elevating the head and so in keeping the anaspid away from the bottom. The terminal

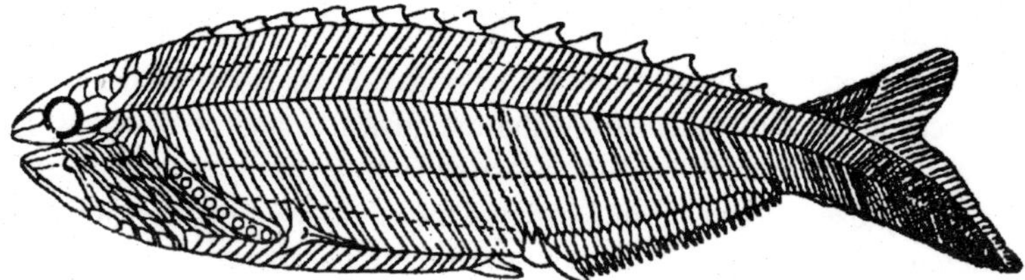

Fig. 12.3. Reconstruction of Rhyncholepis, an anaspid.

position of the mouth suggests that anaspids may have remained just under the surface sucking in algae which floated in the sun-lit water. The body must have been fairly flexible, for fossil remains show it to have been covered by a mosaic of very small plates in front and behind the downward-curving row of gill slits, by rows of narrow, vertically set, bar-shaped scales. If the muscles were as well developed as the shape of the trunk and tail lead one to guess, the anaspids may have been relatively strong and quick in their movements. Among the anaspids there were forms that showed reduction and even disappearance of the dermal plates and scales over much of the body. With what adaptation or change in habits this loss was associated, palaeontologists may never known.

THE HETEROSTRACI

The third group of ostracoderms, in contrast to the first two contained a number of marine forms. In fact, the Heterostraci, as these ostracoderms are called, must have radiated widely and lived successfully in a broad range of watery habitats. Many specimens have been recovered from sediments deposited in shallow seas, yet in every family of these jawless fishes there seem to have been freshwater representatives. To what special abilities the heterostracans owed their success is not clear: they were no better equipped for locomotion than other ostracoderms. Like the anaspids, they relied upon a hypocercal caudal fin for lifting the head in swimming, but their projections from the body to restrict their rolling while moving forward were even less extensive. Some forms possessed a single, long dorsal spine and a pair

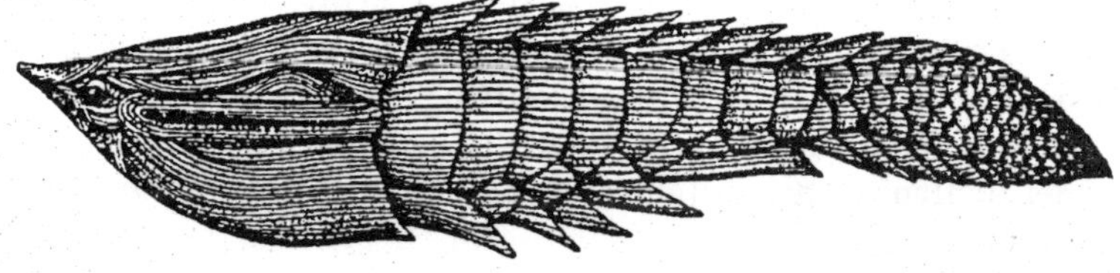

Fig. 12.4. Reconstruction of Anglaspis, a heterostracan.

of lateral ones; other types had not more than ridges of sharp, elevated scales to steady them.

Most heterostracans, from the primitive cyathaspids to the more advanced pteraspids, seem from the cylindrical shape of the body to have been swimmers in open water. Only one subgroup, the Drepanaspidae, prominent in Devonian times, showed the dorsoventral flattening that indicates the bottom-dwelling habit of life. The body in all the Heterostraci was covered in front by a sheathing of dermal plates and in back by large, overlapping or small, closely fitted scales. The presence of a row of scales at the edge of the mouth has caused some palaeontologists to speculate that heterostracans may have been able to nibble at food rather than merely to draw in organic bits with the respiratory current. If that were so, the Heterostraci may have enjoyed some slight advantage over other types of ostracoderms in utilizing materials in the environment as sources of foods.

Palaeontologists agree that the Heterostraci struck out on a separate evolutionary path from that followed by the Osteostraci and the Anaspida. One indication of the divergence is the difference in arrangement of the openings in the head shield. Absent are the close-set eyes with single nostril and pineal aperture between them on the middorsal line.

Instead heterostracans show what may be the first appearance of the pattern that was to characterize later vertebrates: the eyes are situated more laterally, the pineal opening is middorsal but not invariably present, and a single nostril is nowhere apparent. The location of the nasal aperture has not been proved to the satisfaction of all palaeontologists, but most researchers are of the opinion that the nostrils were paired and opened at each side of the mouth. Stensio and his colleague Jarvik insist, however, that the heterostracans retained a single nostril anteriorly on the head. The head shield itself differed in its construction from that of the Osteostraci. The heterostracan covering contained no bone cells and was, in early forms, composed basically of four plates, one dorsal, one ventral, and, between then on each side, a branchial plate covering the gill area. Water passing over the gills emerged from a single opening behind the branchial elements rather than through separate holes in the ventral part of the shield, as in most of the Osteostraci. That there was an evolutionary trend toward reduction in the solidity of the dermal head covering in heterostracans is indicated by the increasing subdivisions of the plates in later forms.

THE COELOLEPIDA

Similar to the Heterostraci in body from and in lack of the osteostracan-anaspid arrangement of eyes and median nostril are the mysterious ostracoderms, which have been classified as the Coelolepida. Palaeontologists know little more about them than that they did exist. Isolated scales have turned up in rocks of late Silurian and early Devonian age as well as a few whole specimens which have revealed nothing beyond, the general shape and size of these forms. They were small, apparently flattened, and covered with tiny spined scales. Their fossils come from deposits laid down in brackish waters. Presumably they swam in company with other forms of ostracoderms in bays where fresh water from rivers diluted the salty sea. Since their structure is poorly known, it is almost impossible to speculate upon their relationship to the other jawless fishes. Although they may simply represent a separate radiation from the ancient base of the ostracoderm stock, some palaeontologists prefer to regard the coelolepid group as a degenerate offshoot from heterostracan stock. They emphasize the likeness in shape and the lateral position of the eyes in the two forms and point out that the evolutionary trend toward fragmentation of the heterostracan dermal elements if continued to an extreme point, could well have resulted in a covering of fine scales like that of the Coelolepida. Westoll, however, suggested that coelolepids might be incompletely ossified young heterostracan individuals.

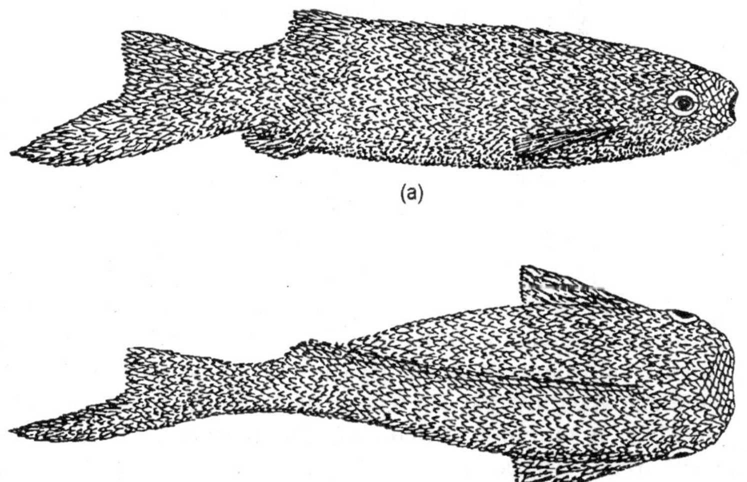

Fig. 12.5. Reconstruction of Phlebolepis, a coelolepid. (a) Lateral view; (b) Dorsal view.

Origin of the Vertebrates an Unsolved Problems

The possibility of a relationship between the Heterostraci and the Coelolepida is but a small part of the problems that palaeontologists would like to solve. What remains to be understood is the early evolution of the entire ostracoderm assemblage and the origin in even more ancient times of the vertebrate line from its non-vertebrate ancestral source. Difficulties of such magnitude exist, however, in the study of these matters that investigators have had to confine themselves to defining the problems and to building theories on the small amount of evidence they do have. Although nonscientists sometimes accept current theories as the answers given by science to these questions of universal interest, students of evolution understand them as starting points for research.

The Earliest Known Ostracoderms

The earliest known evidence of ostracoderm existence consists almost entirely of fragments of dermal bone that show similarities in microscopic structure to heterostracan dermal plates. These plates of bone are present in rocks of Middle Ordovician age in formations exposed in South Dakota, Wyoming, and Colorado. There is one deposit of vertebrate material that is apparently older, a group of small denticles of early Ordovician date found at a site near Leningrad. Since the denticles have no distinctive characteristics relating them to a particular type of ostracoderm, it must remain only an assumption that the vertebrates that produced the dentin and enamel-like material were ostracoderms of some kind. The assumption seems a safe one, however, because the discovery of an impression of a head shield of a heterostracan of Ordovician age has demonstrated the existence of differentiated forms as early as the middle of the period.

All the Ordovician fragments have been made the basis of genera: the Russian fossils are classified as *Paleodus* and *Archodus*, the North American ones as *Astraspis* and *Eriptychius*. These early vertebrate fossils, although they bear witness to the existence of ostracoderms in the Ordovician period, do not afford much help to palaeontologists who are trying to discover the origin of the jawless fishes. Although some of the fragments of dermal bone are at least an inch long, they are not large enough to give any idea of what *Astraspis* and Eriptychius looked like. If one assume from the histological evidence that the animals were recognizable as heterostracans, it follows that this Ordovician material, even should more of it be found, can reveal little about the transition to the ostracoderm level from a preexisting lower line.

Reasons for the Absence of Pre-Ordovician Vertebrate Fossils

What the earlier history of the vertebrate group was and why no vertebrate fossils have been found in pre-Ordovician rocks, palaeontologists can only guess. It may be that the first dermal ossifications were so thin that none has been preserved in recognizable for. Possibly, dermal bone appeared with comparative suddenness, after the organization of the vertebrate body was well established, as the result of a small but crucial change in the chemistry of the deep layers of the skin: research into the question of hard-tissue formation has shown that ossification will take place in connective tissue when a certain chemical equilibrium is reached. The vertebrate pattern of soft organs systems may have been evolving for millions of years before the relatively modest genetic mutation occurred which for the deposition of bone. If, as this theory suggests, the dermal skeleton was among the last of the ostracoderm structures to evolve, the earliest vertebrates would have been soft-bodied and thus rarely preserved after death. A paleontologist might expect, as a stroke of good fortune, to discover an impression of such a soft animal but no more.

Theories of Vertebrate Origin from Invertebrates

In the absence of known vertebrate fossils from rocks older than those of Ordovician time, it has been concluded that the backboned animals were the last of great groups to evolve. Representatives of virtually all the invertebrate phyla appear more than 100 million years earlier in Cambrian deposits, in the oldest strata sufficiently undistorted by the forces of heat and pressure to yield significant amounts of fossil material. Although so large a time gap exists between the first appearance of vertebrate and invertebrate organisms, zoologists and palaeontologists have assumed that the former must have been derived from the latter and have speculated widely about the most logical or probable nonvertebrate ancestor for vertebrate animals.

Some nineteenth-century biologists, observing what they considered to be similarities between the anatomy of living, adult vertebrates and invertebrates, suggested that vertebrates evolved from annelid worms or from arthropods. Some hypotheses entailed turning the lower forms upside down to explain the dorsal location of the vertebrate nerve cord and the ventral position of the heart. The weakness of these idea was soon apparent: aside from the improbability of a dorsoventral reversal, it was obvious that the double, solid invertebrate nerve cord was not translatable into the single, hollow vertebrate structure and that neither the annelid nor the arthropod blood-pumping devices could give rise to the vertebrate heart. In fact, every part of these theories of vertebrate origin disintegrated under examination.

Those biologists who turned for clues to the vertebrate characteristics that appear early in the embryo produced more viable theories. In accordance with the general concept of Ernst Haeckel that the developmental steps in the formation of an individual give some indication of the evolutionary history of the group to which it belongs, these biologists focused their attention upon the establishment of the three primary layers of cells in the embryo (the outer ectoderm, the inner endoderm, and the central mesoderm) and the first structures to arise. The discovery of similarities in the origin of the mesodermal layer in echinoderms and vertebrates and of the distinctly different method of its origin in other multicellular animals gave rise to the possibility of an evolutionary link between these two groups. This idea has proved difficult to pursue, however, because all extant echinoderms develop a radial symmetry and a semisessile habit that are far distant from the vertebrate condition. A connection between echinoderms and vertebrates is not impossible on this account, but to maintain it, one must assume that a protovertebrate group diverged from the echinoderm line in pre-Cambrian times before the development of the specialized, radially symmetrical forms.

The Vertebrates as Chordates

Relationship to two, and possibly three, other groups of nonvertebrate organisms has been postulated on the basis of the appearance in vertebrate embryos of three structures, the rod-like notochord, which develops in the dorsal midline; the tubular medial nerve cord, that forms above it; and a series of paired, laterally placed clefts in the body wall, which allow passage to the outside from the pharyngeal region of the gut. These structures were discovered in the little boneless water-dwelling lancelet *Branchiostoma*, better known as amphioxus, and also in the larvae of tunicate forms. A suggestion of their presence was found in a third group, the acorn worms and their relatives. The common possession of the three anatomical characteristics seemed of such significance that the vertebrates, the lancelet, the tunicates, and the acorn worms were classified together as members of the phylum Chordata. Since no one has suggested a more logical alliance for any of the nonv .ebrate chordates, the group has been maintained in the classification scheme. Although some biologists believe that the acorn worms are misplaced in the phylum Chordata, others see these animals and the other chordates as constituting a

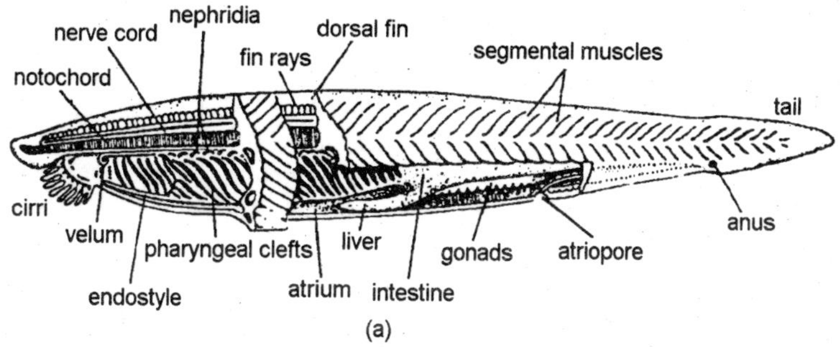

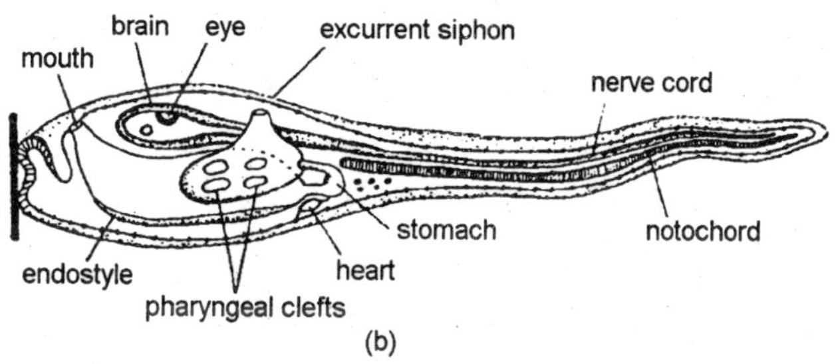

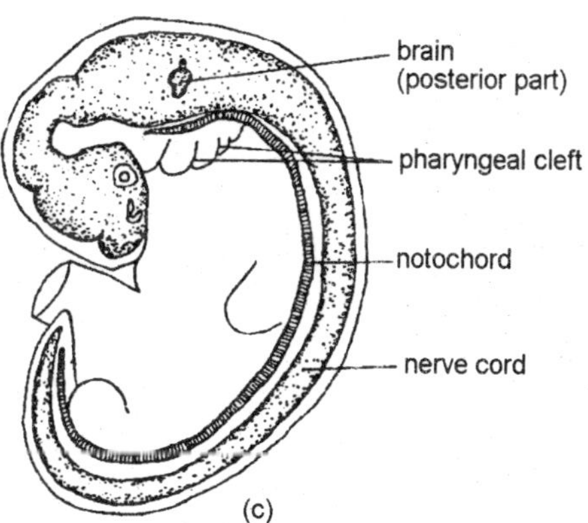

Fig. 12.6. The three chordate characters (dorsal tubular nerve cord, notochord, and pharyngeal clefts) as seen in (a) Branchiostoma (=Amphioxus); (b) larval tunicate; (c) vertebrate embryo.

natural group whose members have evolved from a common stock. The latter workers have formed a theory of vertebrate origin based on descent from primitive forms related to the lower chordates existing today.

The Evolution of the early Chordates

According to this theory, the animals at the base of the chordate line must have been soft-bodied, sessile forms of small size, resembling perhaps the minute pterobranch relatives of *Balanoglossus*, the acorn worm. These hypothetical forerunners of the chordates may have trapped food particles (as pterobranchs still do) by waving soft, unjointed appendages in the surrounding water. One group of pterobranchs exhibits a pair of slits in the body wall, through which excess water, ingested with food material, finds its way out of the gut. The existence of related animals some of which have food-collecting appendages and no openings through the body wall, some of which have both, and some of which (like the acorn worms) have only a series of paired clefts has

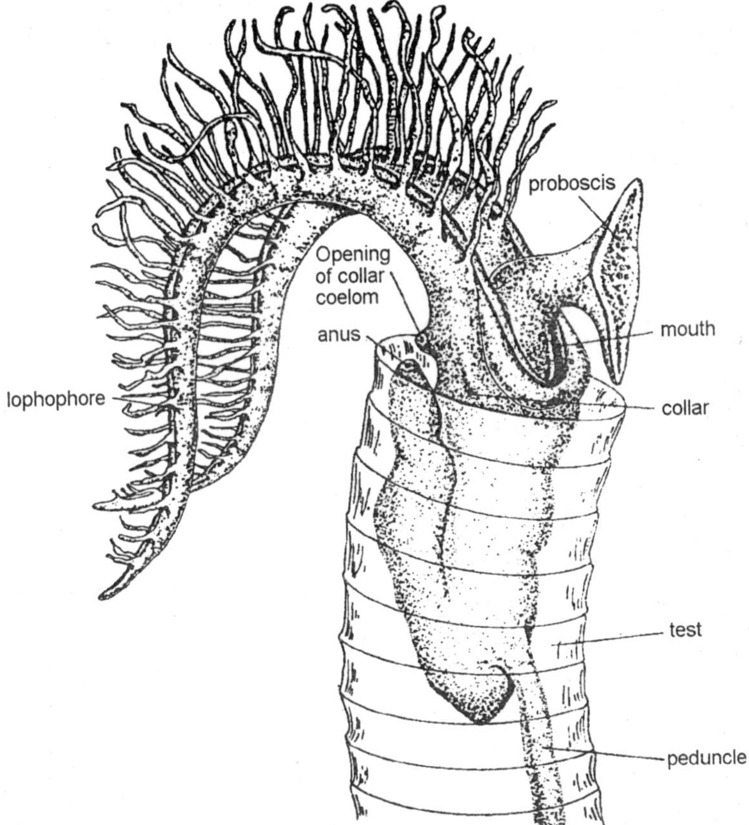

Fig. 12.7. A pterobranch, Rhabdopleura, showing one member of a colony, connected by a test-encased peduncle to other individuals.

given rise to the thought that a transition from one type of structure to the other may have occurred early in the evolution of the chordate line. With the change in structure would have come in the way food particles were collected: in the absence of oral appendages the animals would have had to draw food-laden water into the gut and expel the water, strained of its organic material, through the paired openings in the body wall. This idea is supported by the fact that filter feeding, as this process is called, is characteristic of all the lower chordates and was also used, palaeontologists are quite sure, by the ostracoderms.

If the filter-feeding acorn worms are not degenerate but exhibit a very primitive state of chordate or organization, one must conclude that the paired openings from the gut to the outside, generally called gill slits, were the first of the chordate characteristics to evolve. The other who, the notochord and the dorsal hollow nerve cord, are poorly expressed in acorn worms. A short rod which stiffens only the proboscis of *Balanoglossus* is hesitantly accepted as representing the former and a concentration of nervous tissue in the dorsal region of the anterior collar some development of the latter. The incomplete nature of the typical chordate structures in these animals is of ambiguous significance. Either the acorn worms represent a group that had gone half the distance to full chordate status—and that belief is evidenced in the assignment of the name Hemichordata—or they are actually unrelated forms whose anatomy has developed superficial similarities to that of chordates. If the latter is true, of course, those who suppose the chordates to have arisen from pterobranch-like animals are on the wrong track.

Whether they are correct or not, it does seem possible that gill slits which functioned in the collection of food and so had a vital role in sustaining life may have been the earliest of the chordate structures to become firmly established. In the tunicates, assigned to the subphylum. Urochordata, only the gill slits are present throughout the life of the animal. The notochord and dorsal tubular nerve cord in these marine forms exist during the larval period, disappearing during the metamorphosis which produces the sessile or free-floating adult. They compose the bulk of a muscular tail by means of which the larvae may swim, some distance from the parent animals before assuming their vegetative existence. Those who regard the urochordates as modern representatives of animals ancestral to vertebrates suggest that the role of the notochord and nerve cord in tunicates is primitive and witness to the fact that these structures arose as adaptations for dispersal of the young. If mobility proved to be a decided advantage, allowing individuals to avoid injury or to follow the food supply, it is possible that animals that retained their larval form longest survived more often to reproduce their kind. In time, a population characterized by members having a prolonged juvenile stage and a relatively short, nonmotile adult life would have evolved. If, then, paedogenesis took place, that is, reproduction when the animals were still in the larval state, the degenerate adult form could have disappeared entirely.

Relationship of Amphioxus to Vertebrate Ancestors

Had the evolution of chordates taken place in this way, the earliest of the animals to possess all three of the typical characteristics as well as the segmented axial muscles responsible for locomotion might have resembled the modern amphioxus. Although amphioxus has had a long history of its own, it may have carried into the present slightly modified versions of the primitive notochord and nerve cord, the former extending from the anterior tip of the body into the tail and the latter having at its anterior end a slight enlargement presaging the brain. The common ancestor of amphioxus and the vertebrates, if it existed as this theory suggests, may have inhabited shallow places, filter feeding in quiet water, and swimming away from invertebrates that chanced to disturb it. Beyond this supposition theorists cannot go. Although they have based their reasoning carefully upon the evidence of comparative anatomy and relied upon the phenomena of variation and natural selection as agents of evolutionary change, they cannot offer the fossil evidence that would substantiate their ideas.

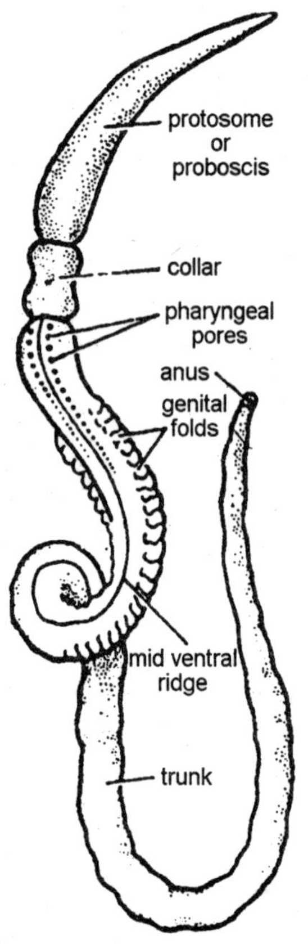

Fig. 12.8. Saccoglossus.

Habitat of the First Vertebrates

As palaeontologists are unable to trace the progression of forms from the invertebrate level through the earliest chordates to the bone and scale-covered ostracoderms, they have tried to determine at least the nature of the watery environment in which the evolution of the jawless vertebrates took place. In this problem, there is some pertinent paleontological and geological evidence. Nevertheless, the question at hand—whether the vertebrates arose in fresh or salt water—is far from settled. By late Silurian time, it appears, ostracoderms were living in both marine and freshwater environments. It is their habitat in the earlier Ordovician period that assumes crucial importance. Among the sedimentary rocks of that age which are now accessible, few were deposited in bodies of fresh water or on floodplains of ancient continents. Water that fill depressions or run through furrows in land masses are relatively shallow: minor changes in level serve to elevate old bottoms and expose the accumulated sediments to the forces of erosion. The small number of Ordovician continental sediments which have survived continue to be searched, but undoubtedly much fossil material from the period has long since disappeared. Marine deposits of Ordovician age and older are more common than continental deposits. Although many of these rocks have proved to be fossiliferous, the absence of vertebrate remains in them is striking. Palaeontologists who maintain that vertebrates originated in fresh water point to the lack of vertebrate fossils in early marine rocks as evidence in favour

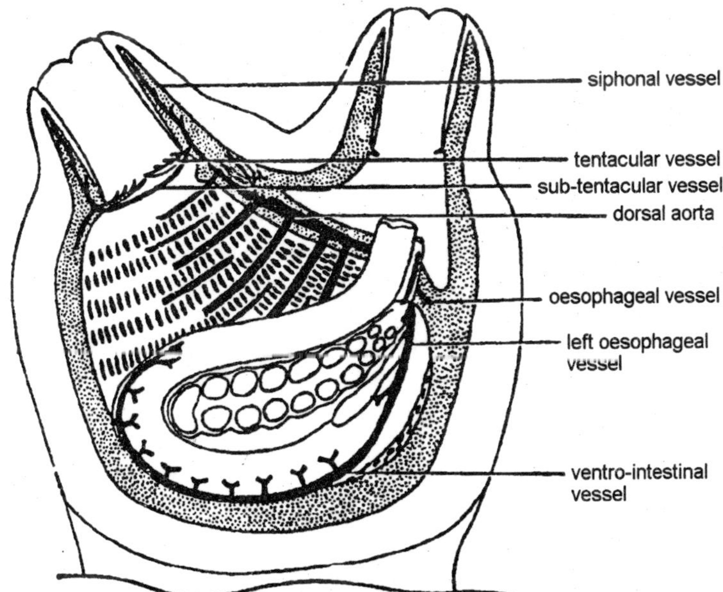

Fig. 12.9. Herdmania. Dorsal aorta and its branches.

of their argument. Those on the opposite side in the debate believe that the fossils exist in such deposits but have not yet been found. The advocates of freshwater origin do not accept this reasoning as valid. Why, they ask, should vertebrate fossils have been completely missed in pre-Silurian marine sediments when they are found easily in post-Silurian ones? Possibly, comes the reply, because the first vertebrates were so few in number that traces of their existence are extremely rare: one might expect such fossils to remain undiscovered. The absence of vertebrate remains in material collected from marine beds of early Paleozoic age constitutes a piece of negative evidence. Those who favour the idea of vertebrate origin in the sea are quick to point out that negative evidence is not conclusive in any argument, and their opponents are forced to agree.

Smith's Argument for the Origin of Vertebrate in Freshwater

Palaeontologists who believe that vertebrates developed in fresh water have received backing from another quarter. Homer Smith, a renal physiologist, argued that the vertebrate kidney is basically a water-excreting, salt-conserving organ that arose in a group of animals establishing itself in a freshwater habitat. In such a situation, the environmental water is hypotonic to the body fluids of the organism. Vital salts are lost from the tissues by diffusion through the semipermeable epidermis, and water tends always to seep into the body by osmosis. Any animal which lives in fresh water must be able to pump out the excess water and prevent the loss of its salts. Vertebrate kidney tubules are built expressly to perform these tasks. The glomerulus at the head of each tubule is a water-expressing structure, and the cells of the tubule walls are specialized to resorb necessary ions from the dilute filtrate passing outward. Smith was sure that kidney tubules with large glomeruli were primitive and arose with the vertebrates body itself because they are widespread throughout the vertebrate line. To Smith it seemed obvious that the earliest vertebrates must have been freshwater forms. Their descendants migrated to salt water and eventually to land as modifications arose in the structure of the kidney tubules, modifications that restricted the volume of water excreted. Marine fishes, unable to live unless they prevent loss of water to the hypertonic environmental fluid, show tubules without glomeruli. Land animals, whose bodies are exposed to the drying air, either, have glomeruli reduced in size or develop kidney tubules without glomeruli. Land animals, whose bodies are exposed to the drying air, either have glomeruli reduced in size or develop kidney tubules with special water-resorbing segments. Smith's ideas do not discourage the workers who continue to postulated a marine cradle for the vertebrate line. They argue that animals living in sea water hypotonic to their body fluids would require, like freshwater forms, kidney tubules which pumped out water and saved salts. They agree with Smith that large glomeruli characterized the primitive kidney but maintain that such a kidney could have evolved in dilute salt water as readily as in fresh.

Rebuttal by Advocates of a Marine Origin

Positive evidence should come from a study of the Ordovician fossils that are available. However, palaeontologists and geologists who have analyzed the sediments in which the oldest pieces of dermal bone were found have produced, not a definitive answer to the question, but testimony which has either been disputed or rationalized by theorists on both sides. The presence in the sediments of kinds of invertebrates which live in shallow marine habitats—brachiopods, pelecypods, and gastropods, for example—has led some investigators to suppose that the Ordovician ostracoderms lived in sea water. Others caution against jumping too quickly to this conclusion on the basis of a general inspection of invertebrate fossils present in the rock formation. Seas advance and ebb within relatively few millions of years, leaving behind marine and continental deposits interlayered in a complicated pattern. One must be sure that the marine invertebrates are discovered at the same location and in the same stratum as the ostracoderm bone.

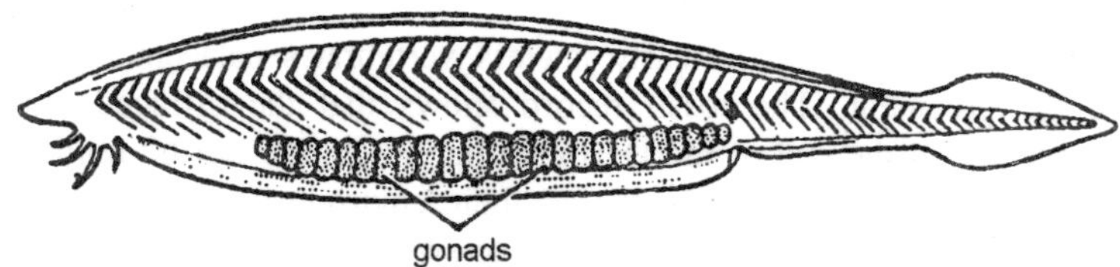

Fig. 12.10. Amphioxus. Diagram showing the location of gonads.

Interpretation of Fossil Evidence for the Habitat of the First Vertebrates

Even supposing arbitrarily that the vertebrate fossils were buried in sand under brackish water at the sea edge, however, does not solve the problem. These remains may represent early forms that were newly adapted to marine life and not at all typical of the larger number of vertebrate animals of the period. It is possible, too, that the assumption that the ostracoderm fossils are remnants of marine forms is itself false. The living animals may have inhabited rivers or streams and been washed seaward after death. Advocates of a freshwater habitat for Ordovician vertebrate have proposed this interpretation. It can be put to the test, because the condition of fossil material gives some clue to whether transportation of the dead animals occurred or not. An animal which dies, sinks, and is quickly buried is often preserved with a minimum of disarticulation of its skeletal elements. Dislocated of partial remains, however, do not prove that transportation from another area took place, because dead animals may be broken apart by scavengers and their bones scattered over the bottom. Fossil fragments that have been transported often show broken edges worn smooth or sorting of materials, as larger pieces settle out of a current sooner than smaller ones. The rocks in which such fossils lie sometimes reveal a pattern of ridges which were formed as water currents flowed in ripples over the mud. With these factors in mind, investigators have attempted to make a judgment concerning the source of the Ordovician ostracoderm fossils. They have found the pieced of dermal bone to be somewhat worn but not distinctly sorted. Some workers believe that the large quantity of bone found in the Harding Formation in Colorado makes it unlikely that the material was transported from a distance. Others think that accumulation of bone fragments in shallow bays where rivers emptied should still be considered as a possibility. The discussion goes on. For the present, at least, the environment of the first vertebrates, like the identity of their ancestors, remains unknown.

The Fate of the Ostracoderms

About the fate of the early jawless vertebrates palaeontologists are in agreement. The fossil record shows that the ostracoderms disappeared at the end of the Devonian period. In the last 60 million years of their existence they shared their environmental niche with new forms of vertebrates whose ability to swim and feed was superior to their own. The more advanced fishes which arose in the Devonian possessed hinged jaws and could engulf relatively large objects. Although it is possible that some of them sustained themselves on plant material and small invertebrates, other must surely have preyed upon the defenseless ostracoderms. The jawless forms which survived in the presence of these powerful natural enemies may have adopted habits of concealment or developed protective colouration. Their numbers dwindled, nevertheless, and extinction followed.

Relationship of the Ostracoderms to the First Jawed Fishes

Since the ostracoderms were the first vertebrates, it is though that they must have given rise to the jawed forms which succeeded

Fig. 12.11. Myxine glutinosa.

them. No series of fossils has appeared, however, to testify how the transition occurred. The known ostracoderms seem to have been too specialized in their structure to have served as ancestral stock for any of the primitive jawed fishes. Most of the jawed forms themselves are far from generalized types when they first appear on the scene. In the absence of fossil evidence, palaeontologists have been cautious in their speculations: they cast the osteostracans and anaspids out of the direct line toward gnathostomes because the peculiar arrangement of eyes and pineal and naso-hypophyseal openings characteristic of these ostracoderms seems unlikely as a basis for the arrangement of the comparable structures in jawed vertebrates. The Heterostraci emerge as a more probable source, in their minds, only because these jawless forms show eyes located more laterally and are generally presumed to have had paired nostrils situated near the mouth. Since all known heterostracans demonstrate peculiar structural features of one sort or another, palaeontologists do not regard any of them as the ancestral form they are seeking. If the ostracoderm ancestor of the jawed fishes ever comes to light, many workers believe that it might prove to be a heterostracan-like form of a very primitive type.

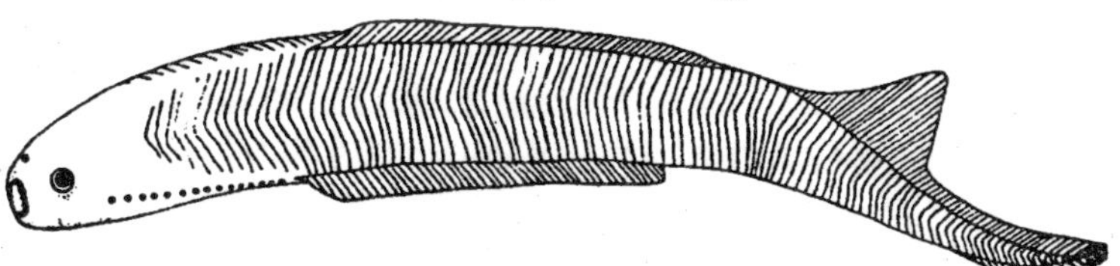

Fig. 12.12. Reconstruction of Jamoytius, a Silurian form which may be an anaspid distantly related to modern cyclostomes.

Modern Descendants of the Ostracoderms

There is some evidence that, besides giving rise to more advanced forms, the ostracoderms may have left behind a group of jawless descendants. The extant cyclostomes–the hagfishes and lampreys–show, side by side with specialized structures, certain characteristics that are similar to those of the ancient jawless fishes. The ammocoete larvae of the lamprey are filter feeders, sucking water from the stream bottom into the mouth and express in it through seven pairs of gill slits. At metamorphosis, modifications in the oral anatomy take place which prepare the animals to assume the parasitic habit peculiar to the adult. If one regards the specialized mouth structures which appear late in the development of the individual as relatively recent evolutionary occurrences connected with degradation to parasitism, it is possible to suppose that lampreys are derived from ostracoderm stock.

Jamoytius

The absence of jaws and paired fins in the cyclostomes is surely primitive: no anatomical or embryological evidence exists to suggest that lampreys or hagfishes are descendants of forms that once had these structures. Between the eyes lampreys have a naso-hypophyseal opening and a pineal organ whose arrangement on the dorsal surface of the head is reminiscent of that of osteostracans and anaspids. In fact, the appearance of these structures and the possession of two rather than three semicircular canals in the ear region have led investigators to suggest that the lampreys are modern representatives of the osteostracan-anaspid line. This reasoning is supported by the appearance of an extraordinary fossil from called Jamoytius. It was found first in Silurian rocks in Lanarkshire, Scotland, by an amateur collector who sold his two specimens to the British Museum (Natural History) in 1914. Both consisted only of a carbonaceous film on the surface of rock fragments, but certain important structures were decipherable: the animal, about 6 inches in length, had large eyes set on either side of a terminal oval mouth, a pair of lateral fin folds, and segmental, vertically arranged markings on the body that were assumed at first to represent well-developed but primitively constructed myomeres. E.I. White, observing these features in Jamoytius and noting the absence of bony dermal plates, stated that the fossil was evidence at last of the primordial ancestor of the higher chordates. He believed that Jamoytius was a conservative descendant of the stock that gave rise to the first vertebrates and the cephalochordate line to which amphioxus belongs.

The finding of new specimens of Jamoytius resulted in interpretations of the fossil that differed from White's. Although no one denied that *Jamoytius* had bulky axial musculature, the surface markings that had seemed to White to be segmental myomeres were recognized in the later specimens as scales similar in arrangement to those of anaspids. They were peculiar in being thin, unossified, and flexible. In each segment of the body, one scale extended from the middorsal to the midventral region without subdivision. The arrangement of the scales, eyes, mouth, and lateral fin folds and the structure of the branchial apparatus (which had not been visible in the first two specimens) led Ritchie to identify *Jamoytius* as an anaspid that was allied to a group transitional to cyclostomes. In support of this conclusion, Ritchie remarks that the mouth of Jamoytius may even have been partially specialized for the type of feeding characteristic of the Recent jawless fishes. He reasoned that the ovoid mouth was too small to have allowed sufficient intake for the filter feeding of so large an animal; he speculated that it might have served instead as a scraper, enabling *Jamoytius* to remove algae from rocks. Late forms might have progressed from this type of feeding to scraping organisms from the surface of other animals and finally to puncturing the skin and subsisting on flesh and blood, as adult cyclostomes do today.

Mayomyzon, a Fossil Cyclostome

Although Ritchie observed that *Jamoytius* shows closer affinity to lampreys than to hagfishes, he concluded that the anaspids were ancestral to both groups of cyclostomes. Other palaeontologists have taken a different point of view. Stensio suggested that the osteostracan-anaspid line of ostracoderms gave rise to the lampreys alone and that the hagfishes were derived independently from the heterostracans. He based his argument in part on his opinion that the heterostracans, like the hagfishes, had a single median nostril. This theory has found little favour with the majority of workers, who think it more likely that the Heterostraci had a pair of nostrils and were closely related to the forebears of the jawed fishes. The single known fossil cyclostome from the Paleozoic era helps little to settle any of the questions that have been raised: *mayomyzon* from the Upper Carboniferous is very similar to *Petromyzon*, the modern lamprey. Bardack and Zangerl, who described *Mayomyzon*, discovered in its structure no new evidence corroborating an anaspid-lamprey transition and nothing that might make clearer the relationship between the ancient lampreys and the hagfishes.

Loss of Bone in the Evolution in Modern Jawless Fishes

In postulating a phylogenetic connection between the ostracoderms and the cyclostomes, palaeontologists are not ignoring the presence of dermal bone in the older animal. The lack of bone and denticles in all living jawless fishes does not preclude their origin from the heavily armored extinct forms. The gradual reduction and loss of bone has occurred in several vertebrate lines as they passed their peak and entered into a decline. The cyclostomes are obviously the last members of a once more numerous group. The basic cartilaginous structures of the braincase and closely associated visceral skeleton characterizing the few existing genera may well be all that remains of the protective skeletal complex housing the brain, sense organs, and branchial pouches within the head shield of the ostracoderms that flourished in early Paleozoic times. Unfortunately, there is no way to verify the theory that has sprung from studies in comparative anatomy, because palaeontologists have found, except for Mayomyzon, no fossilized jawless fishes in rocks formed between the Devonian period and the current age. They have just enough fossil evidence to establish the nature of the ostracoderms at the time of their greatest success but not enough to ascertain either how they gave rise to later forms or even how, originally, they came into being.

13

First Amphibians

The origin of the amphibians was a momentous event, since attainment of the land opened vast new territories of the vertebrate animals and permitted them to evolve their most advanced forms. Despite the importance that terrestrial vertebrates were to have, however, their initial evolution was not in any way unusual or spectacular. The amphibians were not the last survivors of a lesser class but one of a number of new forms produced as the early bony fishes diversified rapidly in the Devonian period. At their first appearance, they gave the impression less of a revolutionary new group than of fishes peculiarly adapted for special habits of life. Outwardly, except for their legs, they resembled the rhipidistian fishes from which they sprang. Very likely, they continued to seem in the shallows, as their sharp-toothed forebears had, preying upon the abundant placoderms and early paleoniscoids to be found there. Palaeontologists are quire certain of the relationship between the rhipidistians and the amphibians even though they have not discovered the animals intermediate between the finned and limbed forms. The remains of the oldest tetrapods in their collections leave no doubt about the derivation of the axial skeleton from fishes of the rhipidistian group. Since the fossil material provides no evidence of other aspects of the transformation from fish to tetrapod, palaeontologists have had to speculate how legs and aerial breathing evolved and why a group of fishes produced forms that habituated themselves little by little to life on land.

Place of Emergence

The three problems which come to mind are the place, the impelling cause, and the time of this important event, and of these the first has been established, for while certain creatures have forsaken the sea and, crossing the strand, become adapted to terrestrial life, such instances are rare and in no case do they embrace all of the members of a class or phylum, but isolated genera or even species only. Such, for example, are the land crabs, *Birgus latro*, etc., several species of which live in damp woods far from all water and, as they are found on islands which, like the Dry Tortugas, have no permanent terrestrial waters, must have had their initial air-breathing adaptation along the strand.

The terrestrial vertebrates, however, apparently did not so emerge, but rather were descendants of inhabitants of land waters, for in such a habitat alone can we find a sufficiently great impelling cause for an event so far-reaching and radical in its ultimate results. Furthermore, the ancestral habitat could not have been within the limits of the tidal zone but was beyond the influence of the sea.

Impelling Cause

Enemies in the Water

Barrell has discussed the probability of several possible causes which may have led to the emergence of the vertebrates. Of these, the first is enemies in the water, which he deems inoperative, for among land-going fishes of to-day those few which crawl on land do not do so to escape their enemies. He also emphasizes the balance which always obtains between carnivorous and herbivorous creatures of a given habitat, and the fact that the amphibian go back to the waters to bring forth their young, and the youngest and therefore the most helpless stages are spent in the waters. Add to this the fact that the earliest amphibia which are known from their skeletons, and their ancestors, are, in many instances, powerful armed carnivores themselves, and their forsaking of the ancient habitat for personal safety seems hardly adequate.

Food on the Lands

Food on the lands is also considered and inadequate cause. Here the argument lies in the rarity of the passage of crustaceans, gastropods, and vertebrates from a truly marine to a truly terrestrial mode of life through the ever present path of the tidal zone, which seems to prove that the unused though increasingly abundant food of the land realm cannot operate as a sufficient cause for this change, nor, so far as this factor is concerned, to the river faunas have an advantage over those of the tidal zone.

Lure of Atmospheric Oxygen

That the lure of atmospheric oxygen is also inoperative is proved by the small direct use made of air for respiration by pelagic marine fishes even when they—the flying fishes, for instance—live an active life near the surface and in frequent contact with the air. It is principally in fresh-ate fishes that accessory respiratory organs are employed and their use is in direct relation to the varying impurity of the waters in which they live. This varying impurity, which often means a reduction of respirable oxygen in the water, is literally impossible in the sea. Streams may locally contaminate the adjacent waters by their load of sediment or other impurities, but they cannot seriously lessen the degree of aeration. Then, too, marine fishes are never confined to such localities, but can migrate should conditions become unsuitable; while with fresh-water fishes this may not be true.

Recurrence of Unfavourable Environment

The real cause, therefore, seems to be not the need of safety or food, nor the desire to breather atmospheric oxygen, but rather an adaptation which has been forced repeatedly to a greater or less degree upon fishes by the recurrence of an unfavourable environment. Hence it appears as though the emergence were compelled by variations in the environment as measured by the amount of dissolved oxygen in the land waters, and the question arises whether such variations occur and, if so, under what climatic conditions. The required climate does occur in various parts of the world, but is neither arid nor humid, but semi-arid—conditions which are found in the tropics especially, where, instead of the fourfold change of seasons whose determining factor is temperature, there are alternate seasons of drought and copious rain, occurring in definite cycle. In such a region during the rainy season the streams would be high and fully aerated, but when the rains ceased the waters would gradually slacken their current, until instead of a flowering river of pure water there would be a succession of stagnant water pools, in which the concentrated plant and animal life would die and by its decay charge the waters with impurities and exhaust their free oxygen. Thus we would get a great fluctuation of oxygen content and so a very variable environment in its ability to support water-breathing life.

Under such conditions, if life existed, a high premium would be placed upon powers of aestivation, or torpidity induced by summer heat and dryness, or of breathing the atmosphere, or both combined; and a rapid eliminating of the unfit, that is, of such as did not possess even the rudiment of this power, would occur. There are among existing fishes a number possessing supplementary respiratory organs which may be one of two structures: either (1) spongy outgrowths of the gills which can retain moisture and, as long as it is retained, utilize the free oxygen of the air for the aeration of the blood; or (2) a modified swim-bladder which may have one of several functions but whose principal purpose is a hydrostatic one—to maintain the fish at a given level in the water.

Teleosts

The first of these structures, that of the accessory organ connected with the gills, is found exclusively in teleostean fishes, a group in which the air-bladder never subserves a respiratory function. Several such teleosts exist, among them the climbing perch, *Anabas scandens*, and the mud skippers, *Peri-ophthalmus* and *Boleophthalmus*, but they are generally fishes which voluntarily leave the waters for migration or in pursuit of food, and rarely is their evolution the result of adaptation to the environmental conditions postulated above.

The swim-bladder, on the other hand, is entirely absent in the two groups known as cyclostomes and elasmobranchs and, as we have seen, is never of respiratory value in the teleosts, so that its function in this direction lies in the groups between, in other words, among ganoids and dipnoans or true lung-fishes, and these fishes are to-day all denizens of semi-arid tropical climates, living under conditions of varied water aeration arising in the way we have described. These air-breathing fishes are of such importance to our argument, and are so few, and represent so ancient a group or groups that some account of the individual species is worthy of record. It should be borne in mind, however, that these are relic forms, representative of Devonian

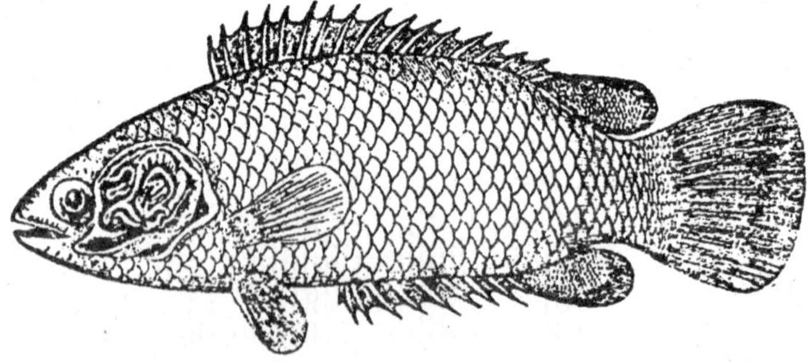

Fig. 13.1. Climbing perch, Anabas scandens, with gill-cover (operculum) removed to show accessory respiratory organ.

time when almost all fresh-water fishes belonged to one or the other of these two groups.

Ganoids

The ganoids of especial note are specifically of the order Crossopterygii, or the lobe-finned ganoids, and include but two related genera, *Polypterus* and *Calamoichthys*, both African in distribution. Of them the better known is *Polypterus*, of which there are several species. *P. bichir* haunts the deeper holes and depressions of the muddy bed of the Nile, although it is not essentially a bottom-liver or mud-fish. It is most active at night when in search of food, and then it may readily be taken by trawl-lines. The lobate pectoral fins are used for progression, but their primary function is to act as balancers, and they exhibit the characteristic trembling movement so often seen in the balancing fins of teleosts. Polypterus does not readily live out of water, rarely longer than three to four hours, and then only when covered with damp grass or weeds. *P. bichir* is said to feed on small teleosts, which it swallows whole, and to these there may be added, in other species, batrachians and crustaceans. The observations of Budgett show that in captivity Polypterus often remains motionless for a long time at the bottom of the water, the anterior part of the body resting upon the tips of the pectoral fins. According to the same observer, the air-bladder is an accessory respiratory organ, supplementary to the gills, rather than a hydrostatic organ". This air-bladder is an out-pushing of the gut, and in the Crossopterygii arises from the ventral side of the gullet and is paired structure exactly as in the amphibian lung. It is not, however, cellular and is thus a very inefficient respiratory organ.

In the genus *Calamoichthys* the body is elongate and eel-like in shape. The pelvic fins are entirely lacking but the pectorals and the series of dorsal finlets are comparable to those of Polypterus except that the latter are relatively fewer. *Calamoichthys* has a more restricted distribution than Polypterus, being confined to certain rivers in West Africa such as Old Calabar River and those of the delta of the Niger on the coat of Kamerun. It is a very agile fish, swimming like a

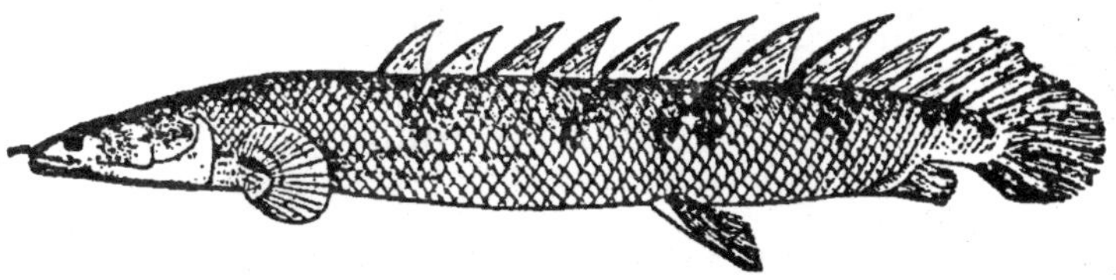

Fig. 13.2. African crosspterygian fish, Polypterus delhezi.

snake and subsisting on insects and crustaceans. The name signifies palm-fish, from its frequenting the roots of the palm trees.

Neither *Polypterus* nor *Calamoichthys* is known fossil, but the group Crossopterygii to which they belong once included a large number of important fishes. Of these *Holoptychius* of the Devonian is interesting because of the intricate infolded structure of the teeth, which has a striking parallel in those of certain amphibia (labyrinthodont). *Undina*, another form from the Upper Jurassic, exhibits a well developed air-bladder in the fossil specimen.

Dipnoans

Of the dipnoans or true lung-fishes three genera only are extant but they never were as numerous as the Crossopterygii. The living forms are, first, the Australian genus *Ceratodus*, or, to be more accurate, *Neoceratodus forsteri*, the barramunda, which is now confined to the Mary and Burnett rivers in Queensland. This form "frequents the comparatively stagnant pools or water-holes which alternate with shallow runs and are usually full of water all the year round. In these pools, filled with a rich growth of aquatic vegetation, and often the favourite haunt of the platypus (*Ornithorhynchus*), the fish is fairly abundant. Inactive and in sluggish in its habits, usually lying motionless on the bottom, the fishes easily captured by the natives with hand-nets or baited hooks. Neoceratodus lives on fresh-water crustaceans, worms, and molluscs, and to obtain them it crops the luxuriant vegetation of the water-holes much in the same way that a polychaet (worm) or a holothurian (sea-cucumber) swallows sand for the sake of the included nutrient particles. Apparently the air-bladder is a functional lung at all times, acting in conjunction with the gills. At irregular intervals the fish rises to the surface and protrudes its snout in order to empty its lung and take in fresh air.

While doing so the animal makes a peculiar grunting noise, 'spouting,' as the local fishermen call it, which may be heard at night

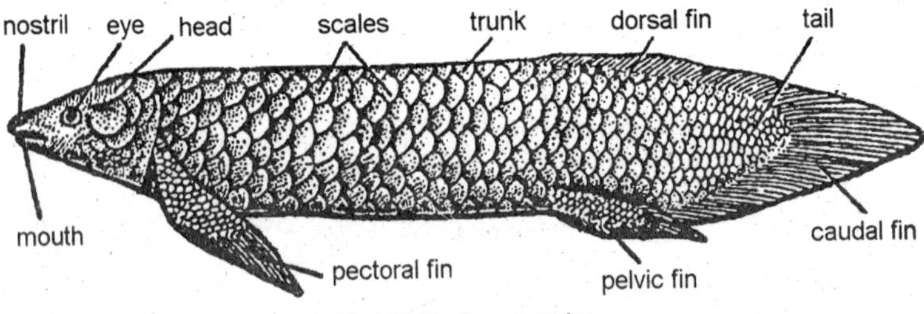

Fig. 13.3. Neoceratodus.

for some distance, and is probably caused by the forcible expulsion of air through the mouth. Useful as the lung is as a breathing organ under normal conditions, there can be little doubt that its value as such is much greater whenever gill-breathing becomes difficult or impossible. This seems to be the case during the hot season, when the water becomes foul from the presence of decomposing animal or vegetable matter. Semon records a striking instance of this in the case of a partially dried-up water-hole, in which the water had become so found that it was full of dead fishes of various kinds. Fatal as these conditions were to ordinary fishes, Neoceratodus not only survived, but seemed to be quite healthy and fresh. Such observations are of exceptional interest. Not only do they afford a clue to the conditions of life which, in the course of time, probably led to lung-breathing Neoceratodus, but they also suggest the possibility that a similar environment has been conducive to the evolution of air-breathing vertebrates from gill-breathing and fish-like progenitors.

"In spite of its pulmonary respiration, Neoceratodus more closely resembles the typical fishes in its habits than any other Dipneusti. It lives all the year round in the water. There is no evidence that it ever becomes dried up in the mud, or passes into a summer sleep in a cocoon, and the well developed condition of its gills suggests that these organs play a more important role in breathing than in either *Protopterus* or *Lepidosiren*. The fish is not known to leave the water, and the paired fins, useful no doubt as paddles, are quite incapable of supporting the bulky body on terra firma. In fact, when *Neoceratodus* is taken out of its natural element it seems to be more helpless than most other fishes, and in spite of its capacity for lung-breathing, soon dies unless kept moist by artificial means" (Bridge). Neoceratodus grows to a length of five to six feet.

Ceratodus, a fossil ally, Mesozoic in age, was very widespread compared with the limited distribution of its living relative. Its very characteristic crushing teeth occur in the Trias of England, Germany, Indian, and South Africa, and also, more rarely, in the Upper Jurassic and Lower Cretaceous strata of England and in Colorado and Wyoming (Morrison formation). Its remains are often found associated with those of carnivorous dinosaurs, but the significance of this, if any, is not apparent.

Protopterus and *Lepidosiren*, which represent a separate family of lung-fishes, the Lepidosirenidae, differ from Neoceratodus in that the air-bladder is a double organ, while in the latter it is single.

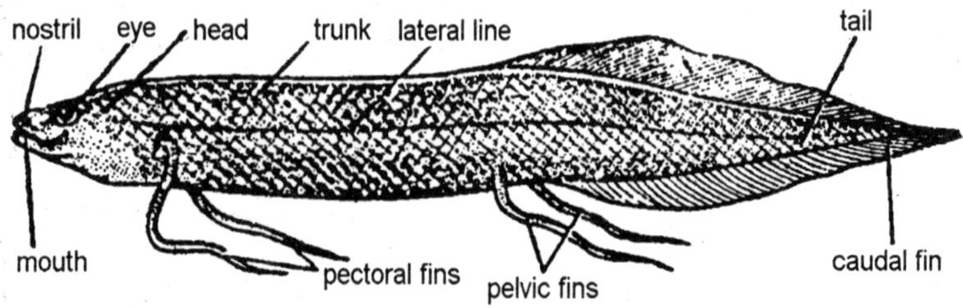

Fig. 13.4. Protopterus.

Protopterus is the African lung-fish, and has a wide distribution ranging from the river Senegal and the White Nile on the north to the Congo basin, Lake Tanganyika, and the Zambesi on the south. The three known species live in marshes in the vicinity of rivers. They are carnivorous, their food consisting mainly of frogs, worms, insects, and crustaceans, but when confined together they are very apt to display cannibalistic traits. The tail is the chief locomotor organ and they are remarkably agile and quick in their movements. The limbs are useless for swimming but are used for crawling over the bottom. Then they show a definite elbow- or knee-like flexure at about their mid-length. Protopterus ascends to the surface from time to time to breath air into its lungs. In the dry season, however, it burrows into the mud to a depth of about 18 inches, where it forms a lining to the cavity in which it lies in the form of a capsule of hardened mucus secreted by skin glands. This capsule has an aperture the margins of which are pulled inward to form a short tube that is inserted between the fish's lips. The fish within the capsule is surrounded by a soft slimy mucus which keeps the skin moist, while respiration is effected by drawing the outer air through burrow and tube into the mouth and thence to the lungs. The lungs are, therefore, the sole means of respiration during the period of aestivation, while the body-fat and muscle-tissues of the fish, as in hibernating mammals, supply it with the necessary food. The dry season varies, but lasts in general from August to December, nearly half the year. When, with the advent of the rainy season, the marshes once more become flooded, the capsule is dissolved, Protopterus emerges from its burrow, and resuming its active life, very soon begins to provide for its coming young. The larvae have much the appearance of a young salamander with four pairs of external cutaneous gills and two pairs of simultaneously developed limbs. It also has chromatophores in the skin whereby its colour may be changed. As in the salamander, the assumption of lung-breathing is marked by a reduction of the cutaneous gills, which takes place about seven weeks after the eggs are deposited. Protopterus attains a length of about six feet.

Lepidosiren, the mud-fish with but a single existing species, is a South American form, occurring along the course of the main Amazon, entering some of its large tributaries and also the Chaco Boreal to the west of the Upper Paraguay River. "The home of the Lepidosiren (or 'Lolach,' as the natives call the fish), of the Chaco country, so to be found in the wide-spreading marshes and swamps, which for a great part of the year are almost choked by a luxuriant growth of their own peculiar vegetation and covered by a floating carpet of surface weeds, with here and there deeper and clearer water and slow-flowing streams. In the dry season the water gradually shrinks and the swamps eventually become dried up. Of sluggish habits, the fish wriggles slowly about at the bottom of the swamp like an eel, using its hind limbs in an irregular bipedal fashion as it wends its way through the dense network of subaqueous plants. Lepidosiren is not exclusively carnivorous. The enormous numbers, seems to be its favourite food; but masses of confervoid algae are also eaten and in its earlier stages it is probable that the fish is more herbivorous than carnivorous". Lepidosiren rises to the surface at intervals to breathe, the rate varying with the degree of impurity of water. It feeds voraciously during the rainy season, storing up a supply of fat against the period of aestivation, which is passed in a deep tubular burrow, much as with protopterus. The entrance to the burrow in this instance, however, is closed by a plug of clay perforated by several holes. On the coming of the waters the plug is pushed out and the fish escapes. Development is quite similar to that of Protopterus and in each case parallels the amphibia very closely. There are many other parallelisms of structure and habits between the two groups, so many in fact that, as Dean said, it is almost impossible to look upon them as of no greater significance than convergences.

Lepidosirenidae are as yet unknown as fossils, but there is reason to believe that their evolution from the order Dipneusti has been in a

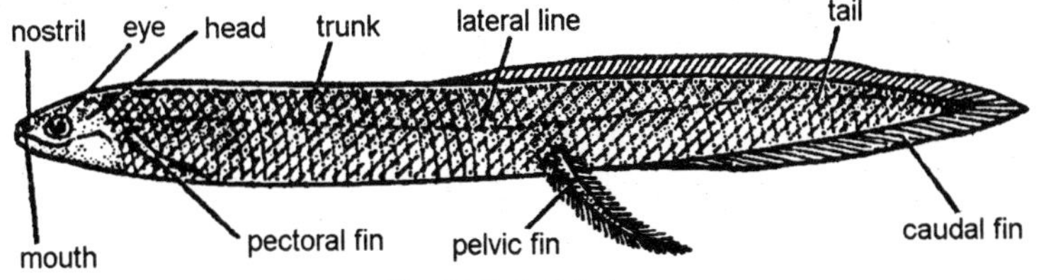

Fig. 13.5. Lepidosiren.

manner retrogressive. *Dipterus*, the most ancient of lung-fishes, may be taken as a starting point and Dollo has selected a remarkable series of genera, *Scaumenacia*, *Phaneropleuron*, *Uronemus*, Ceratodus (Neoceratodus), Protopterus, and Lepidosiren, in which the evolutionary sequence agrees perfectly with their succession in time. Back of Dipterus lies an unknown ancestry, but one which probably falls within the group of crossopterygian ganoids of which Polypterus and Calamoichthys are the living representatives. The trend of evolution among the Dipneusti, if one may judge from Lepidosiren, is leading to an elongated eel-like type, which will be both limb- and scale-less, which may be indications or racial senescence. These two interesting groups of air-breathing fishes, the Crossopterygii and Dipneusti, are both on the eve of their racial passing. The lesson which they teach; however, is of great significance and brings as back once more to the theme of our discussion—that of the most momentous emergence in prehistory; and or inquiry now leads us to a consideration of the probable time of emergence.

Time of Emergence

The evidence here is two fold: first, the fossil record, and second the geologic evidence of climatic conditions such as we have assumed.

The fossil evidence, which will be reviewed in greater detail later, points to a time earlier than Upper Devonian, for it is upon sediment referable to that period that the earliest known footprint of a terrestrial vertebrate has been impressed. The time of emergence therefore cannot be later than the age of this footprint, and form the nature of things must somewhat antedate it, although how much we have no means of knowing, as it was a time of accelerated evolutionary change.

The climate evidence points to the same result, for, as Barrell has shown by a careful study of the sediments and of other phenomena connected with the rocks of Devonian age, these were times of warmth and seasonal rainfall tending toward more marked semiaridity of climate in the Upper Devonian. There is, moreover, a dominance of dipnoans and crossopterygians in the fish fauna. Of these fishes certain could and probably did adapt themselves after the manner of their living descendants to the increasingly long dry seasons, until the latter became so long that the period of activity was not sufficient for the creature's life needs. Then came the emergence for instead of aestivation the animal must adopt some other mode of life which would prolong the time of its activity in spite of its climatic restrictions. Thus the more

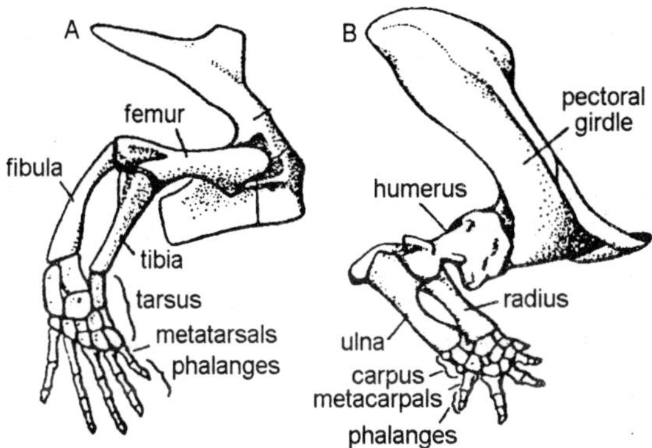

Fig. 13.6. A–Lateral view of the pelvic girdle and limb of a primitive labyrinthodont; B–lateral view of the pectoral girdle and limb of Eryops.

ambitious among the lung-breathers, not content with the limitations imposed upon their lives, emerged from the age-long aquatic home and ventured into the new and untried habitat. Many may have essayed the emergence, but it is probable that relentless nature, weeding out the less fit for so valorous an undertaking, destroyed all but a single sort, for there is not evidence that the ancestry of the amphibia is to be found in more than one evolutionary lineage.

The Devonian Environment

Since the evolution of any group of living things is a progressive response to the demands of the environment, investigators have discussed at length the conditions that prevailed at the time of the rhipidistian-amphibian transition. By pooling interpretations based upon geological, botanical, and zoological data, they have been able to visualize the earth as it was then and to guess what might have driven the incipient amphibians on their new course. In the Devonian period, they reason, primitive green plants were advancing tentatively beyond the dampest ground at the edges of freshwater lakes and streams. Harbingers of the great tree-fern and horsetail forests of Carboniferous time, these early land plants sheltered the many-legged arthropods which were also acclimating themselves to life in dry air. Beyond the water's edge, the land stretched away unprotected by vegetation. Rain fell and ran off in torrents, crumbling igneous and metamorphic rock into pebbles, sand, and clay. In some areas, the wind picked up the finer particles and blew them into dunes, while in other places loads of sediment were washed into the water and layered over the bottom. The red colour of much of this transported material has provided scientists with an important clue to the climatic conditions that existed in the Devonian years. Geologists know that the ruddiness of soils is due to hematite, a compound containing iron in ferric, or highly oxidized, form. Iron in this state is produced when sediments containing ferrous silicates are laid down in warm and most areas where the surface is subject to drying from time to time. Ferric iron is stable once it appears, and deposits containing it are compressed into layers of rock called red-beds.

As long ago as 1916, Barrel pointed out that the existence of extensive Devonian red-beds signified wide spread semiarid conditions during the period when the amphibians arose. He believed that lungs evolved as an adaptation to the droughts that intervened between weeks or months of steamy rainfall. Lull concurred and emphasized the role of natural selection in preserving a population of vertebrates that was developing the ability to sustain itself out of water. Romer described graphically how the earliest tetrapods might have saved themselves from death by leaving a drying pond and hitching themselves overland to a deeper, better-filled pool. Critics of Romer's theory challenged it on several bases. Inger questioned the idea that the Devonian climate was surely semiarid, stating that red soils are being formed today in parts of the world where tropical rains are uninterrupted. Even if dry seasons did exist, he maintained, it was hardly likely that the first amphibians behaved as Romer thought they did. Since they were dependent upon small fish for food, they would have lived in permanent waters rather than in ponds which became periodically uninhabitable. A drought severe enough be destroy the pond or lake in which they had established themselves would certainly evaporate similar bodies of water within a radius of many miles. An attempt to migrate any distance under such conditions would result in the animals dying of desiccation on the bare ground. Inger imagined that lungs were selected for, not because their possessors ventured onto dry terrain, but because these organs enabled fishes in warm, oxygen-poor water to survive by breathing air at the surface. He thought that the animal's first excursions beyond the water's edge were prompted by increasing population pressures in old communities; during humid nights, individuals driven by lack of adequate food or space might have squirmed over the wet earth in search of less crowded ponds much as the fighting fish *Betta* and the catfish *Clarias* do today in tropical Borneo.

Structural Modifications

Why the amphibians left the water and when constitute only a part of the puzzle to be solved by those who are studying the origin of the first tetrapods. Paleontologist tracing amphibian history have also to analyze the structural modifications necessary to transform a fish into an animal viable on land before they can explain the steps by which rhipidistian forms evolved into the earliest terrestrial species. It is possible to draw such guidelines, because observation of living animals makes patent the different demands on the vertebrate body of watery and aerial environments. Underwater, animals are supported by the medium, kept moist, and supplied with dissolved oxygen. Once they come out into the dry air, they are faced with the necessity of holding themselves up and moving in a much less buoyant substance, of extracting oxygen from a gaseous mixture, and of conserving the water that forms the basis of their protoplasms. The physical forces operative in the terrestrial environment make certain mechanical arrangements a necessity and others an impossibility if the animal is to succeed in these tasks. Determining the time and the order in which the required structural changes took place is a more speculative matter. Although many of the modifications were correlated and so must have occurred simultaneously, palaeontologists are not certain of how much alteration took place as protoamphibians were acclimating themselves to land life and how much occurred preadaptively at an earlier time when the animals were still wholly aquatic.

Although several Devonian fishes possessed lungs, it is not all clear that elaboration of these structures was an early even in amphibian, evolution. Ever since its initial appearance as an outgrowth from the pharyngeal region of the gut, the lung seem to have functioned as a respiratory membrane. Its usefulness to aquatic forms confined in waters of low oxygen content surely explains the perpetuation of the origin in several lines of bony fishes. Although much later in vertebrate evolution the lung itself would increase in complexity as advances in the design of internal organs made possible a higher rate of metabolism, a simple internal respiratory sac not too different from that of the fishes would very likely have sufficed for the cold-blooded protoamphibians. A land vertebrate could not survive , however, without some mechanism for assuring the passage of air from mouth to lungs. In fishes this transport may be a passive process: a fish can gulp air and then plunge head downward, causing the bubbles to rise through the pharynx and so pass backward into the lung. A land animal which remains horizontal must have some way of forcing of drawing air through its respiratory tract. Certain structural innovations which appear in early tetrapods could have arisen in conjunction with the requirement for a respiratory current. The increased length and stoutness of the

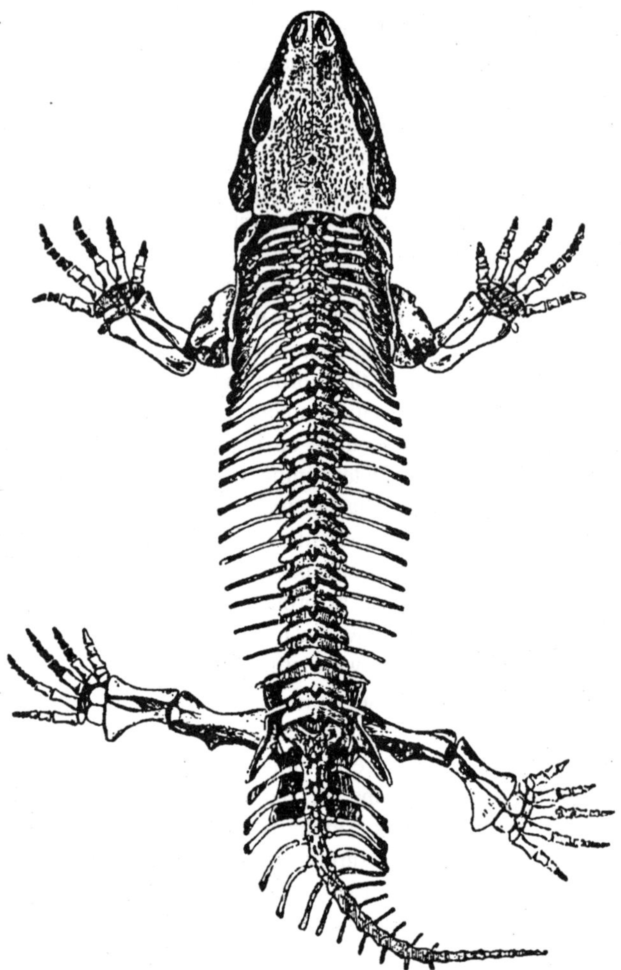

Fig. 13.7. Seymouria in dorsal view; original about 20 inches long.

ribs characteristic of first known amphibians afforded added surface for the attachment of muscles which could have produced a rise and fall in the body wall, rhythmically changing the pressure around the lungs. Air sucked into the respiratory sacs when their internal pressure feel below that of the outside, would have been expressed when the pressure on the lungs rose. This kind of respiratory mechanism, which survived in higher vertebrates, was surely not the only one which evolved in the ancient amphibians. The gradual broadening of the head in many forms can be interpreted as evidence for the existence in

Paleozoic times of the force-pump apparatus that remains in the ribless amphibians of today. In these animals, the floor of the mouth is lowered with the nostrils open to draw air into the mouth, the large its underlying muscles sheet and the greater the efficiency of the mechanism that depends upon their action. The tendency toward flattening and widening of the head evident in several lines of Paleozoic amphibians could be explained by postulating use of the force-pump process in these animals and the continuing selection of types whose skull structure made possible its more effective operation.

In the fossils record there is evidence to suggest that reliance on either once of these respiratory mechanisms developed slowly, both phylogenetically and in the life of the individual. The young of the early amphibians, like the young of modern forms, grew gills first and then developed functional lungs as they matured. During their metamorphosis, the animals almost certainly depended to some extent upon the thin, moist skin (if they were not heavily armored) and the mouth lining as auxiliary respiratory surfaces, as existing amphibians do. Whether the older amphibians were able to continue to respire through the skin as adults is not clear. Although fossil evidence is scanty, it seems that many, if not all, of them retained a covering of bony scales that would have limited gaseous exchange at the surface of the body. The scales over the back and flanks were thinner than those of the ancestral rhipidistians, however, and possibly not present everywhere, so that some cutaneous respiration might have persisted throughout life in these animals.

One paleontologist believes that the appearance of legs facilitated respiration as well as locomotion in terrestrial vertebrates. I.I. Schmalhausen points out that a fish lying on its side on the ground bears the weight of the body wall upon its viscera despite the presence of ventral ribs. When the internal organs are compressed in this way, the fish can force air through the pharynx only with great difficulty. For air to pass easily in other lungs, the trunk has to be propped up so that the respiratory organs hang suspended in the body cavity. A rhipidistian fish could have used its muscular fins to lift its body from the surface temporarily, but in the absence of legs the body would have dragged or flopped when the animal moved. Although there might have been a rhipidistian that could expend the large amounts of energy necessary to make the lungs work under these conditions, a form with a fish-like trunk would have been better adapted for living on land if the body was kept clear of the ground.

Changes in the Skeleton Associated with Terrestrial Life

Elevating the body was also desirable as preparation for easy locomotion over dry ground. A fish moves against the water which buoys it up, generating a minimum of friction, but a tetrapod in contact with the substratum would be considerably hundred by the scraping of its body along the uneven surface unless, like the snakes, it was specially modified for that kind of locomotion. Raising the trunk and tail eliminates at once the necessity of slithering along every rise and fall in the terrain and the danger of scrapping the epidermis to shreds in doing so. The transformation of the fins from steering devices to piers for the suspension of the body involved changes in every part often appendicular skeleton. Besides reorienting at least a portion of the limb in a vertical direction and exchanging for the fin rays feet that could be planted flat on the ground, the relationships of the girdles to the axial skeleton had to be substantially modified. In fishes, the pelvic girdle always consisted of a pair of bony plates embedded in the ventral body wall. A tetrapod limb, pressing upward against such a structure, would cause it to sink inward and compress the soft organs in the posterior part of the body cavity. If the hind legs were to hold up the animal and push it forward, there had to emerge a rigid connection between the pelvic girdle and the vertebral column. With such an arrangement, the thrusting force of the foot against the ground could be transmitted with little loss to the main axis of the body.

A union of the appendicular and axial parts of the skeleton in the pectoral region already existed in fishes: from earliest times, the dermal shoulder shield had articulated firmly with the back of the skull. This association, while it was undoubtedly advantageous for fishes, was unsuitable for a tetrapod. The forelegs are a sort of landing gear

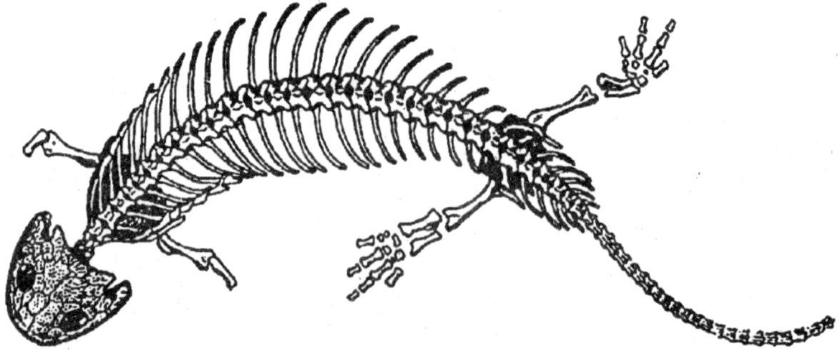

Fig. 13.8. The skeleton of the Upper Permian seymouriamorph kotlassia.

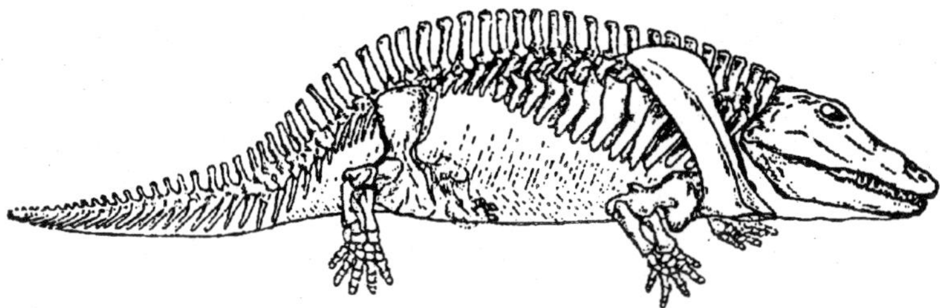

Fig. 13.9. Eryops, a large Lower Permian rhachitomous amphibian about 5 feet in length.

in a terrestrial animal: they receive the with of the body as it is propelled forward by the hind limbs. To absorb the shock generated by the descent of the trunk, the pectoral girdle has to be disconnected from the head. Even if an amphibian had been able to withstand the tension on the skull that would have been created at every step so long as the articulation between girdle and skull remained, another difficulty would have ensured. Because the scapulocor-acoid element which receives the limb bone abuts the cleithrum of the shoulder shield along its vertically oriented medical surface, pressure from the leg would set up a shearing stress between the endochondral part of the girdle and the rigidly held dermal portion. Since the skeleton is most vulnerable to shearing forces, the resulting weakness of the suspensory apparatus would limit the weight, and thus the size of terrestrial vertebrates. Transformation of the girdle into one strong enough to support a walking animal of any mass demanded a loss of the connection of the dermal bones with the skull and a concomitant expansion of the endochondral part to which the legs are attached.

Freeing the skull from the pectoral girdle surely brought with it other advantages for an animal making its way on land. For the first time, a vertebrate would be able to turn its head and to lift it above the level of the body without tilting its tail downward. This ability made it possible to see beyond small obstacles in the immediate environment and so to chose the most convenient path or to discover enemies lurking nearby. Since the skull, separated form he girdle, remained cantilevered from the front of the vertebral column, its elevation above the ground required its development as an inflexible structure with space available on the posterior surface for attachment of the strong neck muscles that hold it up. Specifically, the braincase would have to become more solid than it was in the ancestral fishes and the occipital region broader.

Although many fishes flourished with little ossification of the vertebral column, supporting a body in the aerial environment necessitated increasing the amount of bone in the pat of the skeleton. While the long-bodied early tetrapods still needed the kind of flexibility conferred by the unjointed but pliant notochord, that structure, reinforced as it was inmost ancient fishes by cartilaginous elements, never would have sustained the weight of the whole animal out of water. Evolution of stronger column was made possible by the appearance in the rhipidistians of small bones that hugged and constricted the notochord: anteriorly in each segment an intercentrum embraced the notochord from below, and, just behind, a pair of nubbins called pleurocentra occupied a dorsolateral position. Selected for at first, perhaps, because the intercentrum provided secure footing for the high-spined neural arch, the bones could later expand to share and then entirely accept the stresses projected through the vertebral column of the land animal. Contact between these central elements was apparently insufficient to produce stability, however, for no tetrapod evolved without an accessory articulation more dorsally between the neural arches. The outgrowth fore and aft of paired zygapophyseal, or yoking, processes from the roof of each arch resulted in the development of gliding joints which added strength to the column but interfered little with its flexibility.

Protecting the Body from Desiccation

The structural modification which enable an animal to support itself and to respire in an airy medium would be of no benefit unless the animal could protect its body from desiccation. Since vertebrates, like all living things, are composed of watery protoplasm, this is a formidable task. Eventually, in the tetrapod line, it was to be accomplished in a way similar in principle to the utilized by all organisms adapted for terrestrial life: the body became enclosed in a nonliving coat impervious to water, which exposed moist tissues in as few places as possible. Whereas other forms of life manufactured cork and cuticle, shell, slime, and exoskeleton to retard the loss of water, tetrapods evolved a layer of dead, keratin-filled cells at the surface of he epidermis which confined internal fluid reasonably well. Gills, which have to remain moist and exposed if they are to function, had to be abandoned, and respiratory membranes developed in more protected parts of the body. The formation of internal nares was preadaptive for animals the had to restrict contact between he dry air of the environment and their respiratory organs. With a pair of narrow passages that led from the exterior to the front of the oral cavity, protoamphibians could admit in two fine streams the air necessary for respiration rather than exposing the moist interior of the mouth broadly by parting the

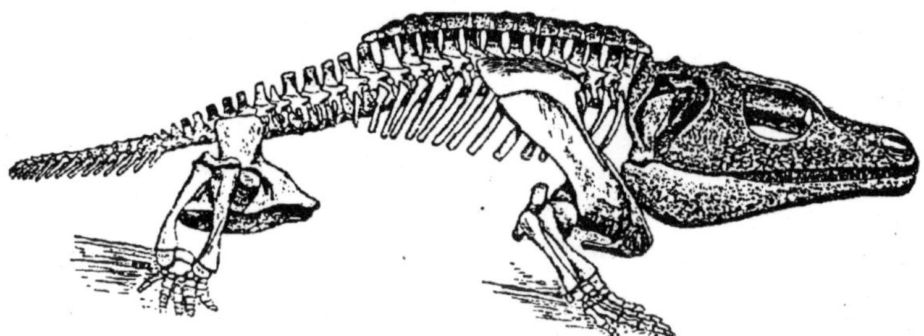

Fig. 13.10. Cacops, a small rhachitomous amphibian from the Lower Permian.

jaws. In higher tetrapods, protection of the internal tissues would be increased by further isolating the respiratory stream within the oral cavity and lengthening the distance through which air had to travel before the lungs.

If the condition of modern amphibians is representative of the level to which the first tetrapods had brought their water-conserving abilities, it is obvious that the bare beginnings of protection against desiccation were sufficient to allow vertebrates to gain the land. In living amphibians, the skin exhibit a keratinized layer but one thin enough in most species to be permeable to water under certain circumstances. The delicacy of the protective layer enables it to serve a double function: covered with mucus, it slows the loss of water when the animal is exposed to the air but permits fluid to enter when the animal submerges itself after a terrestrial excursion. The increase in cutaneous permeability shown on the latter occasion is under hormonal control. Although it is not known whether endocrine activity of this sort is a specialization in modern amphibians or a legacy from ancient forms, it is safe to suppose the physiological mechanisms must have appeared early to supplement the structural alteration as which

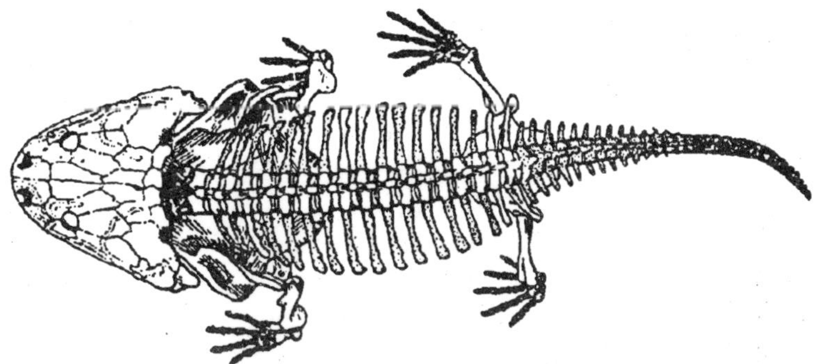

Fig. 13.11. Dorsal view of the skeleton of the Triassic stereospondyl Buttneria.

adapted vertebrates for land life. If hormonal regulation of cutaneous permeability did exist in the first amphibians, it would have been effective combined (as it is in extant forms) with endocrine control of the kidney. If the emergent tetrapods secreted hormones that acted to conserve water by inhibiting filtration of fluid form the bloodstream into the urinary tubules or by encouraging the recovery of water by resorption through the tubule wall, the ability of these animals to survive in the atmosphere would have been enhanced.

Scales disappeared, and other structural defenses against drying out remained limited among the amphibians. These animals assured their viability by behavioural adaptations rather than by paralleling the higher tetrapods in the development of increased physical resistance to desiccation. They frequented moist areas adjacent to streams or lived in marshes, where the heat of the sun made the air steamy rather than dry. When drought was inescapable, they burrowed or lay dormant underground. Undoubtedly, the earliest amphibians fertilized their eggs externally, as many living forms do, laying them in the water, where they were safe from drying if not form the depredations of hungry fishes. Surviving embryos, unprotected by extraembryonic membranes, became gilled larvae that passed through an obligatory aquatic stage before they could step out on land. Those later amphibians which became more terrestrial evolved a variety of special reproductive habits designed to accommodate or abbreviate the larval requirement for water. Their eggs are laid in damp moss, gelatinous froth, or temporary puddles or fertilized and held for a time within the body; the embryos may mature with unusual speed, acquiring adult characteristics before they leave the egg. None of these peculiar inventions permitted the degree of independence from fresh water enjoyed by higher vertebrates, which enclose their embryos in a fluid-filled sac and protect them further with a shell or by housing them internally until they can survive in dry air.

Fossil Record

Footprints

The earliest record of a terrestrial vertebrate is the single footprint of *Thinopus antiquus* mentioned above. This is impressed upon a slab of sandstone and is form the uppermost Devonian (Chemung). It was found in 1896 by the late Professor Beecher of Yale and by him presented to the Museum where it is now treasured. These same beds contain ripple-marks and mud-cracks, and impressions of rain-drops and land plants also come from the same general horizon. A characteristic marine mollusc (*Nuculana*) is preserved on the under side of the footprint slab. The associated strata show dominant delta conditions on the outer margin of which the sea had contributed to the material, for in the wide oscillations of the strand-line characteristic of delta fronts, deposition under shore conditions and deposition under river conditions alternate (Barrell).

This Devonian is directly overlain by Lower Carboniferous (Mississippian) Coal Measures, represented in Nova Scotia and New Brunswick by the Horton series. These contain the remains of plants and crustaceans and the footprints of amphibians. No bones have been found in these beds, but the footprints indicate, at the beginning of the Carboniferous period and before the deposition of the Lower Carboniferous limestones, the presence of both large and small species similar to those of the coal formation. One interesting type, *Hylopus hardingi*, found in the Lower Carboniferous shales of Parrsboro, Nova Scotia, shows a stride five times the length of the foot and twice the width of the trackway as though the creature which made it stood high on its legs like an ordinary mammal. This looks very much like a cursorial adaptation; if so, it is the earliest on record.

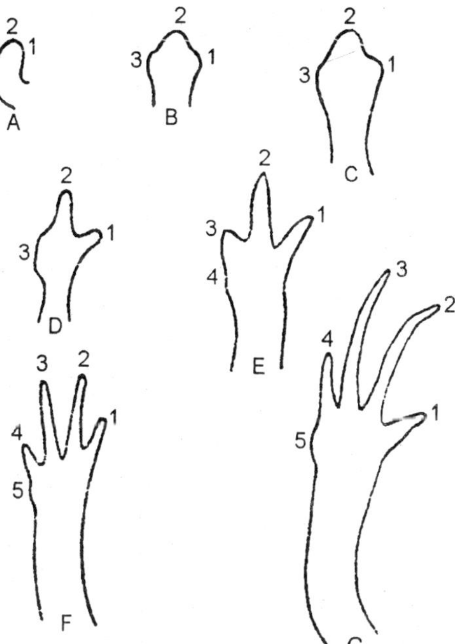

Fig. 13.12. Development of the hind foot of a salamander, Triton taeniatus.

The next higher level to record the passing feet of these primal terrestrial forms is the Mauch Chunk series of Pennsylvania, assigned by geologists to the upper half of the Lower Carboniferous. Here has been recorded *Palaeosauropus primaevus*, a five-toed track of considerable size as these early forms run, and more careful search of these same beds at Pottsville has brought to light several other species, some very small and delicately impressed. Other tracks have come from Virginia and are referred to the same general age (Hinton formation).

An amazing series of footprints of Permian age representing number of genera and species has recently come to light in the Grand Canyon, where they may be seen traveling up the strata until they disappear under the overhanging cliff. Some of these are preserved in the Yale Museum and some in the National Museum at Washington.

First-Skeletal Remains

The first amphibian bones were found in the Upper Devonian of Greenland and are comparable in age to the of the Thinopus footprint. Other skeletons come form the Edinburgh Coal Measures of Scotland which have been referred to the Lower Carboniferous. There are in no sense transitional forms, but are fully developed amphibians. Above the Lower Carboniferous Coal Measures we have red shales and sandstone in which bones are invariably rare and footprints abundant, and so it is with the Scottish record.

It was during Permian and Carboniferous times especially that the great deployment of amphibia occurred, and we have from various places, notably in Europe and in Nova Scotai and the various places, notably in Europe and in Nova Scotia and the United states, the remains of a varied assemblage of forms, some small, others huge, heavily armored types with complex vertebrate, still others with complexly infolded teeth; some with well developed crawling limbs, yet others limbless, elongate, indicating that already the condition which we have

Fig. 13.13. Permian stegocephalian, Cacops aspidephorus, from Texas, allied to the primitive reptiles.

called racial old age, or senility, with its attendant degenerative specialization was upon them. One and all were alike in this—they went, presumably, back to the waters to lay their eggs, and their young were therefore aquatic and breathed by means of gills. But there are among them many of which this cannot be proved and some may actually have been transitional, not between amphibians and fishes, but between amphibians and the succeeding class, the Reptilia. Certain of the forms, such as *Cacops*, discovered in the Permian of Texas by professor Williston, show such a combination of characters pertaining both to the amphibia and reptiles that, as the distinguished discover says, it may become necessary to revise our definition of the former group.

14

GIANT REPTILES

The world 'dinosaur' was-invented by Sir Richard Owen more than a century ago to designate certain large fossil reptiles that were then being recognized and described for the first time. Dinosaurs are often though exclusively as immense animals, but some were no bigger than a hen. In each-order existed both large and small reptiles. The biggest were the most massive animals ever to exist on land. Some dinosaurs were quadrupedal - walking on all fours, others were bipedal and walked on the hind legs. In. these, the weight of the body was pivoted on the hip bone. Which became fused to the backbone, giving more rigid support. The fore limbs in many forms were short and adapted far grasping. They arose in early Triassic times from some primitive thecodontstock, steadily progressed in size and specialization throughout practically the whole Mesozoic and became extinct at the close of the Cretaceous. During this long period they were lords of the terresterial world. For 150 million years, from the late Triassic to the end of the Cretaceous, dinosaurs dominated the terrestrial environment. They not only inhabited the land but also invaded the sea and air. They were unquestionably the most successful of all reptiles and rivaled the Cenozoic mammals in their structural diversity and wide distribution. Dinosaurs included the largest terrestrial animals that have ever lived, as well as species that may have equaled many animals in their high metabolic rate and relative brain size.

HISTORICAL ACCOUNT

The remains of dinosaurs were discovered in the Jurassic and Cretaceous rocks of southern half of England and were described by *Buckman* 1824 and *Nuctell*1925. *Sir Richard Owen* established for them the order Dinosauria and much additional light was thrown on their structures and relations by *Huxley, Philips* and *Coke* prior to the year 1875. During subsequent years the discoveries of complete skeleton of Ignadon in Wedon (Belguim), and many skeletons of primitive genera in the Triassic rocks of Germany and N. America have provided a very good knowledge of this order. It was eventually sub-divided into three orders namely Pteropoda, Sauropoda and Predenota by *O.E MerSip. H.G. Seeley* placed them into two great assemblagean - the Saurischia (reptile like dinosaur) and Ornithischia (bird like dinosaurs).

Period

The known records of dinosaurs extends from Middle Triassic to the very close of Cretaceous period. This is particularly true of the saurichian dinosaurs. The herbivorous ornithischians on the other hand appeared in the late Triassic period.

Distribution

Dinosaurs were first found in Germany, at Gagolin. It is however believed that the dinosaurs originated somewhere in North Atlantic basin which connected lower end of N. America and Europe. From here they began their world wide march of conquest. They extended all over the world and their fossils have been collected from United states, Canada, Brazil, England, Belgium, France, Germany, India, Australia and Africa. Thus they have a world wide distribution excepting Newzealand.

Ancestral stock

Triassic thecodonts were the direct ancestors of the dinosaurs. Most of the thecodonts were small reptiles. They had a narrow diapsid skulls that lacked a pineal opening. Many of them adopted a bipedal type of locomotion. The forelimbs, freed from locomotor duties, were available for use in grasping and handling. The thecodont body plan is, as *Colbert* has said, "the blue print to dinosaurian body forum." Doctor *Friedrich Von Huene*, derives the dinosaurs from primite cotylosaurian stock which arose in Carboniferous time.

Habits and Habitats

In habits the dinosaurs were nearly as varied as the mammals are today. Some of them were carnivorous, others were herbivorous while still others were of very peculiar nature as far as their feeding nature was concerned. Initially they lived in semi-aridity climates mainly in the ter-resterial environments. Some of them developed aquatic characters, some aerial or volant and some cursorial characters.

Classification of Dinosaurs

The dinosaurs were diapsid reptiles characterized by two openings on each side of the skull behind the eyes. The word dinosaur does not denote a single and natural group of reptiles but rather two distinct reptilian orders: the Ornithischia and the Saurischia. The key character that distinguishes the two orders concerns the pelvis. The Saurischia had a pelvic much like that of the ancestral thecodonts, described as triradiate with three prongs. The Ornithischia had a tetra-radiate pelvis much like that of birds.

Order Saurischia

The saurishians were the dinosaurs most like their thecodont ancestors. These were generally carnivores, except the sauropoda. The jaws were armed with rows of pointed teeth, some of them six inches long. Pelvic tri-radiate in which pubis and ischia were large and projected downwards and met in a ventral symphysis. There were two orbital vacuities in the skull. The group includes the bipedal carnivores and the quadrupedal, amphibious herbivorous sauropods.

It is further divided into two sub-orders:

Theropoda

Dinosaurs belonging to this sub-order existed during Triassic to Cretaceous period, and retained their ancestral carnivous habit. The principal evolutionary changes which they show were a gradual increase in size of body and proportionate decrease in that of forelimbs. (bipedal locomotion). These creatures therefore walked or ran entirely on the hind legs. When walking, left bird-like tracks made by three clawed toes. A further toe, which did not reach the ground, pointed backwards. They showed following features:

1. Skull small.
2. Cutting type of teeth.
3. Completely ossified cranium with pre-orbital vacuities.
4. Short forelimbs resulting bipedal locomotion.
5. 3-5 digits with prehensile claws.
6. Degitigrade in locomotion.

Some of the better known theropods are:

Ex. *Allosaurus, Tyranosaurus.*

Allosaurs

They had relatively large and very strong limbs to carry their great weight. They reached 12 meters in length and were among the

Fig. 14.1. Allosaurus.

most powerful camivores of their time. The skull was high and laterally compressed; the orbital opening was triangular and smaller than the principal anterior orbital fenestra. The arms and clawed hands, though relatively small, were still capable of being used as an aid to feeding. Mouth was armed with dagger like long, laterally compressed teeth. The tail was long, and the posterior prezygopophysis were considerably elongated.

Tyranosaurus

This was the largest carnivorous dinosaurs and called the *King of Dinosaurs*. It was about 47 ft. long and 20 ft. fall. Its weight was about 8 to 10 tons. It was the most formidable animate engine of destruction that ever lived. The hind limbs were tree like in size and adapted for running. Feet with three powerful toes, each armed with a massively curved claw, 6 to 8 inches in length. The head was about 4 feet long and rather broad. The large jaws were armed with numerous flattened saw edged, dagger like teeth. The forelimbs were reduced to such a degree that they must not have been very useful. Long jaws and large teeth gave an effective bite for dealing with large prey, which constituted the food of these dinosaurs. Strong muscles to operate the big and heavy jaws. Because of the weight of big skull the neck was shortened which served to avoid adverse leverages. *Chatterjee* (1985) proposes that they may have evolved directly from Triassic thecodonts.

Fig. 14.2. Tyrannosaurs.

Sauropoda

The term Sauropoda was coined by *Prof. Marsh.* The sauropods were the herbivorous dinosaurs. They were characterized by their enormous size with weights as great as 80,000 kilograms (*Colbert,* 1962) and body lengths approaching 30 meters. They were all plantigrade and quadrupedal. The limb bones were extremely massive with very rugose (ridged) ends. In almost all of them the hind legs remained proportionately long. The heads of Sauropods were absurdly small for animals of such great size. With such small heads they must have spent most of their lives eating. The teeth were reduced in number and size and they had serrated margins with more or less spoon shaped. The nostrils of some of them were located high up on the head, seemingly to make breathing possible while the mouth was engaged in under water feeing. It is thought that these giants spent much of their lives in lagoons and swamps. The bulk was so great that it is difficult to see how the legs could have furnished adequate support, for protracted periods of time, without the aid of buoyancy provided by surrounding water.

Ex. *Brontosaurus* and *Diplodocus*.

Brontosaurus

One of the largest dinosaurs, reached a length of 67 feet and weight about 38 tones. Its complete original specimen is preserved at Yole and similar in the American museum of Natural History. This specimen is from the Jurassic rocks. Much of the length of this animal is attributable to the long neck and tail. The small head contained a brain disproportionately small even for a reptile. The brain was hardly larger disproportionately small even for a reptile. The brain was hardly

Fig. 14.3. Brontosaurs.

larger than that of a cat. The back-bone was extremely massive but the vertebrae were much hollowed out on the sides and these hollows may have been filled with water air sacs as in birds. In this specimen, there was a single large claw on the inner toe of each front foot, and claws on the inner three digits of the hind foot. The pelvis was a huge structure for strong support and muscle attachments and the shoulder blades were likewise very long and heavy. The vertebrae of both neck and tail had strong spines that allowed large surfaces for muscle attachments.

Diplodocus

It is the another well known sauropod, differs from *Brontosaurus* in the more slender form, so that even with a length of 87 feet with the skull about 55 centimeters, long at the-end of a very long neck. It was by no means so weighty as the latter. In this species the terminal ten feet of the tail was like a whiplash, as the contained vertebrae did not decrease further in size. This may have proved a very efficient weapon of defence. A complete skeleton of *Diplodocus* from sheep Creek, Wyoming, about 15 miles from Bone Cabin quarry, is now mounted in the Carnegie Museum at Pittsburgh. In *Diplodocus* the teeth seem surprisingly weak for so large an animal.

Fig. 14.4. Diplodocus.

Order Ornithischia

The first known record of Ornithischia is that their fossil tracks upon the Connecticut valley's late Triassic rocks. They were derived in common with the Saurischia from the Parasaurischian stock. But the Ornithischia were on the whole more specialized than were the Saurischia. One indication is seen in the fact that the Ornithischia departed from the thecodont patterns of pelvic structure, while the Saurischia retained this pattern.

All of the Ornithischia were herbivores. Their teeth were somewhat leaf shaped, with serrate edges. Most of the Ornithischia lacked teeth in front of the mouth. Presumably this toothless region was covered with a horny beak somewhat like that possessed by turtles.

Four sub-orders of Ornithischia are recognized:

Ornithopoda

This group includes all the bipedal ornithischia. Best known among them are the *duckbilled dinosaurs*. They were the most primitive of bird like dinosaurs and had the typical tetra radiate pelvis. They had long, powerful hind legs and somewhat reduced forelegs. The bill was broad and flat and there was a prominent crest on top of the head. They are believed to have been amphibious. The Ornithopoda, although a large and varied sub-order that contains genera ranging from the upper Triassic through the Cretaceous periods, does include other primitive forms like *Camptosaurs*. It represents a somewhat generalized morphologic type from which more specialized Ornithischians evolved. The best known of the Ornithopods are: *Camptosaurus* and *Iguanodon*.

Camptosaurus

It was a small and medium sized dinosaur, no more than 6 or 7 feet in length, sometimes as much as 20 feet long. It was predominantly bipedal animal, with strong hind limbs but forelimbs were small. Perhaps the quadrupedal method of locomotion was utilized when *Camtrosaurus* was moving about slowly to feed. In its general build *Camptosaurus* was a heavier animal than the theropod dinosaurs of

Fig. 14.5. Camptosaurus.

similar size. It was an in-offensive plant eater. The skull was comparatively low and rather long. The temporal openings were large. The lower jaw was somewhat shorter than the length of the skull. Teeth were broad and leaflike and arranged in a single row. Long neck with opisthocolous vertebrae.

Iguanodon

The skeleton was found in a coal mine at Bernissart in Belgium. *Iguanodon* was about 34 feet in length. It was a bipedal vegetarian that probably roamed Europe in herds during the Cretaceous. In this reptile the thumb was enlarged into a sharp spike that may have been used as a weapon for defence. It was strongly built with a heavy built head and horny beak. Teeth were found in the hind part of the jaws Hind limbs were much longer and stronger than the forelimbs. Neck was long and thick. The tail was heavy and long. The feet had only four toes each. Pelvic girdle was remarkably 4-radiate with a prepubic bone pointing forward. It had affinities with crocodile and birds in having long hollow bones.

Fig. 14.6. Iguanodon.

Anatosaurus

Anatosaurus is the best known genus as mounted specimens from Wyoming, Montana, and United States, National Museum. It grew to great size, attaining lengths of thirty or forty feet and weights during life of several tons. Inspite of their size, these dinosaurs were predominantly bipedal, the hind limbs being very heavy and the feet broad. It was a fine cursorial type. It also possessed a powerful tail which judging from the vertical expansion implied by the bones, was admirably adapted for swimming. The skin was utterly without defensive armour. The hands, which possessed four fingers, were webbed, as were probably also the feet.

The skull was elongated, and in front both skull and lower jaw were very broad and flat in form rather similar to the bill of a duct. The dagger-shaped teeth in the sides of the jaws were enormously multiplied in number. There were five hundred or more teeth. The premaxillary and nasal bones were pulled back over the top of the skull to form a hollow crest.

Stegosauria

The stegosauria were slow moving, essentially Jurassic reptiles which grow to a length of about 25 feet. They were quadrupedal, but the forelimbs were shorter. The skull was very small, and the brain tiny and considerably smaller. They developed peculiar plates on the back and spikes on the tail. The well known example of this order is the *Stegosaurus*.

Stegosaurus

It was found in the late Jurassic period and was the most grotesque of dinosaurs. Back was provided with two rows of huge upstarding plates. The plates bear traces of numerous blood vessels. This implies that they possibly acted as a heat exchange device.

Fig. 14.7. Stegosauria.

The tail was armed with fearful horn-like spices, about 62 cm or more in length. The animals were about as large as elephants, but the head was extremely reduced. Brain was hardly larger than the terminal joint of one's finger, with a cranial capacity of about 56 c.c. According to Prof. *Williston* the *Stegosaurus* intelligence was not greater than that of a three week old kitten. *Stegosaurus* is famous for having a brain much smaller than the enlargement of the spinal nerve cord in the sacrum which has given rise to the popular belief that this animal had two "brains".

The *Stegosaurus* was permanently quadrupedal and the hind legs were much larger than the forelegs, making the pelvic region the highest point of the body. The shoulders were low. The reptile had great muscular power especially in the tail and hind limbs.

A few scattered remains of this group are known as the lower Cretaceous with possibly attributable fragments from the upper Cretaceous of India.

Ankylosaurus

They were heavily armoured dinosaurs somewhat reininiscent of turtles or of armadillos in the completeness of their armour plate. They have been called the '*tanks*' of the *Mesozoic battlefield*. They were bulky, quadrupedal reptile, some twenty feet in length, not very high at the back, and very broad. Their legs were heavy, hind limbs much longer than the forelimbs. The skull was very broad as compared with its length. The top of the head and the entire back were completely covered by a continuous armour of heavy polygonal, bony scutes, and along the sides of the body there were long, bony spikes. In the neck region, the armour was formed in half rings. The armoured tail terminated in a great mass of bone. That obviously formed a club or bludgeon. The teeth were singularly small and weak, indicating that the *Ankylosaurs* must have fed upon soft plants.

Fig. 14.8. Ankylosaurs.

Ankylosaurs were widespread in North America and, Asia in the upper Cretaceous, but only a single genus appeared in the latest Mesozoic beds. *Galton* (1980) recognised several fragmentary specimens from the Middle through Upper Jurassic of Europe as probable ancestors.

Ceratopsians or horned dinosaurs

The last of the dinosaurs to evolve were the *Creatopsians*, appearing in late Cretaceous times. Their entire evolution was confined to the second half of the Cretaceous period. They possessed a horn over each eye and a horn on the nose. They possessed a parrot like beak and a great frill of bones projecting backward over the neck. The head in giant *Ceratopsians* constituted an unusually large proportion of the body. It appears to be exceedingly large, making up a third or so of the total length of the body. The molar teeth were broad and flat, indicating a herbivorous diet. The horned dinosaurs were animals of moderatesize, not truely giants. They were very numerous during late Cretaceous times, but faded rapidly from the picture towards the end of that period. The excellent example of this group is the *Triceratopus*. In *Triceratopus* there were three huge horns above the eyes hence called three horned dinosaur. In *Monoclonius*, the nasal horn is very large and the brow horns under-developed.

Adaptations of Dinosaurs

As the great dinosaurs increased, there were many adaptations as a result of stresses and strains placed upon the bones, muscles and ligaments. In *Tyranosaurs* there were several tonnes of weight to be carried around so that the legs became very heavy and strong and feet broadened to form a good support and to furnish fraction against the ground. The connection between the hip bones and back bones was strengthened by the lengthening of sacrum. The dinosaurs lived in all Kinds of surroundings. Some of them were uplanded forms which were rather active and others were low landed animals living on marshy places near rivers and lakes. Forms like *Brontosaurs* were probably sluggish. Some dinosaurs were adapted for aquatic life. In *Diplodocus* the nostrils were placed on the top of the head.

There were various adaptations for herbivorous and carnivorous dinosaurs. In some a sharp bird like beak was found for the purpose of tearing green leaves. For grinding and chewing the Ornithischians were provided with batteries of teeth complexities.

The dinosaurs developed many bodily weapons for defence which were mainly of offensive variety. Almost all of them had spikes or clubs. On the end of the tail lethal weapons of great value to attack.

Affinities

With Rhynchocephalia

Both are diapsid, but in Rhynchocephalia the quadrate is firmly fixed while in dinosaurs quadrate is large and fixed or slightly movable. Intensive aspects of skull are alive in both, have no teeth on palate in Rhynchocephalia. Pubic bones enter into the acetabulum in Rhynchocephalia, however, is not performed as in dinosaurs.

With Crocodiles

Both are diapsid and the quadrate is firmly fixed in crocodiles. Pubic is extended from the perforated acetabulum by an ischial process in crocodiles. This is not the case in dinosaurs. Pineal foramen is absent in both. Thecodont teeth on palate present. Both atlas and axis show resemblance in both the cases. Crocodiles have false palate device for eating underwater (not in dinosaurs).

Avian affinities

1. Close function of astragalus calcaneum with bola which shows tendency to form tibio-tarsus.
2. Quadrate seems to have been slightly movable as inbirds.
3. Sometimes the scapula and coracoid are fused at their proximal ends as in birds.
4. Pelvic arches and the bones of hind limb shared simultaneously and ilium is depressed posteriorly.
5. The long ischian extends backwards and usually joins a median neutral symphysis.

Decune in Dinosaurs

The ancient reptiles began to decline in the late Cretaceous period about 100 million years ago. The dinosaurs were "lords of all they survived". Then suddenly they all became extinct. Why this mass extinction of creatures that had been successful for so long? This is one of the great unanswered qucstion of palaeontology. The possible causes of extinction are tried to justify this in the following lines.

1. The major cause of decline was supposed to be the changing climate. At that time North America rockes were being formed and the climate was growing cooler and drier. Swamps also became dry as the land masses arose.
2. Extinct reptiles were very huge and massive, therefore, they could not cope with the changing time.

3. During Mesozoic time, food was ample. The dinosaurs required plenty of food, which was not available due to the climate changes.
4. Epidemics, the eating of dinosaur eggs by mammals, and harmful effects of radiation are among the suggested reasons for the extinction.
5. Better protection from enemies was also not possible due to the large size of animals.
6. The animals like *Brontosaurs* and *Diplodocus*, had tonnes of flesh, small heads and feeble teeth were unfit to succeed in any struggle for existence.
7. Another factors also played in part. The larger the animal the fewer young ones does it bring forth at a time and longer are the intervals between the births and slower the growth of the young ones. Thus larger animals were unable to breed profusely.
8. There was no single sufficient cause of extinction viz., *Ichthyosaurs* and *Plesiosaurs* were marine reptiles, well adapted to an aquate life, but they became extinct without any cause. Simple explanation given by Zoologists for it is the *law of Progress*, and the trend of life towards the replacement of larger lower animals by those smaller and intellectual better developed.

Very recently Dr. *T. Swain* studied the massive extinction of dinosaurs. According to him they were poisoned and murdered by the rising Angiosperms, a class of flowering plants containing alkaloids. There were 14 families of dinosaurs. Out of which 12 were purely vegetarian. Thus, they were tricked by the chemistry of new plants and died of alkaloid poisoning. The rapid decline of the remaining two families of carnivorous reptiles followed because the principal diet of these reptiles was their vegetarian cousins.

15

REPTILES OF MESOZOIC

Reptiles are a primarily terrestrial assemblage, but during the Mesozoic several groups also dominated the marine environment. Among the most conspicuous marine reptiles were the ichthyosaurs, which were comparable in size and shape to large pelagic sharks and dolphins, and the plesiosaurs, which are characterized by their long necks, short trunk, and large paddlelike limbs. Ichthyosaurs appeared in the early Triassic, were most numerous in the early Jurassic, but died out well before the end of the Cretaceous. The plesiosaurs were diverse throughout the Jurassic and Cretaceous. Less well known are the nothosaurs and placodonts of the Triassic, which show lesser degrees of aquatic adaptation but are already clearly distinct from all groups of earlier terrestrial reptiles.

The specific ancestry and relationships of these groups have long been subject to dispute. There is still little evidence to establish the affinities of the ichthyosaurs, which are distinguished as a separate reptilian subclass, the lcthyopterygia. The placodonts appear to be derived from among early diapsids, but we cannot be more specific about their ancestry at present. Nothosaurs and plesiosaurs are united by unique derived features of the skull and shoulder girdle that support their inclusion in a distinct order, Sauropterygia, which may be a sister group of the lepidosaurs.

Aquatic adaptation is a common phenomenon among reptiles, as exemplified by the mesosaurs among the Captorhinida; the turtle families Plesiochelyidae, Chelonidae, and Dermochelyidae; the pleurosaurs; and several families among the Squamata. Seymour (1982) pointed out that aquatic adaptation is particularly easy among primitive amniotes because of their low metabolic rate, tolerance of anoxia and low body temperature, and great capacity to make use of fermentative metabolism for muscle activity. If we judge from modern marine iguanids, we see that adaptation to locomotion and feeding in the water does not necessarily require any structural or physiological specialization. Aquatic locomotion requires only one-fourth the metabolic expenditure of terrestrial locomotion in the iguana. Reptiles may have become secondarily aquatic whenever the balance between terrestrial and aquatic food sources and predators favoured life in the water.

Sauropterygians

The sauropterygians provide the most complete evidence of the sequence of events that leads to a specialized aquatic way of life. The radiation of primitive diapsids includes a number of aquatic lineages. Among the younginoids, the families Younginidae and Tangasauridae are similar in most skeletal features, but the tail in tangasaurids is laterally compressed with long neural and haemal spines to form an effective swimming structure. The abdominal cavity is filled with small stones that would have served for ballast. Tangasaurids are precluded from the ancestry of later aquatic diapsids by the specialization of the carpus, in which the lateral centrale is separated from contact with the fourth distal carpal by the large medial centrale. However, the remainder of their anatomy is close to that of primitive sauropterygians. Neither the front nor hind limbs are specialized for aquatic propulsion, nor are they reduced in size. As in modern aquatic lizards and crocodiles, the forelimbs were probably held against the side of the body to reduce drag.

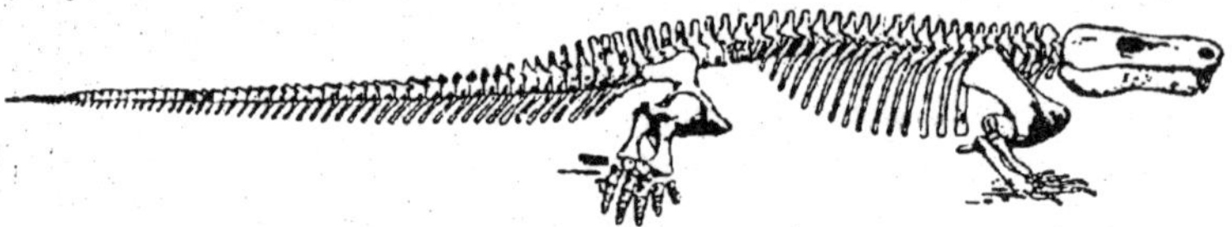

Fig. 15.1. The skeleton of Limnoscelis.

Claudiosaurus

Contemporary with these eosuchians in the Upper Permian of Madagascar are the remains of a further aquatic genus, *Claudiosaurus*, which may be closer to the ancestry of the later sauropterygians. Most skeletal features of Claudiosaurus resemble those of early terrestrial eosuchians, although the carpus retains the primitive diapsid pattern in which the lateral centrale reaches the fourth distal tarsal. The most conspicuous difference is the presence of a long neck, which results from the posterior displacement of the shoulder girdle. The presacral vertebral count remains 25. The third, rather than the fourth, digit of the manus is the longest, which gives the hand a more paddlelike appearance.

The most important derived feature that *Claudiosaurus* shares with nothosaurs and plesiosaurs is the loss of the lower temporal bar. The remainder of the skull closely resembles that of the early eosuchian *Youngina*. The general similarity of the skull of eosuchians, *Claudiosaurus*, and nothosaurs demonstrates that the configuration of the cheek that is typical of sauropterygians evolved from that of primitive diapsids through the loss of the lower temporal bar rather than from genera that primitively lacked a lateral temporal opening. In contrast with squamates, which have also lost the lower temporal bar, the quadrate is solidly supported by the pterygoid in *Claudiosaurus* and later sauropterygians.

The palate of *Claudiosaurus* also resembles that of nothosaurs and plesiosaurs in the loss of the transverse flange of the pterygoid and the reduction in the size of the suborbital fenestrae and interpterygoid vacuities. In further contrast with terrestrial lepidosaurs, the sternum is not calcified or ossified in this genus, although an impression of an apparently cartilaginous structure is evident in most specimens between the gastralia and the posterior margin of the coracoids. The loss of ossification of the sternum may be attributed to a change in function of the forelimbs. In lizards, the sternum serves as a surface for the rotation of the coracoid to extend the stride. However, the alternating movements of the forelimbs are not so important in terrestrial locomotion would be disadvantageous in the water, since they would force the head and front of the trunk from side to side. If such alternating movements of the limbs were not advantageous in aquatic reptiles, selection would quickly act to reduce the sternum.

In contrast with tangasaurids, the tail of *Claudiosaurus* is slender and shows no specialization for aquatic locomotion.

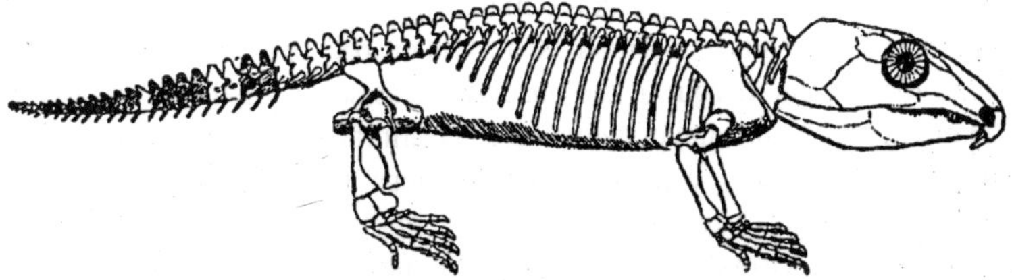

Fig. 15.2. Labidosaurus, a Lower Permian cotylosaur.

To judge by proportions of the trunk, tail, and limbs, this genus probably swam like modern aquatic lizards and crocodiles, with the front limbs folded alongside the body and the propulsive force produced by lateral undulation of the rear portion of the trunk and tail. The configuration of the joints of the girdles and limbs indicates that *Claudiosaurus* was probably still capable of terrestrial locomotion, but the large amount of cartilage suggests that it may have dependent on the buoyancy of the water for support most of the time.

Nothosaurs

The nothosaurs, which are known primarily from the Middle Triassic of Europe and China, represent a more advanced stage in aquatic adaptation. Their limbs are reduced relative to primitive terrestrial reptiles but not highly modified for aquatic propulsion. Ossification of the girdles, carpals, and tarsals is greatly reduced, and the structure of the joints shows no features that would facilitate support and movement on land.

Nothosaurs range from fewer than 20 centimeters to more than 4 meters long, and show considerable variation in the proportions of the head and neck. However, the structure and proportions of the remainder of the body are relatively uniform, which indicates a similar manner of aquatic propulsion throughout the group.

Pachypleurosaurus from the Middle Triassic of the Alpine region is a well-known genus that belongs to the family Pachypleurosauridae. It reached a little more than 1 meter in length, including a moderately long neck and long tail. The skull is particularly small and its relative size decreases significantly during growth.

Apart from its small size, the skull is typical of primitive nothosaurs. As in eosuchians and *Claudiosaurus*, the upper temporal opening is smaller than the orbit. There is a strong upper temporal bar that is important in supporting the cheek in the absence of a lower temporal bar. The quadrate of *pachypleurosaurids* is embayed posteriorly like that in lizards and may have supported a tympanum. The presence of an impedance-matching ear is confirmed by the structure of the stapes, which is a slender rod that extends directly lateral toward the margin of the quadrate. It is suprising that a primarily aquatic group would have specialized the middle ear to be sensitive to airborne vibrations, especially since there is no evidence for an impedance-matching middle ear in primitive eosuchians and *Claudiosaurus*. However, most primarily aquatic modern frogs have a well-developed middle ear.

The palate of *Pachypleurosaurus* is typical of other nothosaurs in the broad extension of the pterygoids posteriorly and medially, so that the interpterygoid vacuities are completely closed and the base of the braincase is covered ventrally. This can interpreted as an extension of the changes already observed in *Claudiosaurus*.

The neck is long in all nothosaurs. *Pachypleurosaurus* has approximately 18 cervical vertebrae. The number in the trunk range from 20 to 21. Intercentra are lost, except anterior to the atlas and axis and at the base of the tail. There are three sacral vertebrae—one more than in eosuchians. Other nothosaurs may have as many as six. The sacral ribs have very small surfaces for attachment to the ilium, which is itself very reduced. The sacral attachment appears very weak in all nothosaurs, and it is difficult to explain why the number of sacral ribs has increased.

In *Pachypleurosaurus*, the neural and haemal spines near the base of the tail are expanded dorsally and ventrally to form an effective surface for sculling.

In many nothosaurs, the ribs and vertebrae are greatly thickened or pachyostotic, as were those of mesosaurs. This thickening would have increased their specific gravity, enabling them to remain submerged with a minimum of effort. This characteristic is particularly evident is smaller genera, including *Neusticosaurus*.

The configuration of the pectoral girdles in nothosaurs is unique among reptiles. The bones are relatively poorly ossified, with little definition of the glenoid. The base of scapulae, clavicular blades, and interclavicle form a strong anteroventral bar. Behind this bar is very large medial opening that is bordered posteriorly by very elongated coracoids. The blade of the scapula is greatly reduced relative to terrestrial reptiles and located well anterior to the glenoid. It would have provided little area for the attachment of muscles from the trunk that support the body on the girdle in terrestrial reptiles. The stem of the clavicle is fused to the scapula. Ventrally, the scapula expands beneath the clavicle, in contrast with the relationship in most reptiles, in which the blade of the clavicle is external to the scapula.

In all nothosaurs, the distal limb elements are reduced relative to the humerus, which is a particularly stout element in the larger species. The elbow joint is very poorly defined. The olecranon of the ulna is not ossified, and articulating surfaces for the ulna and the radius on the ventral surface of the humerus are absent. The lower limb was probably capable of being extended directly laterally, in contrast with

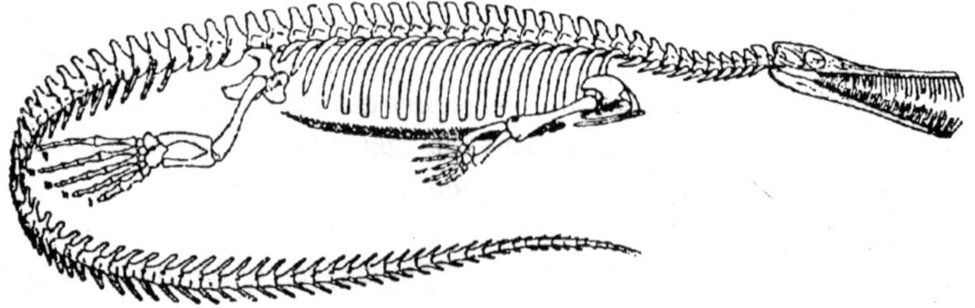

Fig. 15.3. The skeleton of Mesosaurs, a late Carboniferous or early Permian aquatic reptile.

its habitually flexed position in quadrupedal reptiles. All nothosaurs retain five distinct digits. In most genera the primitive phalangeal formula of 2, 3, 4, 5, 3 is retained, but *Pachypleurosaurus* shows a loss of phalanges while *Ceresiosaurus* exhibits hyperphalangy.

The pubis and ischium resemble those of primitive reptiles except for the development of a thyroid fenestra. The blade of the ilium is very narrow and provides little area of support for the sacral ribs. In *Pachypleurosaurus*, the rear limb is considerably shorter than the anterior, but the foot is broad and may have been important in steering.

Aquatic Locomotion in Nothosaurs

Nothosaurs have long been thought to be intermediate in their degree of aquatic locomotion between the pattern or primitive terrestrial reptiles and plesiosaurs. The limbs appear to be suitable for paddling but not for more sophisticated aquatic locomotion.

The restoration of *Pachypleurosaurus* is lateral view shows that the vertical expansion at the base of the tail would have provided the largest surface for aquatic propulsion. The primitive pattern of nothosaur locomotion was probably an extension of that seen in aquatic lizards and crocodiles, in which lateral undulation of the trunk and tail played the primary role in propulsion. The limbs in all nothosaurs are reduced relative to the length of the trunk and may initially have been held to the sides of the body to reduce drag. The reduction of the size and degree of ossification of the limbs in nothosaurs probably dates from a primitive stage in their adaptation to life in the water. In some nothosaurs, the ventral portion of the shoulder girdle is extremely reduced, with neither the clavicles nor the scapulae expanded ventrally. This reduction may reflect the most primitive condition in the group.

The configuration in other genera, in which the ventral portion of the scapula and the clavicular blade are expanded but have a relationship that is reversed from that of most reptiles, is certainly a specialization of nothosaurs. It may be associated with selection for more active use of the forelimbs in propulsion. All nothosaurs are characterized by large coracoids that are greatly extended posteriorly. We can associate this extension with the elaboration of muscles that move the forelimbs posteriorly. We can attribute the redevelopment of the anterior portion of the ventral surface of the pectoral girdle to a later stage in which more powerful protractors of the humerus evolved. The elaboration of the forelimbs in locomotion presumes a change in behaviour and perhaps in the central nervous system control that emphasizes symmetrical, rather than asymmetrical, movement.

The pelvic girdle and rear limbs in nothosaurs are reduced and show little specialization indicative of an active role in propulsion.

The pachypleurosaurids are primitive in the structure of the skull roof, with small, upper temporal openings like those of early eosuchians. Most Pachypleurosaurids are less than 1 meter long, and the low neural spines and broad flat neural arches indicate that lateral undulation of the region may have remained the most important force in aquatic locomotion.

Other nothosaurs, including *Nothosaurus*, are much larger—up to 4 meters long—and the skull is much wider posteriorly, with the upper temporal openings larger than the orbits. *Nothosaurus* has much taller neural spines and narrower neural arches, which suggest a more rigid trunk. The forelimbs are much more robust and more strongly differentiated from the rear limbs than in the smaller Pachypleurosaurids, which suggests that they were more important in propulsion.

The Ancestry of Pleiosaurs

Nothosaurs appear to be plausible ancestors of the plesiosaurs in most skeletal features. However, such a relationship seems to be contradicted by the structure of the palate and the shoulder girdle. The palate of plesiosaurs is *less* specialized than that of nothosaurs in the retention of interptery-goid vacuities and exposure of the base of the braincase between the pterygoids. The configuration suggest that plesiosaurs may have evolved from more primitive diapsids rather than from any of the well-known nothosaurs.

One genus form the Middle Triassic, *Pistosaurus*, which was originally described as a nothosaur, retains a more primitive pattern of the palate. The skull is similar to that of nothosaurs in retaining nasal bones that are the lost in typical plesiosaurs, and it has been associated with postcranial remains that are similar to those of nothosaurs. Unfort-unately, the phylogenetic significance of this material

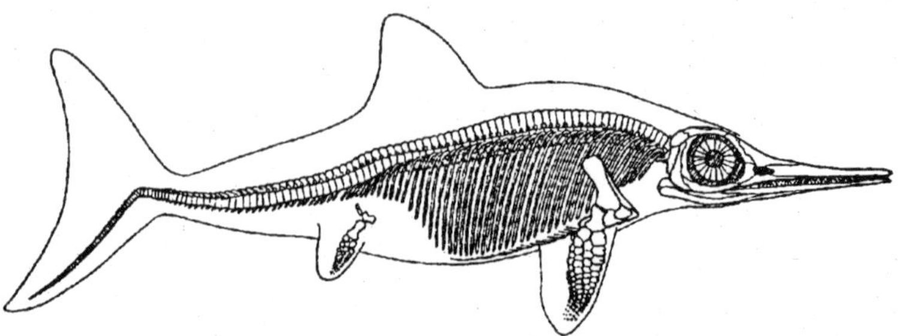

Fig. 15.4. A Jurassic ichthyosaur.

is difficult to assess, since we are not certain that the skull and postcranial material belong to the same species. If they do *Pistosaurus* provides a strong link between nothosaurs and plesiosaurs. The pattern of the palate of *Pistosaurus* may represent retention of a primitive condition and so indicate that the lineage leading to plesiosaurs had diverged earlier than the appearance of the oldest known typical nothosaurs. Alternately, it might reflect a later reversion to a primitive condition. Such a reversion could result from a change in developmental patterns, with the retention of an early onotogenetic stage in which the pterygoids had not yet extended to the mid-line beneath the braincase.

The other problem in accepting a nothosaurian origin for plesiosaurs is the great difference in the configuration of the shoulder girdle. Typical plesiosaurs have what appears to be a more primitive pattern, with both the scapula and coracoid greatly expanded ventrally, somewhat like those in primitive eosuchians. Andrews (1910) descried growth stages in advanced plesiosaurs that show that the ventral portion of the girdle is initially open, somewhat similar sequence phylogenetically from the nothosaurs through the most primitive Lower Jurassic plesiosaurs to the more advanced plesiosaurs. Unfortunately, the anatomy of the earliest plesiosaurs remains poorly known.

More detailed knowledge the transition between nothosaurs and plesiosaurs is limited by the absence of articulated remains of either group in the late Triassic. Disarticulated vertebrae, girdles, and limb elements hint at a transition between the two groups, but so far there is no evidence of the manner of change between the markedly different limb proportions and pattern of the pelvic girdle that differentiate the two groups.

PLESIOSAURS

As a group, the plesiosaurs are distinguished from nothosaurs by the much greater size and increased similarity of the forelimbs and hind limbs. The pectoral and pelvic girdles are both greatly expanded ventrally, and there is only a short distance between them. The gastralia are limited in extent, with no more than nine rows, but they are extremely massive.

The means of locomotion among plesiosaurs has long been in dispute. Watson (1924) suggested that they rowed through the water, while Tarlo (1958) and Robinson (1975) proposed that plesiosaurs employed subaqueous flight in the manner of sea turtles and penguins.

Godfrey (1984) argued that the structure of the girdles differs from that of sea turtles and penguins in lacking any specialized accommodation for dorso-ventral movement of the limbs. The great anterior-posterior extent of both the pectoral and pelvic girdles in plesiosaurs would have permitted strong anterior and posterior strokes like those that are use for aquatic propulsion in the sea lion. The sea lion is propelled primarily by horizontal retraction of the forelimbs but English (1976) demonstrated that the recovery stroke provides some

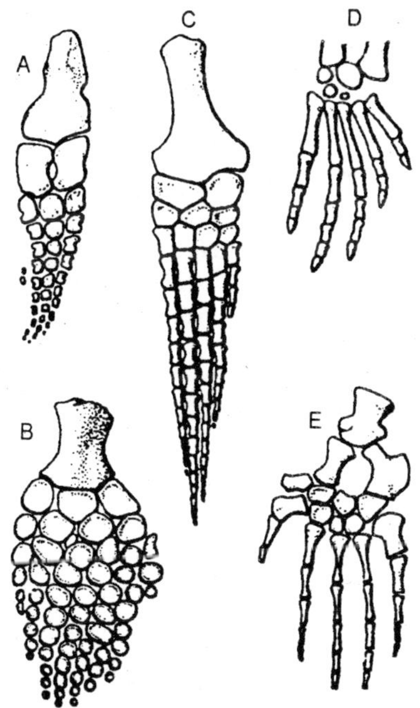

Fig. 15.5. Limbs of various aquatic reptiles. A–Pectoral limb of the Triassic ichthyosaur Merriamia; B–same of the Jurassic Ophthalmosaurs, C–same of the Cretaceous plesiosaur Elasmosaurus; D–pes of the nothosaur Lariosaurs; E–pectoral limb of the mesosaur Clidastes.

anteriorly directed lift as well. Although the recovery stroke would not provide as much thrust as the dorso-ventral movements of the flippers in penguins and sea turtles, the great inertia of the large pleiosaurs would have allowed them to continue forward during the recovery stroke as is the case for large penguins between wing beats.

The lengthening of both girdles and the flattening of the puboischiadic plate into the horizontal plane, together with the increased bulk of the gastralia, would have contributed greatly to the rigidity of

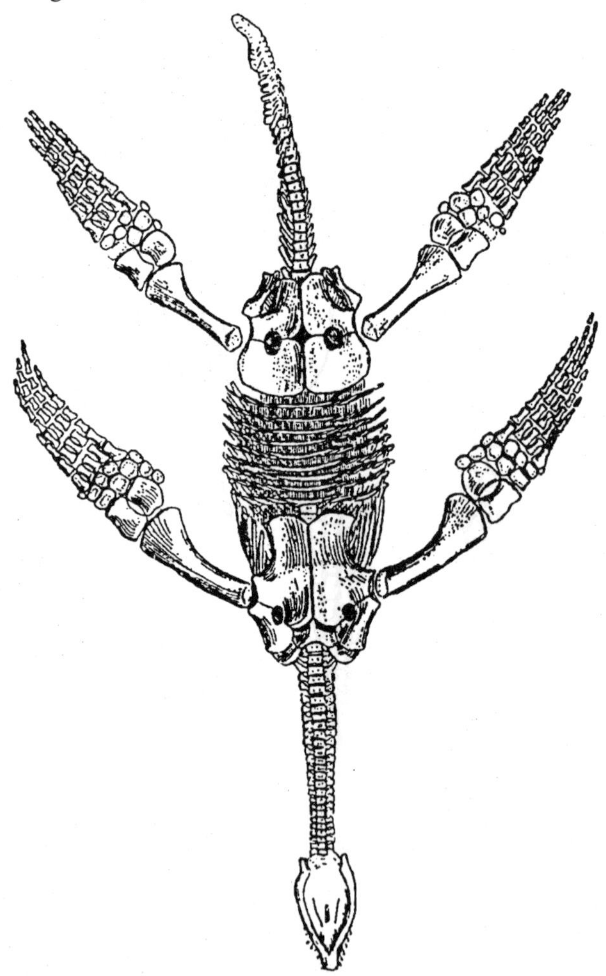

Fig. 15.6. Thaumatosaurs, a Jurassic Plesiosaur, ventral view, to show the abdominal ribs and expanded girdles.

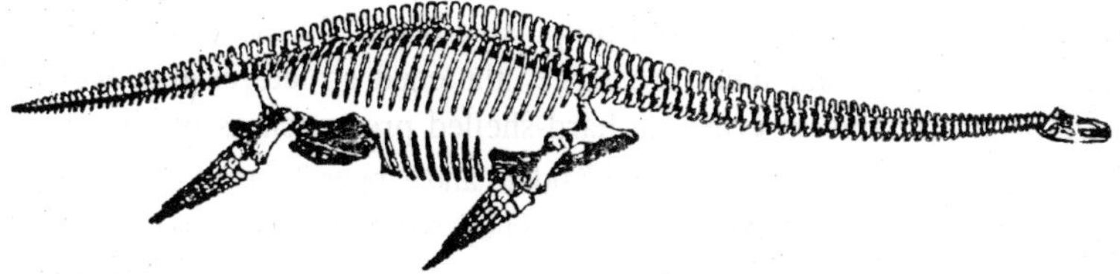

Fig. 15.7. Muraenosaurus, a long-necked Jurassic plesiosaur.

the trunk region in plesiosaurs. This rigidity was probably necessary to permit the effective use of both the forelimbs and hind limbs as paddles. The tail is short and probably served as a rudder rather than as an important element in propulsion. The stiffening of the trunk and the use of the rear limbs as paddles may be the most important changes between nothosaurs and pleiosaurs.

The proportions of the trunk and limbs remain relatively constant among plesiosaurs from the Lower Jurassic to the end of the Cretaceous. On the other hand, proportions of the head and neck vary extensively and progressive changes are evident in the details of the vertebrae, girdles, and limbs. In a recent review, Brown (1981) recognized approximately 40 plesiosaur genera. He divides them into four families. Like most other authors, including Welles (1962) and Persson (1963), Brown recognized two major groups of plesiosaurs, the Plesiosauroidea and the Pliosauroidea.

Among the Plesiosauroidea, the head is relatively small, as among the nothosaurs. The neck is relatively long, with up to 76 vertebrae. Pliosaurs have much larger skulls (in some genera they exceed 3 meters in length) with a much longer symphysis between the lower jaws, but the neck may have as few as 13 cervical vertebrae. All pliosaurs are included in a single family. The plesiosauroids are typically divided into two groups, the more primitive Plesiosauridae, which may be close to the origin of the entire assemblage and have as few as 28 cervical vertebrae, and the more specialized elasmosaurs, which have 32 to 76 cervical vertebrae.

All advanced plesiosaurs, regardless of the families to which they belong, increase the number of phalanges in the paddles and develop single-headed, rather than double-headed, cervical ribs. Both elasmosaurs and pliosaurs persisted until the end of the Mesozoic.

Placodonts

The placodonts are a distinct aquatic group that was nothosaurs. Placodonts have short, stout bodies, with the limbs only moderately specialized as paddles. The tail is long and laterally compressed but not in the form of a specialized propulsive organ. In most placodonts, the cheek and palatal teeth are large, flattened structures that would have been effective in crushing hard-shelled prey such as molluscs.

We know placodonts from the Middle and Upper Triassic of Europe, North Africa, and the Middle East. They were probably confined to shallow, coastral waters. Placodonts are quite diverse when they first appear in the lower beds of the Middle Triassic, and they are already so specialized that we have not been able to establish their ancestry.

Placodus, which Broili (1912) and Drevermann (1933) described, is characteristic of the group. The skull is massive and well consolidated for support of the large, flattened teeth on the palate, maxilla, and lower jaw. The anterior teeth of the dentary and premaxilla are procumbent and spatulate at the tip. They may have served to dislodge attached prey. In contrast with most early reptiles, the dentary bone is elevated posteriorly as a large coronoid process for the insertion of jaw muscles. The palate is strengthened by the sutural attachment of the palatine and pterygoid at the midline, and there is no trace of interpterygoid vacuities. The suborbital fenestra is reduced to a narrow slit. The internal nares opens by a single passage behind the median vomer. The base of the braincase and the basipterygoid processes are still visible behind the palate, in contrast with the condition in nothosaurs. In *Placodus*, the braincase and dermal bones are integrated to form a nearly solid occipital surface, with greatly reduced posttemporal fenestrae. As in nothosaurs, the large upper temporal openings are surrounded by the parietal, squamosal, and postorbital. The cheek is solid, without a trace of a lower temporal opening, and shows little ventral emargination.

No features of the skull support close affinities with other aquatic reptiles or with any particular group of primitive terrestrial amniotes.

Placodus has approximately 28 presacral vertebrae, only slightly more than in most terrestrial reptiles. The neural spines are tall and the transverse processes extremely elongate. However, the centra are primitive in being deeply amphicoelous. This unusual combination of vertebral features is diagnostic of the placodonts. There are three sacral vertebrae and a long slender tail. The endoskeletal element of the pectoral girdle are poorly ossified, like those in other members of the group; the coracoid is oval, and the scapula is short and narrow. As in nothosaurs, the interclavicle underlies the blades of the clavicle.

The pelvic girdle is primitive in the platelike nature of the pubis and ischium, with no thyroid fenestration. The limbs, especially the carpals and tarsals, are poorly ossified and they have little specialization for aquatic propulsion.

The gastralia are well developed in all placodonts. *Placodus* is characterized by the presence of a single row of dermal ossifications above the neural spines. However, the generally more primitive genus *Paraplacodus* lacks such ossification. In a closely related, but divergent, lineage (which is represented by the genera *Cyamodus*, *Placochelys*, and *Henodus*), the entire trunk region is covered by a dermal carapace that superficially resembles that of turtles. It is composed of a large number of polygonal ossicles. They are not at all comparable with the large plates of turtles, but in *Henodus* they were apparently covered by epidermal scutes. The vertebrae are fused to the carapace and reduced in numbers. Most members of this assemblage of armored placodonts have cheek and palatal teeth like those of *Placodus*, but in some genera the anterior teeth are lost and may have been replaced by a horny beak. Most of the teeth are lost in *Henodus*. This genus is also peculiar in the closure of the upper temporal opening.

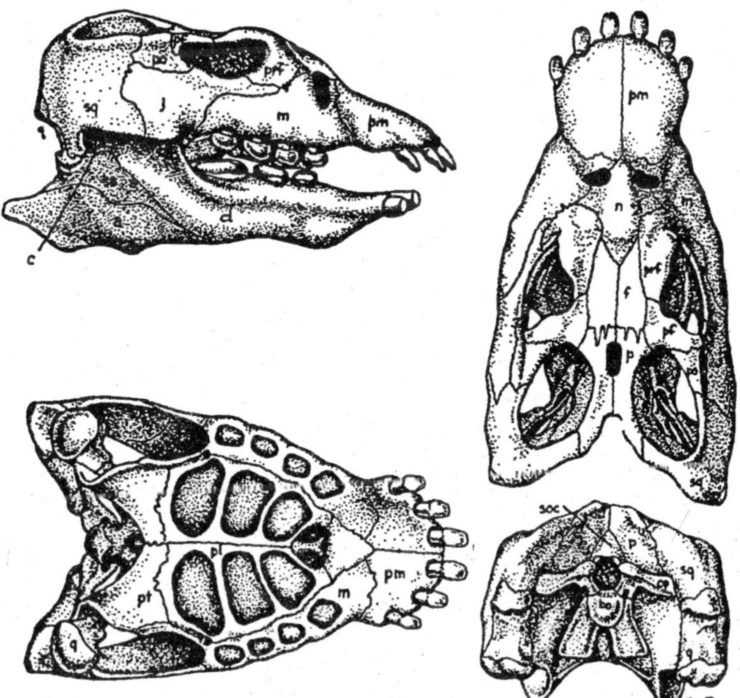

Fig. 15.8. The skull of the placodont reptile Placodus.

Helveticosaurus apparently represents an early offshoot from the remainder of the placodonts. The vertebrae are comparable to those in *Placodus* in being deeply amphicoelous and in having very long transverse processes. The limbs and girdles show a similarly low degree of ossification. However, the trunk region is much longer, with approximately 43 presacral vertebrae. The cheek teeth are not flattened but resemble the more anterior teeth of other placodonts in being long and somewhat recurved. Palatal teeth are not evident. In the only two speciments we know, the skull is badly disarticulated, which precludes

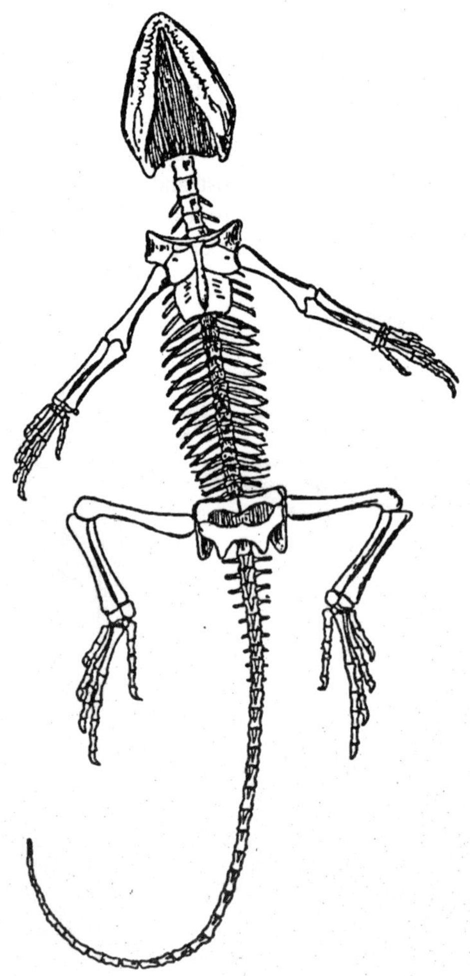

Fig. 15.9. Homoeosaurus, a Jurassic rhynchocephalian.

restoration of tis original configuration. The interclavicle retains a long stem. The margins of the anterior portion of the interclavicle may be covered by the clavicular blades, in common with most primitive terrestrial reptiles. Despite some primitive features, *Helveticosaurus* does not help to establish the origin of placodonts of their relationships with other groups of aquatic reptiles.

One day argue that the absence of lateral temporal opening is specialized character of placodonts that evolved within the group of strengthen the skull in relationship to the crushing dentition. However, there is no direct evidence for this hypothesis. Other reptiles, including Araeoscelis and *Trilophosaurus* appear to have closed a primitively open cheek in relationship to a crushing dentition, but they show no other features in common with placodonts. The establishment of the affinities of this group must await the discovery of new fossils from the Lower Triassic or Upper Permian.

ICHTHYOSAURS

Throughout their known fossil record, ichthyosaurs were the most highly specialized of all marine reptiles. In advanced genera from the Jurassic and Cretaceous, the body is spindle shaped, the limbs are reduced to small steering fins, and the caudal fin is a large, lunate structure. The body form corresponds very closely with that of the modern teleost family, Carangidae, which includes the fastest swimming of all fish, the mackerals and tuna, which achieve speeds in excess of 40 kilometers an hour. The vertebrae are highly specialized, with the centra in the form of very short, deeply biconcave disks. The neural arches are separated from them by cartilage and do not bear transverse processes. The configuration of the skull is greatly modified, with a large orbit, a greatly reduced cheek, and a long snout. The teeth are of uniform shape and set in a long groove instead of distinct alveoli. The ichthyosaurs gave birth in the water to live young, thus avoiding the critical problem of going on land to lay eggs as must modern sea turtles, a practice that may have been necessary for mosasaurs and plesiosaurs as well.

Since, specimens are present in many museums, it might be expected that the evolutionary history of ichthyosaurs would be well known. As McGowan (1983) emphasized, most fossils have come from a few extremely productive localities that represent only very limited periods of geological time; the remainder of ichthyosaur history is very poorly documented. Relatively few detailed anatomical studies have been undertaken during the past 50 years, and most of our knowledge still rests on preliminary work published in the nineteenth and early twentieth century.

Early Triassic Ichthyosaurs

The earliest ichthyosaurs are known from deposits at the top of the Lower Triassic in Spitsbergen and Japan and at the base of the Triassic sequence in China. We know most of the skeleton from numerous specimens of the Japanese genus *Utatsusaurus*. The body form is broadly similar to Jurassic and Cretaceous ichthyosaurs, but most skeletal elements are notably more primitive. The entire skeleton is approximately 1½ meters long. The skull is poorly known, but the lower jaws indicate that it was long and slender. The teeth are long and very slender, and the enamel is infolded at the tip. The neck is short, but the restored presacral column has approximately 40 presacral vertebrae. In contrast with later ichthyo-saurs, the individual centra are approximately as long as they are high and not deeply amphicoelous. The neural arches appear to be only loosely attached to the cen-tra. The tail is long and little specialized from the pattern of terrestrial reptiles, alth-ough the neural spines are somewhat modified and may have supported a low fin. The ribs in the trunk region are very long and slightly exp-anded distally. They are double headed, but both head apparently attached to the centra—as later ichthyosaurs—without the arch possessing a transverse process.

The scapula and cor-acoid are separately ossified, like those of the slightly modified aquatic diapsid *Askeptosaurus*. The inter-clavicle is a small triangular element, as in later Triassic ichthyosaurs. The clavicles are long narrow elements that resemble their counterparts in primitive terrestrial reptiles. The pubis and ischium are greatly reduced, as in the ilium, which was apparently not attached to the vertebral

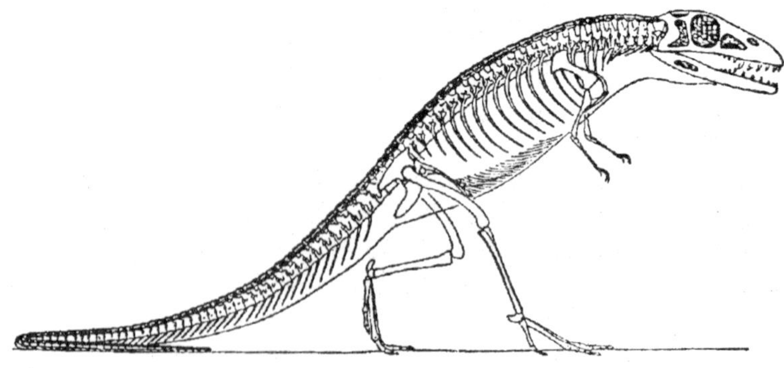

Fig. 15.10: Saltoposuchus, a lightly built Triassic thecodont.

column. The forelimb and, even more so, the hind limb are reduced in size and degree of ossification. The humerus is short and broad. The ulna, radius, and metacarpals, in contrast, are little modified from the pattern in terrestrial reptiles. The carpus appears to have four elements in the proximal row and five in the distal. The manus has five distinct digits. There are four very short phalanges in each of the first three digits, with the hand as a whole giving the appearance of a fin.

The femur, tibia, and fibula are even more reduced than the forelimb, but the tarsus and pes are not preserved. A pattern of gastralia like that of primitive diapsids is retained.

Grippia, which is found in the uppermost beds of the Lower Triassic of Spitsbergen, provides the earliest evidence of the cranial structure in ichthyosaurs. The skull has a large upper temporal openings, but the cheek shows no evidence of a lateral fenestra, although it is slightly emarginated ventrally. The orbit is larger, but the pre- and postfrontals meet above it. The snout, which is incompletely known, is somewhat elongate but probably not as greatly as in later ichthyosaurs. The pineal openings lies between the parietals in primitive fashion, in contrast with its position in later ichthyosaurs at the border of the frontals. The parietals, frontals, and nasals retain a relatively primitive configuration, whereas their pattern is greatly modified in later ichthyosaurs. The quadratojugal makes up a large portion of the cheek, as in later ichthyosaurs but in contrast with primitive amniotes. The squamosal forms much of the lateral border of the upper temporal opening. The external nares are high on the snout and relatively close to the orbit, like those in other aquatic forms.

Surprisingly, the teeth of the maxilla and back of the dentary are hemispherical, like those of the mosasaur *Globidens*, and set in two rows. The more anterior teeth are sharply pointed cones. All the teeth are set in distinct sockets. The Lower and Middle Triassic genus *Omphalosaurus*, which we find in California as well as Spitsbergen, has several rows of similar teeth in the cheek region but a much more advanced structure of the limbs than *Grippia*. Heterodonty and multiple rows of cheek teeth may be specializations of this group of early ichthyosaurs. We do not known the general body form of *Grippia*, but the individual vertebrae and elements of the girdles and limbs appear nearly as primitive as those of *Utatsusaurus*.

The earliest known ichthyosaur from China, Chaohusaurus, shares primitive features with Utatsusaurus and *Grippia*. The snout is somewhat elongated, the vertebrae are longer than high, and the forelimb is modified as a paddle, but the individual elements retain a configuration that is reminiscent of terrestrial reptiles, with a phalangeal count of 2, 3, 4, 4, 2. However, this limb is much larger than that of *Utatsusaurus*.

The Problem of Ichthyosaur Origin

These early Triassic genera probably exemplify in a general way the pattern from which the later Mesozoic ichthyosaurs arose. In

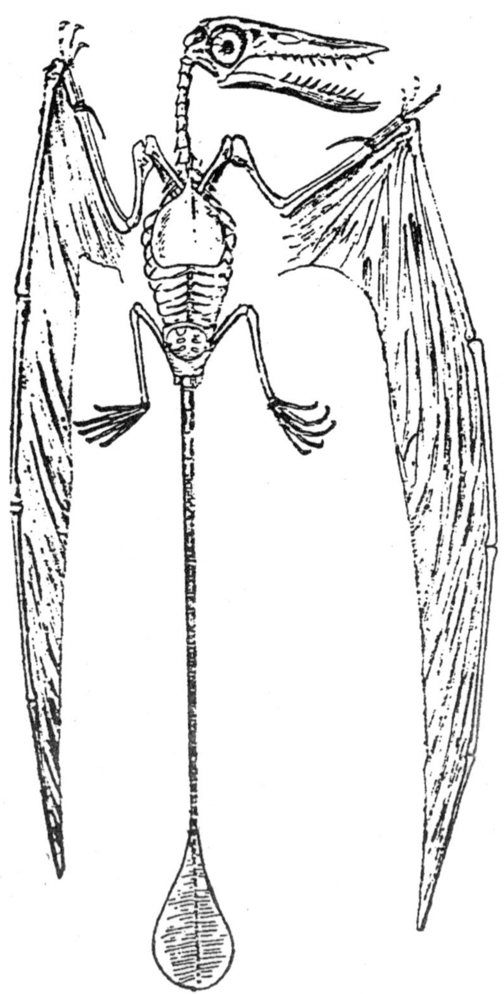

Fig. 15.11. Rhamphorhynchus, a long-tailed Jurassic pterosaur.

contrast with the sauropterygians, the early ichthyosaurs were already highly adapted to an aquatic way of life, although the anatomy of the skull, vertebrae, girdles, and limbs of the Lower Triassic genera is much more primitive than that of their Jurassic and Cretaceous counterparts. However, not even the most primitive features link them to any particular group of terrestrial or aquatic reptiles, although the presence of a dorsal temporal opening suggests affinities with early diapsids.

It was once thought that the presence of an upper temporal opening and a solid cheek constituted the basis for a large taxonomic group termed the Parapsida or Euryapsida that included (at various times in the history of the concept) ichthyosaurs, plesiosaurs, nothosaurs, placodonts, and such terrestrial forms as *Araeoscelis*, *Trilophosaurus*, *Protorosaurus*, *Prolacerta*, and even lizards. It is now clear that lizards, *Protorosaurus*, and *Prolacerta* evolved from the primitive diapsid stock by loss of the lower temporal bar. *Araeoscelis* and *Trilophosaurus*, in which the cheek is solid, may have a similar origin. The postcranial skeleton of these strictly terrestrial forms shows no significant specialization in common with either ichthyosaurs of placodonts, and the common presence of a dorsal openings was either retained from an early diapsid condition or achieved convergently. In the later case, either or both of these groups might have arisen separately from primitive anapsid reptiles.

The anatomy of the skeleton of ichthyosaurs does point to a common origin within primitive amniotes, with marked similarities of the braincase to that of early captorhinomorphs. There is no support for von Huene's (1949) suggestion that ichthyosaurs arose directly from embolomerous labyrinthodonts.

Nanchangosaurus

Another early form that may have affinities with the primitive ichthyosaurs is *Nanchangosaurus* (*Hupehsuchus*) from China. The trunk is fusiform and the limbs are reduced but retain the characteristics of terrestrial forms. The skull has a long, toothless snout and an upper temporal opening. According to Young and Dong (1972), the skull also shows a lateral temporal opening and an antorbital fenestra. There is a row of dermal plates along the vertebral column. These features are characteristic of thecodont archosaurs. Young and Dong place this genus among thecodonts, which suggests that the ichthyosaur habitus was achieved convergently in the two groups. Alternatively, one could interpret this form as a primitive ichthyosaur, showing evidence of the relationship of this group to the archosaurs. Much more remains to be learned of all the early Triassic forms before we can understand the origin of this group.

Middle and Upper Triassic Ichthyosaurs

Middle and Upper Triassic ichthyosaurs are even more widespread geographically and anatomically diverse than those of the Lower Triassic. Two very distinct forms are known from the Middle Triassic, Cymbospondylus and *Mixosaurus*. Cymbospondylus from Nevada was last described by Merriam in 1908. This genus was much larger than the Lower Triassic forms, reaching 10 meters in length. The trunk was very ling, with approximately 60 presacral vertebrae. The tail was also very long and apparently straight, and the skull had a very long snout. The teeth are long, sharp, and set in deep sockets.

The proximal limb bones do not appear greatly modified from the terrestrial pattern. The carpals and tarsals are rounded elements: the distal bones are unknown. The pelvic girdle is striking primitive, with a platelike pubis and ischium; the ilium has a narrow blade. The bones of the pectoral girdle resemble those of *Utatasusaurus*.

Mixosaurus, is the most widely known of Middle Triassic ichthyosaurs, with fossils from Spitsbergen, the Alpine region, the western United States, China, and Indonesia. Superficially, *Mixosaurus* appears close to the pattern of later ichthyosaurs. The body is somewhat more than 1 meter long. The limbs are modified as paddles with five principle digits, which show considerable hyperphalangy. The pectoral fin is substantially larger than the pelvic. The tail is modified as a caudal fin, with the neural and haemal arches considerably elongated, but the end of the tail is not sharply down-turned.

There are 45 to 55 presacral vertebrae. As in *Grippia*, the cheek teeth have wide, blunt tips, although the anterior teeth are sharply pointed. Unlike either earlier or later ichthyosaurs, the dorsal temporal opening is quite narrow.

The only Upper Triassic ichthyosaurs for which most of the skeleton is known, is *Shonisaurus*, which is represented by the remains of many specimens from Nevada that Camp (1980) recently described. *Shonisaurus* is the largest of all ichthyosaurs, reaching a length of 15 meters. In marked contrast with other genera, the manus and pes are both greatly elongated but there are only three rows of phalanges, which gives the appearance of paddles rather than fins. The ribs are expanded distally. The teeth are restricted to the front of the jaw, and the end of the tail is bent ventrally. In contrast with other ichthyosaurs, two vertebrae bear sacral ribs for attachment of the ilia. The other vertebrae approach the pattern of advanced ichthyosaurs. The centra are short, biconcave disks. They bear two rounded articulating surfaces for the attachment of ribs heads.

Merriam (1902, 1904) named several genera of ichthyosaurs from the Upper Triassic of California, included *Shastasaurus*, *Delphinosaurus*, and *Toretocnemus*, none of which is represented by a complete skeleton. In an additional genus, *Merriamia*, the forelimb is much longer than the hind limb, as in *Mixosaurus* and Jurassic ichthyosaurs, but it has only three principle digits; the hind limb has four digits. The skull is advanced over that of most other Triassic forms in having the teeth set in a continuous groove rather than in separate sockets.

Jurassic and Cretaceous Ichthyosaurs

Ichthyosaurs reached their greatest diversity in the early Jurassic, as documented by remains from England and Germany. Post-Triassic ichthyosaurs appear to represent a new radiation from a single preexisting lineage, rather than being a continuation of the early Triassic radiation. Jurassic and Cretaceous ichthyosaurs appear more similar to one another than do the Triassic genera, and all show the same advanced characters.

The most important features relate to the mode of locomotion. Among the adequately known Triassic genera, the tail is straight or only gently curved. From the base of the Jurassic, all advanced ichthyosaurs show a sharp ventral bend in the tail that supports the lower portion of a high, lunate caudal fin. The caudal zygapophyses form a ball-and-socket joint to facilitate lateral flexion but limit mobility in a dorso-ventral direction. In Triassic genera, the zygapophyses are paired and relatively flat, but they become confluent and form a single, broadly curved surface by the early Jurassic.

A comparison with modern fish indicates that the change in body form between Triassic and Jurassic ichthyosaurs can be associated with a change in swimming mechanics from a moderately fast anguilliform mode to the much more effective advanced carangiform mode of modern tunas and the fastest-swimming sharks.

Other skeletal changes also distinguish the post-Triassic ichthyosaurs. In most Triassic genera, the ribs attach by a single head except in the cervical region. In Jurassic genera, the trunk ribs are double headed as well. The pectoral girdles of Jurassic and Cretaceous

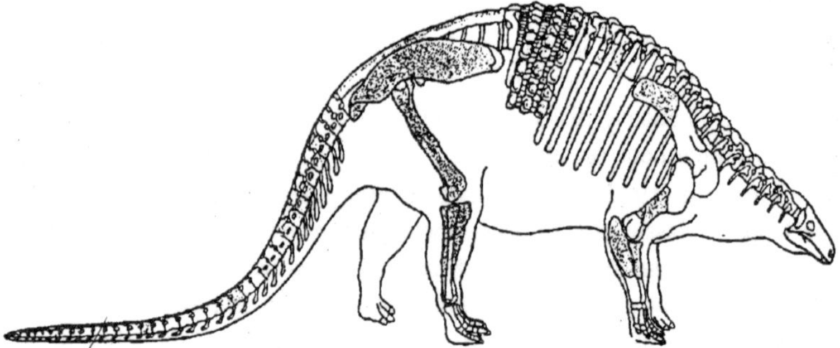

Fig. 15.12. Nodosaurus, a Cretaceous armored dinasaur.

ichthyosaurs are characterized by a T-shaped rather than a triangular interclavicle. The pelvic girdle is greatly reduced and the ischium and pubis are fused to one another in the Upper Jurassic genus *Ophtalmosaurus*.

The paired fins are both more highly specialized. The ulna, radius, tibia, and fibula are so modified that they form an unbroken series with the metapodials and phalanges. There is considerable variety in the configuration of the paddles in Jurassic and Cretaceous ichthyosaurs, and the number and arrangement of the carpals and phalanges has long been used to separate two taxonomic groups. In one group, termed the *longipinnates*, there are three distal carpals and three primary digits; in the other group, the latipinnates, there are four distal carpals and four or more primary digits.

McGowan (1972) extended these groups back into the Triassic and recognized Mixo-saurus as a latipinnate and Cymbospondylus, Merriamia and Shonisaurus as longipin-nates. Appleby (1979) demons-trated that several early Jurassic species exhibited an inter-mediate pattern and suggested that the divergence of these two patterns began only after the end of the Triassic. He sugg-ested that all the Jurassic ichthyosaurs arose from a single lineage of Triassic longipinnates rather than from the latipinnate *Mixosaurus*, which was thought to have such an ancestral position. The poorly known genus *Merriamia* might be the most closely related of the upper Triassic genera.

Detailed descriptions of the skull of early Jurassic ichthyo-saurs have only recently been published. They are based on specimens that have been acid etched from nodules in which the original three-dimensional character of the bone is retained. It was long thought that the supratemporal, bone lost in many reptilian groups formed the lateral margin of the temporal opening in Jurassic and Cretaceous genera and that the remainder of the cheek included both a squamosal and a large quadratojugal separated by a posterior extension of the postorbital bone. More complete preparation demonstrates that the areas identified as squamosal and quadratojugal are parts of a single bone that is simply overlapped by the postorbital. If the single bone is identified as the quadratojugal, the more dorsal element bordering the temporal opening can be recognized as the squamosal.

The occiput exhibits a common pattern in genera from the Upper Triassic through the Jurassic. The elements are not well integrates with one another but were presumably united by cartilage. The pattern broadly resembles that of primitive amniotes with large supraoccipital, but the exoccipitals are small and do not contribute to the occipital condyle. The otic capsule are large and supported laterally against the squamosal like those in early archosaurs.

In the Jurassic genus *Ichthyosaurs*, the stapes is strikingly massive, unlike that of Mesozoic diapsids but resembling that of Carboniferous amniotes. It abuts ventrolaterally against the quadrate. It certainly would not have contributed to an impedance-matching system but may have been effective in transmitting vibrations from the water to the inner ear. The quadrate extends fairly far dorsally but was not embayed posteriorly for support of a tympanum.

The palate is specialized in the loss of the ectopterygoid and the transverse flange of the pterygoid. The basicranial articulation between the braincase and the palate is still retained in the early Jurassic genera, and the interpterygoid vacuities are open.

The Upper Liassic (the lower division of the Jurassic) localities near Holzmaden in southern Germany provide the most striking record of ichthyosaur anatomy. In addition to several hundred completely articulated skeletons, many of the specimens show the outline of the body preserved as a carbonaceous film. These specimens demonstrate the configuration of the fleshy portion of the tail and a large, shark like dorsal fin. Many specimens show the young in *utero* or in the process of emerging from the birth canal.

Despite the diversity of ichthyosaurs in the Lower Jurassic, the group diminishes rapidly in the later Mesozoic. We know little of ichthyosaurs in the middle Jurassic. In the late Jurassic and early Cretaceous, one of the most common and widespread genera is *Ophthalmosaurus*, which is characterized by the enormous size of the orbits. The elements of the paddles are rounded rather than polygonal. The last ichthyosaur *Playpterygius*, is known from scattered remains throughout the world that cover a span of approximately 45 million years and extend to the base of the Upper Cretaceous.

The last of the ichthyosaurs disappeared from the fossil record long before the late Mesozoic extinction of the mosasaurs and plesiosaurs. Despite then earlier appearance, more rapid diversification, and higher degree of aquatic specialization the ichthyosaurs appear to have failed in competition with the other marine reptiles and the advanced sharks that were becoming dominant in the late Mesozoic.

16

First Aves

Many characteristics of birds show close resemblance to those of reptiles and in particular to the archosaurian diapsids. Already in the early Triassic period the small pseudosuchians such as *Saltoposuchus* showed the essential characteristics of the bird group, especially those associated with a bipedal habit. From some such form the birds have almost certainly been derived, by a series of changes parallel in many cases of those found in other descendants of the pseudosuchians, such as the crocodiles, dinosaurs, and pterosaurs.

Jurassic Birds

Modern birds show many reptilian features; but even closer to the archosaurians were the earliest birds, known only from two skeletons of animals about the size of a crow from the Jurassic lithographic stone of Germany. The remains, which pertain to two related general—*Archaeopteryx* and *Archaeornis*—so closely resemble those of some of the smaller bipedal dinosaurs that they might well have been taken for reptiles, were it not for the impressions of feathers which surround them on the stone slabs on which they are preserved. The skull, as far as it can be seen, was already rather birdlike, with an expanded braincase; and, as in modern birds, the sutures were mostly closed. However, well-developed teeth, implanted in sockets, were present in the jaws. None of the bones was pneumatic. The backbone was quite simple in structure, for the centra had the primitive amphicoelous character, and all were free as far as the sacrum, which included only five or six vertebrae. The tail still had the long structure of the typical dinosaur; and the feathers were arranged in two rows along its sides. As would be expected, the hind legs were quite like those of

Fig. 16.1. Fossil Archaeopteryx.

archosaurs, for they are but little changed even in modern birds. The wings had not yet completed their transformation. The sternum was not preserved but must have been small and could not have afforded much support for pectoral muscles (incidentally, ventral ribs were still present). The front limb was still a hand rather than a wing, for the three fingers which it possessed were very similar to those of the carnivorous dinosaurs; the metacarpals were quite separate, the number of joints was complete, and each finger was clawed. It would hardly be suspected that this appendage was used for flight, had not the impression of wing feathers been found back of the ulna and manus. The flying powers, however, must have been slight, for the spread of the wing was much less than that of rather poor fliers, such as the pheasant, among modern birds. These early forms probably were forest types and did little more than plane from tree to tree or to the ground.

Archaeopteryx and *Archaeornis* were already definitely birds but were still very close to the archosaurian reptiles in most of their structures, and it is obvious that they were descended from that group. Their exact ancestors among the small thecodonts which include the ancestors of all the later archosaurians are unknown. The structure of the pelvis suggests that they arose from a stock close to that which gave rise to the Ornithischia. There were, however, in the limbs many features suggestive of a parallelism, if not an actual relationship, to the saurischian dinosaurs.

There has been much discussion as to the origin of flight. The most generally held theory is that the ancestral bird was a tree dweller and that flight started as a slight parachute effect as the "proavis" jumped from branch to branch, the developing wings breaking the fall on landing. A second theory is that the ancestors were ground types and that the feathered arms and tail helped increase running speed by acting as planes.

The early birds of the Jurassic were obviously sharply marked off from later types by several features which are not met with in birds of later geological times. These include the absence of pneumaticity, the primitive reptilian structure of the wing bones, and especially the long reptilian tail. They are the only known forms included in the subclass Archaeornithes.

Cretaceous Toothed Birds

All other birds may be included in the subclass Neornithes, characterized by such features as a reduced tail with a fan of feathers, a well-developed sternum usually with a good keel, and the reduction and fusion of the metacarpals to the typical bird condition.

Our next glimpse of bird life is gained from the chalk beds of the Cretaceous, the only complete fossils begin from the marine Niobrara formation of Kansas. This is, in sense, unfortunate; for we know, in consequence, only sea birds and the more generalized land types are almost unknown. Two forms are known from adequate skeletons. One, Ichthyornis, was a small form about the size of a tern, seemingly rather similar to a tern in habits and not unlike it in many structural features. In Ichthyornis the vertebrae were still amphicoelous, but the wings were highly developed and quite like those of later birds and the sternum was large and keeled. Here, as in Archaeopteryx, teeth arranged in separate sockets were still present; and consequently we are dealing with a type intermediate between the Jurassic forms and modernized birds. A second and considerably larger bird from these deposits in *Hesperornis*. This remarkable form was probably much like the modern loon in habits, for it seems to have been a diving bird with powerful hind legs which spread out sidewise at the ankles to give a vigorous swimming stroke. But *Hesperornis* had been almost completely lost, and of the arm bones there remains only a slender humerus, while the sternum was unkeeled. This is the first of the many types of birds which have taken up a water life. As in *Ichthyornis*, teeth were present, although here they were placed in a groove rather than in sockets; and, also as in that form, there were no teeth in the premaxilla; probably the horny beak was already in process of formation.

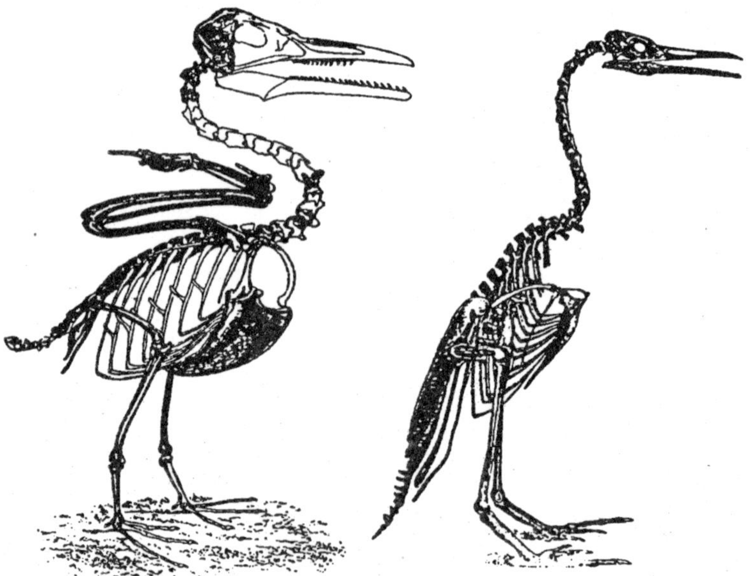

Fig. 16.2. Cretaceous toothed birds. Left, Ichthyornis, a small form with well-developed wings, about 8 inches in height. Right, Hesperornis, a flightless diver.

These two birds may be gathered together in superorder Odontoganathae, in contrast with all other post-Jurassic birds; and the two known types are certainly far enough apart, judging by avian standards, to be placed in two separate orders. The incomplete glimpse which they give us the Cretaceous bird life at least shows us that the radiation of the group must have been far under way at this time to have already produced such dissimilar forms.

Tertiary Birds

All later birds may be distinguished from those already discussed by several characters: loss of teeth; the fact that the two halves of the jaw, separate until then, are fused; and a union of the posterior end of the ischium with the ilium. Such birds, we know, had come into existence by the end of the Cretaceous. A large jaw from the Upper Cretaceous of Canada (*Caenagnathus*) is suggestive of a giant bird rather than a toothless dinosaur. There are, in addition, a few other Cretaceous fragments which may be those of toothless birds; but our remains of them are very poor, and it is not until the beginning of the Tertiary that we have much information about them.

And even here, it must be confessed, our knowledge is none too good. Birds are mostly small forms, and their bones and skulls are very fragile; consequently, we have but few remains, mostly of water birds. With mammals and reptiles we know more fossil forms than living ones; but among birds, as against about twenty thousand living species, we have but four or five hundred fossil forms. Even these are, for the most part known only from a fragmentary bone or so. This leads to great difficulties in attempting to work out the paleontological history of the group, for the birds of today, despite their varied plumage, are very similar to one another in their structure. They are divided into many orders; but the differences, for example, between a humming bird and an albatross are much less than those between a seal and a cat, or between a stegosaur and a duck-billed dinosaur, forms which are commonly placed in a single order. The different orders have, in general, no more differences between them than exist between families in other classes of vertebrates, and, anatomically, generic differences are so slight that fossils are very hard to place.

Palaeognathous Birds

Of all the post-Mesozoic birds, the most interesting from the paleontological point of view are the flightless forms represented today by the ostrich, rhea, cassowary, emu, and kiwi and including a large

Fig. 16.3. Restoration of Aepyorns, the elephant bird, which survived in Madagascar beyond the end of the Pleistocene.

number of interesting fossil types. These forms, here grouped together with the tinamous as a superorder, Palaeognathae, have a number of common features. Most of them, however, are associated with the fact that they have lost the power of flight; the wing skeleton in flightless forms is naturally reduced; the sternum, with weak pectoral musculature, lacks a keel, and the hind legs are powerful running organs. Since the tail is not used for flight, there is usually no pygostyle. The feathers, in correlation with the loss of flight, are usually soft and curly. But, in addition, some presumably primitive features are found in the lack of the distal fusion pubis and ischium found in most modern birds and in the palaeognathous palatal structure described above.

Perhaps representing the central stock from which the ostrich-like birds have arisen are the tinamous (Tinamiformes) living in South America today an almost unknown as fossils. These forms have the appearance of quail or grouse but are quite different from those game birds in their structure. They are essentially ground dwellers and take to the air only in an emergency. They are exceedingly poor fliers, having the little control over their flight and seldom going more than a hundred yards or so. Tail feathers are reduced or absent, but, unlike other ostrich-like birds, there is a keeled sternum.

Ratites

From such a stock may have come the remaining, flightless members of the superorder, often grouped as ratites. Most familiar are the ostriches (order Struthioniformes), the largest of living birds. Now found only in Africa and Arabia, they are known to have existed in Eurasia in the Pliocene, while eggs of a gigantic ostrich-like bird have been found in the Pleistocene in that region. A distinguishing feature of the ostrich, in contrast to other flightless types, is the fact that there are only two toes—the third and fourth.

Fig. 16.4. Rhea americana.

On the South American pampas the place of the ostrich is taken by the rhea (order Rheiformes), with a somewhat different structure (including a three-toed foot) and smaller size. Rheas are known from the Pleistocene, but farther back their history is quite unknown.

In Australia and New Guinea are the emus and cassowaries (order Casuariiformes). Among their distinguishing features is the fact that the wing is more reduced than in either of the last two orders, there being only one digit preserved, and this projecting but little from the body. These forms were present in the Pleistocene of Australia.

Fig. 16.5. Struthio camelus (Ostrich).

Still more interesting were certain fossil types present in the Pleistocene of two island regions. In Madagascar lived a number of species of "elephant birds" (*Aepyornis*, order Aepyornithiformes), some of them as large as an ostrich, with greatly reduced with and heavy legs. There have been found numerous eggshells, some with a capacity of two gallons. It is possible that these large birds lived until quite recent times, and their extermination may have been due to man.

New Zealand is remarkable today for flightless birds, for there are found only the kiwi but numerous flightless rails and other ground dwellers belonging to higher bird groups. In the past there were much larger ratites, the moas, which ranged in size from species no larger than a turkey to creatures 10–11 feet in height. Numerous remains of these birds have been found, and even feathers have been preserved. There is proof from native camp sites that these forms were still alive when man reached the island and that they were used for food.

Quite different from the ostrich-like groups in size and general appearance are the kiwis of New Zealand (order Apterygiformes). These small birds are also practically wingless ground types with a very long bill an small eyes. They are known as fossils from the Pleistocene of New Zealand and from Australia as well.

Fig. 16.6. Restoration of Phororhacos, a large, flightless bird that inhabited South America during the Miocene.

Origin of the Flightless Types

It has been suggested that these various groups of flightless birds are really primitive forms that have descended from types in which flight had never been developed. But upon reflection it is obvious that such a theory is highly improbable. Even in the Cretaceous the toothed *Icthyornis* had already developed powerful wings, and such a form as Hesperornis was obviously specialized rather than primitive. Further, we know that in many other higher groups of birds certain types have also lost the power of flight; the dodo, the great auk, and the flightless rail of New Zealand are all types which have reduced wings and cannot fly but yet are surely descended from flying forms.

The clue to the matter seems to be that, for many birds which seek their food on the ground, flight is necessary mainly as a protection from enemies. With freedom from carnivores, there is no reason why

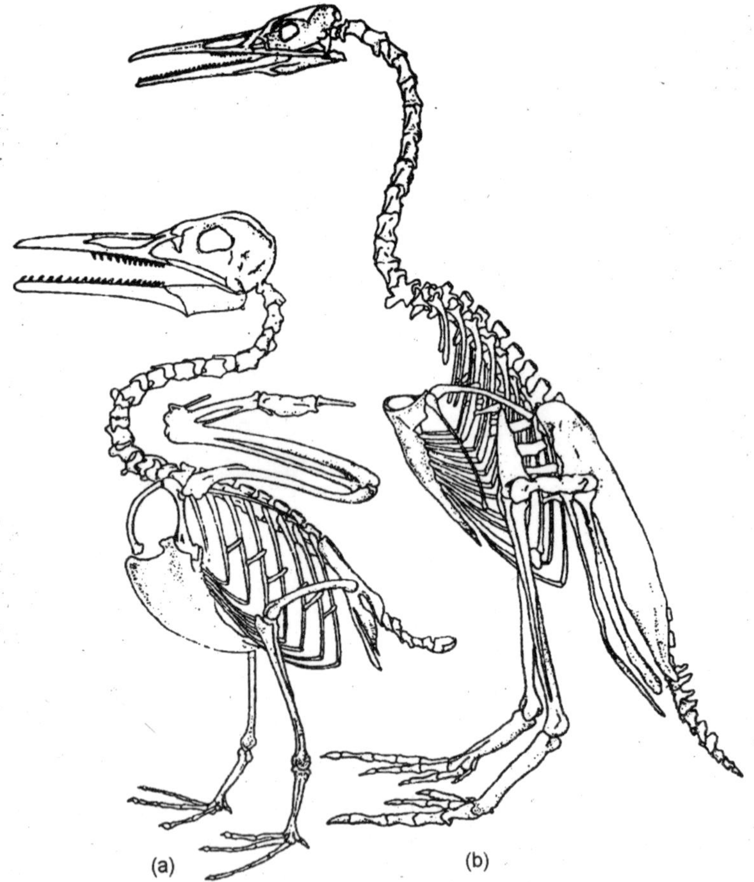

Fig. 16.7. Skeleton of Cretaceous birds. (a) Ichthyornis, with heavy, toothed mandible. (b) Hesperonis, a primitive toothed bird with reduced wings, apparently specialized for swimming and diving.

flight should be maintained. Almost all these flightless forms are so situated as to be practically free from enemies. There are no carnivorous mammals in New Zealand; until the arrival of man there was no ground-living enemy of any kind for the moas and kiwi. In Australia the only carnivores before the arrival of man and his dog were comparatively harmless primitive mammalian types (marsupials). The island of Madagascar is almost devoid of carnivores. The ostrich and rhea are in a worse situation, but they live in open country where carnivores cannot approach without being seen, and the speed of these forms enables them to outrun their enemies. It would thus seem that birds are likely to return to a ground life wherever conditions permit. When flight is thus abandoned, most of the common adaptations seen in these forms, such as reduction of wings and tail and development of powerful legs, would naturally follow. The resemblances between the various ostrich-like types need not, then, really indicate relationship; and below are mentioned other larger birds, now extinct, which in the past evolved along similar lines.

However, certain anatomical features, such as the palate, tend to show that these forms are really related. A probable suggestion is that there existed in the early Tertiary a group of birds in which teeth had been lost but the palate had remained primitive, and which had complete wings and a keeled sternum but were comparatively poor fliers. The living tinamous may be a comparatively unchanged relict of this primitive type. In parts of the world where favourable conditions existed, these forms might have tended to stay on the ground, lose the power of flight, and develop into ostrich-like forms; in other regions they would have lost out in completion with better flying types and have disappeared.

There is little fossil evidence for such a history; but, as has been said, our knowledge of early fossil birds is very inadequate.

Neognathous Birds

All remaining birds may be grouped together as the superorder Neognathae. In these forms there is but seldom any tendency toward a flightless condition; almost always the wings and tail are well developed.

Avian Flight

Flapping flight has evolved four different times during the course of evolution. The wings of different vertebrates represent examples of convergent evolution, and the actual structural details of the wing design are quite different in the three groups. Only birds employ a complicated series of overlapping epidermal derivatives (feathers) as the main wing surface.

Evolutionary Origin of Flight in Birds

Archaeopteryx was fully feathered, and this fact means that feathers themselves must have evolved much earlier in the nonflying ancestors of birds, probably under the influence of selective pressures associated with the acquisition of endothermy in this clade of archosaurs. It seems reasonable, then, to assume that the evolution of avian wings and flight began in a line of well-feathered, insulated, homeothermic, highly active, bipedal animals.

Archaeopteryx is usually considered to have been an arboreal climber, jumper, and glider with limited powers of flapping flight allowing it to extend the distance it could travel through the air between trees. It scampered about bipedally on the branches of trees, aided by grasping toes and reversed hallux, by the grasping claws on the leading edges of its wings, and by along, balancing tail. Most reconstructions and pictures of *Archaeopteryx* are based on this conception.

The arboreal theory of the origin of avian flight has been the most commonly held one. According to this theory, the proavian ancestors *Archaeopteryx* were tree-climbers, which jumped from branch to branch and from tree to tree, much like some squirrels, lizards, and monkeys do. Under selection pressures favouring adaptations that increase the distance and accuracy of travel between trees, structures what would aid in gliding by providing some surface area for life would be favoured. As a result, contour feathers on the arms elongated into primitive flight feathers, and the evolution of volant forms passed from gliding stages through intermediate stages, such as *Archaeopteryx*, in which gliding was aided by weak flapping flight, to fully airborne flapping fliers.

If, as seems reasonable form the fossil record of the thecodont and the structure of *Archaeopteryx* itself, the ancestral lineage giving rise to birds consisted of bipedal, cursorial forms, it is necessary to invoke arboreal selection pressures at all for the evolution of avian flight? Another much criticized theory but one as old as the arboreal theory, in fact, postulates that flapping flight evolved directly from ground-dwelling, bipedal runners.

According to this "*cursorial theory*" the proavians were fast, bipedal runners that used their primitive wings as planes to increase lift and lighten the load for running. In a later development, the wings were flapped as the animal ran to provide additional forward propulsion,

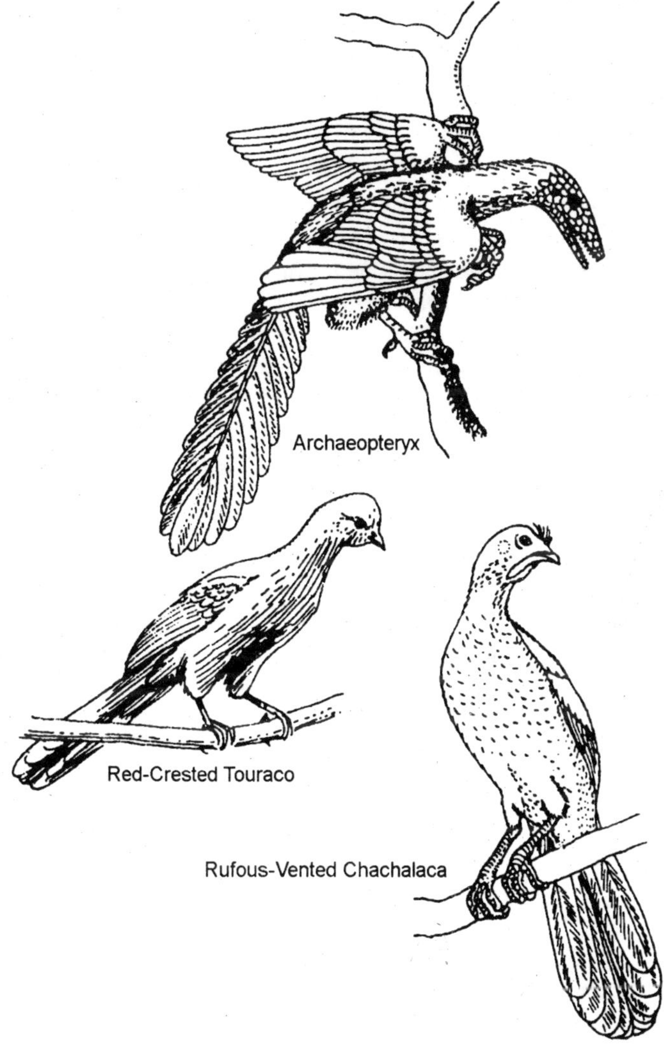

Fig. 16.8. Reconstruction of "arboreal" Archaeopteryx compared with chachalaca, and touraco.

much as a chicken flaps across the barnyard to escape from a dog. Finally, the pectoral muscles and flight feathers became sufficiently developed for full powered flight through the air.

The cursorial theory in its original form fails, however, in physical and mechanical terms as an explanation, because flapping is not an effective mechanism to increase running speed. Maximum traction on the ground is required for acceleration or high speed from leg movements, and this requirement can only be met by solid contact with the feet on a firm substrate. Planning with primitive wings would have reduced traction, and the push from the small surfaces of the protowings probably would not have compensated the loss in speed from the hindlegs, much less added to acceleration.

Gliding Flight

All birds can stretch their wings and glide, and some species depend primarily on this mode of flight to soar on rising air currents. The conditions of a stable glide, one in which air speed remains constant in still air, are like those for level flight, except that the flight path is inclined downward at some angle that is determined by the lift to drag ratio. The lift is inclined ahead of the vertical axis of the air-foil to produce a force in the direction of flight that is equal to the total drag. Obviously a glide can continue without loss of air speed only by loss of altitude, in which case the drag multiplied by distance traveled along the glide path will equal the loss of potential energy, which is the weight multiplied by the loss in height:

$$\frac{\text{weight}}{\text{rate}} = \frac{\text{distance moved}}{\text{height loss}}$$

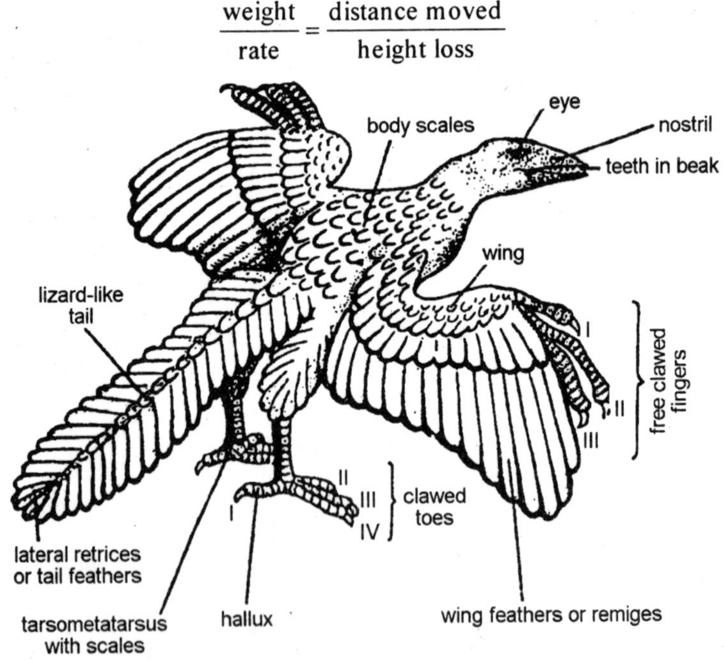

Fig. 16.9. Archaeopteryx. Restroration showing detailed structure.

The higher the L/D ratio, the smaller the glide angle has to be for a given air speed, and the lower the *sinking speed*, the height lost per unit time, and consequently the farther a bird can move horizontally per distance dropped. Since the *L/D* ratio is importantly influenced by the aspect ratio of the wing and by size and weight of the flying body, it can be expected that different species of birds will have vastly different gliding performances, depending on their sizes and the shapes of their wings.

Some studies of different kinds of birds have been carried out in wind tunnels or by observing the actions of wild birds in free flight in comparison with a sailplane flying under the same conditions, in order to determine the basic performance of birds during gliding. The ***glide polar*** is a curve showing the relationship between sinking speed and forward speed through the air and is a convenient way of summarizing the gliding performance of a bird or aircraft. Figure shows the main characteristics of the glide polar. The forward speed at which loss of height is minimal, the speed for minimum sink (V_{ms}), is somewhat higher than Vmin. Still higher is the speed for best glide ration (V_{bg}), which can be determined by drawing a tangent to the curve from the origin of the graph. By flying at V_{bg} a bird or aircraft covers the greatest horizontal distance for a given loss of height. Figure presents generalized glide polars for several species of birds and compares them with the performances of the astomite model airplane and the SHK sailplane.

Because of its very much larger size and wingspan, the sailplane has a far superior *L/D* ratio than any bird that has been studied, approaching 40:1 at air speeds of 20 to 25 m/sec. Some vultures approach the *L/D* ratio of the sailplane, and it can be estimated on the basis of known aspect ratios and flying speed that albatrosses probably have *L/D* ratios in the range of 30 : 1 to 40 : 1. Again, because of its size and weight, the sailplane has to maintain a faster forward air speed than the birds, and its minimum gliding speed is around 18 m/sec, will above the speed for best glide ratios of most birds. Birds have much lower stalling speeds and optimum air speeds for gliding because of their lighter wing loading (again primarily a function of size). In general, soaring birds, such as vultures and albatrosses, have gliding performances superior to birds like pigeons and falcons that rely mainly on flapping flight.

As a mode of locomotion through air, gliding flight can function in two ways. One is to cover distance over ground to a given destination along a single glide path, as when a vulture launches from the top of a cliff in still air and glides to some point on the ground below. The other is to remain aloft by soaring, that is, by gliding in air that is rising faster than the sinking speed of the bird. Any aircraft or bird accomplishes these two goals to the degree that is determined by its design and prevailing atmospheric conditions. A design that is optimal for one goal is not optimal for the other, and this is the basic meaning of the different glide polars.

Birds have lower *L/D* ratios and lower velocities than the sailplane, and therefore their performances are inferior when the goal is to achieve maximum distance along a glide path in still air, although again birds that are primarily adapted for soaring have better performances than birds that rely on powered flight. When static soaring is the goal, however, then some birds that rely on powered flight. When static soaring is the goal, however, then some birds are as good or better than the sailplane. The best avian soarers have lighter wing loadings, lower stalling speeds, and lower sinking speeds than the sailplane, and for these reasons birds can frequently rise in upward moving air when a sailplane cannot do so. The sailplane has two disadvantages compared with the birds; because of its heavier wing loading, it can only gain altitude in air that is rising considerably faster than that required to life a bird; because of its larger size, it has a much larger turning radius than a bird and consequently it cannot maneuver as easily to take advantage of small volumes of rising air (thermals).

Flapping Flight

Flapping flight is remarkable for its automatic, unlearned performance. A young bird on its maiden flight uses a form of locomotion so complex that it defies precise analysis in physical and aerodynamic terms. The nestlings of some species; such as hornbills, some owls, swifts, swallows, some parrots, wood-peckers, petrels, and other develop in confined spaces (burrows, cavities) in which it is impossible for them to spread out their wings and practice flapping before, they leave the nest. Despite this seeming handicap, many of them are capable of flying considerable distances on their first flights. A young African bank martin (swallow), for example, flew continually for 6 min the first time it left its nest. Young whale birds and diving petrels (Procellariiformes) having been observed to fly as far as 10 km the first time out of their burrows. On the other hand, young birds reared in more open nests frequently flap their wings vigorously in the wind for several days before flying—especially large birds such as albatrosses, storks, vultures, and eagles. Such flapping may help to develop musculature, but it is doubtful whether these birds are "learning how to fly"; however, a bird's flying abilities do improve with practice for a period after it leaves the nest.

There are so many variables involved in flapping flight that it becomes difficult to understand exactly how it works. A beating wing is flexible and porous and yields to air pressure, unlike the fixed wing of an airplane. Its shape, expense, wing loading, camber, sweepback, and the position of the individual feathers all change remarkably as a wing moves through its cycle of locomotion. Furthermore, the frequency and amplitude of the beats change in velocity and angle of attack during a single cycle. This is indeed a formidable list of variables, and it is no wonder that flapping flight has not yet fully yielded to explanation in aerodynamic terms; however, the general properties of a flapping wing can be described.

We can begin by considering the flapping cycle of a small bird in flight. According to the forces diagrammed in Figure 16–28, a bird cannot continue to fly straight and level unless it can develop a force or "thrust" to balance the drag operating against forward momentum. The flapping of the wings, especially the wing tips (primaries), produces this thrust, while the inner wings (secondaries) do not move very fast and continue to act as an airfoil generating lift as when the bird is gliding. It is easiest to consider the forces operating on the inner and outer wing separately. Most of the lift, but also most of the drag, comes from the forces acting on the inner wing and body of a flying bird. The forces of the wing tips derive from two motions that have to be added together. The tips are moving forward with the bird, but at the same time they are also moving roughly downward relative to the bird. The wing tip would have a very large angle of attack and would stall if it were not flexible. As it is, the forces on the tip cause the individual primaries to twist as the wing is flapped downward and to produce the forces. The forces acting on the two parts of the bird combine to produce the condition for equilibrium flight.

As the wings move downward and forward on the downstroke, which is the power stroke, the trailing edges of the primaries bend upward air pressure, and each feather acts as an individual propeller biting into the air and generating thrust. During this downbeat, the thrust is greater than the total drag, and the bird accelerates. In small birds, the return stroke, which is upward and backward, provides little or no thrust and is mainly a passive recovery stroke. The bird decelerates during the recovery stroke.

Contraction of the pectoralis major, the large breast muscle, produces the downstroke during level flapping flight, whereas the upstroke results from the reaction of the air on the wing as modified by the relaxing, and controlling movements of the pectoralis major. During the upstroke, also, the primaries twist and open in a venetian-blind manner that allows them to slip through the air with little resistance. This twisting of the primaries in the down stroke and upstroke results from the asymmetry of the leading and trailing vanes.

In larger birds with slower wing actions, the time of the upstroke is too long to spend in a state of deceleration. A similar situation exists when any bird takes off: it needs thrust on both the downstroke and the upstroke. Thrust on the upstroke is produced by bending the wings slightly at the writs and elbow and by rotating the humerus. This movement causes the upper surfaces of the twisted primaries to push against the air and to produce thrust, as their lower surfaces did in the downstroke. In this type of flight, the wing tip described a rough figure-of-eight through the air.

This powered upstroke results from the contraction of the *supracoradoideus*, a deep muscle underlying the pectoralis major and

Fig. 16.10. Drawing from high speed photograph of a chickades, showing the twisting and opening of the primaries during the upstroke in flapping flight. Further forward rotation of the wrist would give the twisted primaries a positive angle of attack on the upstroke with a resultant production of thrust.

attached directly to the carina of the sternum. In most species of birds the supracoracoideus is a relatively small, pale muscle with low myoglobin content, easily fatigues. In species that rely more on a powered upstroke, either for fast, steep take off or for hovering, as in humming birds, or for fast aerial pursuit as in swifts, the supracoracoideus is relatively larger. The ratio of weights between the pectoralis major and the supracoracoideus is a good indication of a birds's reliance on a powered upstroke; such ratios vary from 3 : 1 to 20 : 1. The total weight of the flight muscles is also indicative of the extent to which a bird depends upon powered flight. Strong fliers like pigeons and falcons have breast muscles comprising more than 30 percent of body weight, whereas in some owls, which have very light wing loadings, the flight muscles make up less than 10 percent of total weight.

17

First Mammals

The idea that the mammals include the highest of animals can easily be ridiculed as a product of human vanity; we shall find, however, that in a many aspects of their structure and activities they do indeed stand apart as animals that are 'higher' than others, in the sense in which we have used the word throughout this study. The mammals and the birds are the vertebrates that have become most fully suited for life on land; among them are many species in which the processes of life are carried on under conditions for remote from those in which life first arose. The information in their DNA (deoxyribonucleic acid) provides them with numerous special adaptive devices, not the least being brain capacities that enable them to collect further information during their individual lifetimes. The mammalian organization includes a great number of special features that together enable life to be supported under conditions that seem to be extravagantly improbable or 'difficult'. For example, the surface of the body is waterproofed, and elaborate devices for obtaining water are developed. A camel and the men he is carrying through the desert may perhaps contain more water than is to be found in the air and sandy wastes around. This is only an extreme example of the 'improbability' of mammalian life, which is one of its most characteristic features.

Characteristics of Mammals

The faculty of maintaining a high and constant temperature has opened to the birds and mammals many habitats that were closed to the reptiles. Besides making life possible under extremely cold conditions, such as those of the polar regions, the warm blood vastly extends the opportunities for life in more temperature climates. Mammalian populations, which can feed at night and all through the winter, can of course expand more rapidly than their reptilian cousins, which must hibernate for much of the year, during which period they consume rather than produce living matter.

The success of the mammals is maintaining life in the strange environments is largely due to the remarkable powers they possess of keeping their own composition constant. All living things tend to do this, but it seems probable that the mammals maintain a great constancy than any other animals, except perhaps the birds. Claude Bernard's famous dictum 'La fixite du milieu interieur c'est la condition de la vie libre' (1878) may be doubtful of vertebrates in general, but it can certainly be applied to mammals. They have a life that is more free than that of other groups, in the sense that they can exist and grow in circumstances that other forms of life would not tolerate, and they can do this because of the elaborate mechanisms by which their composition is kept constant. Besides the regulation of temperature there is also a regulation of nearly all the components of the blood, which are kept constant within narrow limits. Barcroft (1932) has pointed out that the achievement of this constancy has enabled the mammals to develop some parts of their organization in ways not possible in lower forms. For instance, an elaborate pattern of cerebral activities requires that there shall be no disturbances by sudden fluctuations in the pressure and composition of the blood.

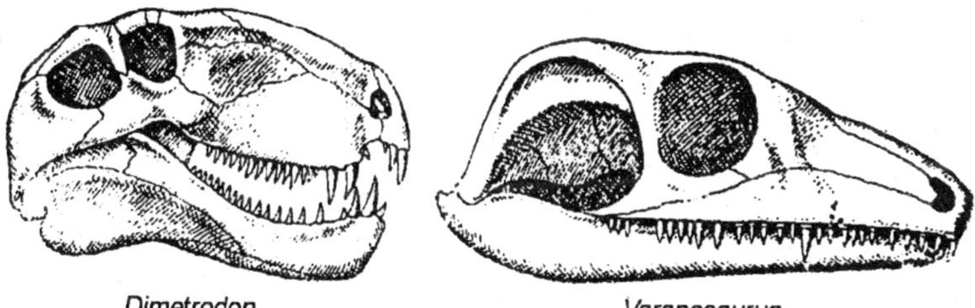

Fig. 17.1. Skull of two early synapsis, that of Dimetrodon was about 40 cm long and of Varanosaurus 23 cm.

We shall expect to find in the mammals, therefore, even more devices for correcting the possible effects of external change than are found in other groups. Besides means for regulating such features as those mentioned above we shall find that the receptors are especially sensitive and the motor mechanisms able to produce remarkable adjustments of the environment to suit the organism, culminating in man with his astonishing perception of the 'World' around him and his powers of altering the whole fabric of the surface of large parts of the earth to suit his needs.

Such devices for maintaining stability often take peculiar and specialized forms in particular cases. The activity and 'enterprise' of mammals has led many of them to make use of particular structures and tendencies in order to develop very odd specializations, which enable them to occupy peculiar niches. What could be more bizarre than the development to the muscles of the nose until a huge mobile trunk appears, so that the heaviest of four-footed beasts, while using its legs for support, can also 'handle' objects more delicately than almost any other animal?

The mammals have developed along many special lines and many of these have already become extinct; others, especially among rodents and primates, remain among the dominant land-animals today. Warmth, enterprise, ingenuity, and care of the young have been the basis of mammalian success throughout their history. The most characteristic features of the modern mammals are thus seen to be largely in their behaviour and soft structures. Mammalian life is above all see active and exploratory. Mammals might well be defined as highly percipient and mobile animals, with large brains, spiral cochlea, warm blood, left aortic arch, and a waterproofed, usually hairy skin, whose young are born alive, and are nourished by milk. Since it is difficult to recognize such characters as these in fossils we cannot say exactly when they arose, and our technical definition of a mammal must be made on the basis of hard parts. These include tribosphenic molars, a dentary - squamosal articulation and three ossicles in the middle ear. The side wall of the skull is formed by the alisphenoid bone. There is an axis–atlas complex and seven cervical vertebrae. The pelvis has an anterior iliac blade and small pubis. Some of these features were not yet present in the earliest mammals, the Prototheria.

The whole series of mammal-like forms from Carboniferous anapsids onwards forms that we separate the reptilian subclass Synapsida from the class Mammalia. The present day mammals form a distinct group of animals, which we identify superficially by their possession of hair. For instance, the duck-billed platypus is immediately referred because of its hair to the mammals. We class it apart from reptiles or birds, even though its internal organization and the fact that it lays eggs show it to have many similarities with reptiles, and its bill is like that of a duck. The technical characteristic of the class Mammalia is conventionally given by the presence of a single dentary bone in the lower jaw, the articular and other bones forming, with the quadrate, part of the mechanism of the middle ear. However, fossils showing intermediate conditions are now known from the upper Triassic, say 180 million years ago, and all stages can thus be traced in the jaws. The full mammal-like condition was established by the middle Jurassic period, about 150 million years ago. The reduction of the jaw bones may perhaps have been associated with the habit of chewing the food, in order to obtain the large amounts necessary to maintain a high temperature.

Origin of Mammals

Stock

At least two views have been held as to the origin of mammals. The older one, that advocated by Huxley in 1880, would derive them from the Amphibia. Much in Huxley's theory is undoubtedly correct, except that authorities do not now believe that the Amphibia represent the next lower stage, but that there was an intervening condition, one in which gill-breathing had been lost but truly mammalian characters had not yet appeared, although some of them were already foreshadowed.

There are found in Triassic rocks in South Africa a group of reptiles known as the Cynodontia, in allusion to their dog-like teeth, and there seems to be a large body of evidence in favour of the view that out of this group, although from no known member of it, the mammals have been derived. The cynodonts resemble the mammals in the possession of a heterodont dentition, in that the teeth are clearly divisible into incisors, canines, and molars, and in the paired occipital condyles, in addition to which there are similarities of construction. Structurally the cynodonts bridge the gap between reptiles and mammals because while the dentary, the single bone of the mammalian lower jaw, is large and important, the jaw is nevertheless complex in that it possesses the several bones typical of the reptile. There are other reptilian characters as well, all of which may reasonably be expected in the remote ancestors of the Mammalia. These cynodont reptiles are low in the reptilian scale, far removed from the dinosaurs and birds with which we have been concerned, although capable of as high a degree of specialization along other lines.

Place of Origin

While we find members of this order in both North America and Africa, although they must have been much more widely diffused, it was possibly in the latter place that the mammals arose, probably due

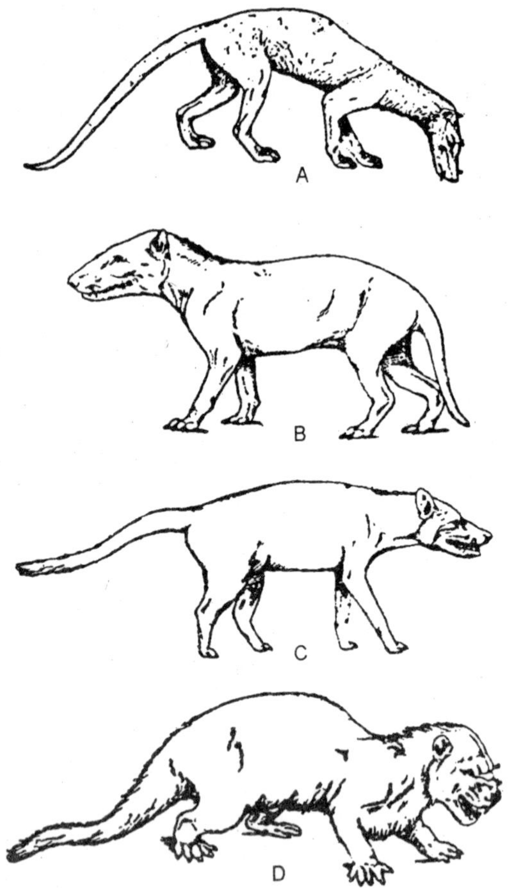

Fig. 17.2. Creodonts. A–Tritemnodon, a primitive hyaenodont, Middle Eocene, North America. B–Hyaenodon, the last survivor of the archaic carnivores, Lower Oligocene, North America and Old World. C–The dog-like Dromocyon, Middle Eocene, North America. D–Patriofelis, Middle Eocene, North America.

not so much to the potentiality possessed by the reptiles of one place over the other as to a happy combination in Africa of a potent stock and an impelling cause.

Cause

According to Broom that all of the characters where in a mammal differs from a reptile are the result of increased activity, for he says that when the cynodont took to walking with feet underneath and body off the ground, it first became possible for it to become a warm-blooded animal. But back of this lay impelling geologic cause which

Fig. 17.3. Coryphodon, a swamp-dwelling amblypod, Lower Eocene, North America.

Broom does not even hint at. These were, first, increasing aridity of climate, which from Permian time well into the Jurassic seems to have been characteristic of all lands, as the extensive series of red sediments may imply. And aridity has been found to be a great stimulative to speed. Add to this the evidence, especially in the southern hemisphere between latitudes 20° and 35° of extensive glaciation—greater even than in the more familiar Pleistocene Glacial period-and we have a high incentive to the retention of bodily heat. For it is well known that cold more than any other factor limits the activity of reptiles and effectively prevents their distribution into the high latitudes. For a while perhaps, there were recurring warm seasons, sufficiently long and frequent so that a normal reptilian life could still be led, the creature hibernating when the weather became too serve. But, as in the case of the aestivating lung-breeding fishes, sufficient time must still be had for the active portion of the creature's career, so that, although in the origin of terrestrial forms premium would be placed upon the capability of atmospheric respiration, here it would be upon ability to withstand the cold and yet remain active, and the acquisition of warm blood and a heat retentive clothing is the only possible means to this end. Hence as immediate geologic causes of mammalian evolution we have, first, aridity, the incentive for speed, rendering possible the development of warm blood and second, the increasing cold to place a premium upon such as did develop it and to eliminate those which did not.

Time

The time of mammalian evolution can only be fixed within certain limits. Geological evidence, if we have read the cause a right, points to its inreption in Permian time for "there is now undeniable evidence that in middle Permian times most of the lands in the southern hemisphere were subjected to as severe a glacial climate as that of

Fig. 17.4. Horned amblypod, Uintatherium, the culmination of its rare, Upper Eocene Wyoming.

the present polar areas, and that this Permian glaciation was likewise interrupted by times of warmth" (Schuchert).

On the other hand, the geologic record seems to point to the Triassic at any rate as the time of the *culmination* of the evolutionary movement, for here are found for the first-time the cynodont reptiles and the actual relies of the mammals themselves. But the cause must always precede the effect and it may be that the known cynodonts were persistent reptilian survivors of the group out of which the mammals actually sprang, and that the earliest known mammals themselves had already had a long transitional period. It seems reasonable to suppose therefore, that the time of mammalian origin was not later than Middle Triassic nor earlier than Lower Permian.

Mammals of the Mesozoic

In spite of all the uncertainties of the fossil record it is now possible to follow the history of the Mammalia back to their origin from cotylosaurian reptiles of the Permian, more than 225 Ma ago. Sufficient information is available for us to be quite sure that some population of early anapsid reptiles, such as *Solenodonsaurus* of the Carboniferous and Permian times, besides giving rise to all the modern reptiles, to the dinosaurs, and to the birds, also produced the mammals. The evidence for this connection rests on a most interesting series of fossils, together with some 'living fossils', the monotremes of Australia. The fossil history is not at all times equally clear. The mammalian stock first became distinct in late Carboniferous times, as a special type of cotylosaurian reptile, with a tympanum placed behind the jaw and later a lateral temporal aperture. This stock quickly became very abundant and successful as the pelycosaurs and therapsids of Permian and Triassic times, but then nearly died out in the Jurassic, during which period we know of the mammals only from fragmentary remains of a few small animals. Then, in the Cretaceous, some of these small forms became more numerous and from them arose, before the Eocene, a variety of different types of mammals, from which the histories of the modern orders can be followed in some detail.

We have, therefore, abundant evidence of the earliest stages of mammalian evolution, say 200 Ma ago from fossils in the Permian Red Beds of Texas. For the next following 40 Ma or so we have also rich material from the Upper Permian and Triassic Karroo beds of South Africa. In these early times the mammal line was a flourishing one, more so indeed than the diapsids or any other of the descendants of the cotylosaurs. The animals were mostly carnivorous, though there were also herbivorous types. Many of these early mammal-like forms dominated the land scene in the Permian. In the later Triassic, however, if our fossil evidence is a safe guide, their numbers became reduced, perhaps the carnivorous dinosaurs took their place. Already at this time many of the essential features of mammalian organization had been developed, so far as these can be judged from the bones. Possibly the soft parts of these Permian forms were also mammal-like, the animals may have been active and 'intelligent', have possessed hair and warm blood. But their brains were small and reptilian in structure.

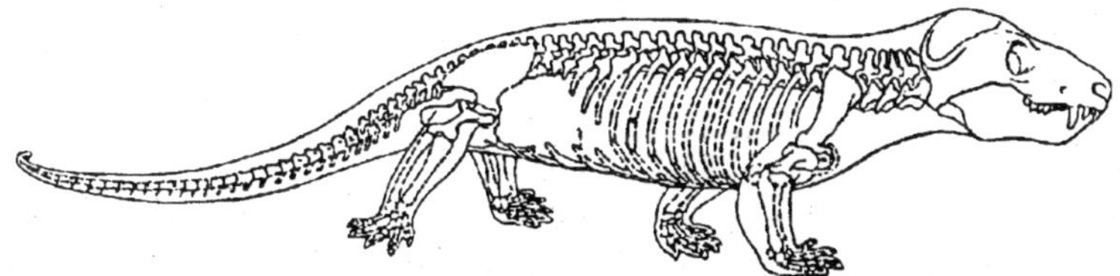

Fig. 17.5. The skeleton of the Lower Triassic cynodont cynognathus.

Throughout the succeeding 100 Ma of Jurassic and Cretaceous times our knowledge of the mammalian stock is dim. Mammals of a somewhat rodent-like type, the multituberculates, became quite numerous and produced some large forms, but then became extinct in the Eocene. Unless we have been singularly unlucky in the preservation of the mammalian remains, no other mammal-like animals larger than a polecat existed throughout this long period. Since we possess detailed information, based on numerous fossils, about scores of large and small diapsid reptilian types, it can hardly be only an accident that mammal-like fossils are so rare. We must conclude that the mammalian organization, after an initial success in the Permian and Triassic, was almost supplanted in the Jurassic and Cretaceous by the various diapsid creatures. Only a few are fossils, mostly of teeth and lower jaw bones, to give us some idea of the nature of the animals that carried our type of organization through the Jurassic period. Then from the top of the Cretaceous, about 70 Ma ago come the first fossil remains of mammals similar to those alive today; rare, insectivorous creatures, not widely different from modern hedgehogs. This was the time of the beginning of an astonishing revolution, which completely altered the life on the face of the earth. The descendants of these shrew-like animals multiplied considerably in the new conditions, and by the earliest Palaeocene times, 65 million years ago, they had produced recognizable ancestors of most of the modern orders of mammals.

Our knowledge of the origins of mammals derived from fossils in supplemented by certain surviving mammals, whose structure shows them to have diverged rather early from the main stock, especially the monotremes (duck-billed platypus and spiny anteater) and the marsupials (kangaroos, opossums, etc.). Unfortunately it is still not clear exactly how these survivors are related to the main stocks as revealed by the fossils. The monotremes are almost unknown except as Pleistocene fossils; so, tantalizingly, we know only little about the affinities of this ancient group. The characteristics of their body skeleton show that they must have diverged from other mammals well back in Mesozoic times. For instance, the pelvic and pectoral girdles are very 'reptilian' in structure. It is therefore not wild speculation to use the characters of the soft parts of monotremes to deduce those of the mammalian stock in late Triassic or early Jurassic times, say 195 Ma ago. The marsupials appears as fossils in the late Cretaceous. Some of the earliest forms are very like modern opossums, and we may conclude that the condition of modern marsupials throws some light on the probable condition of the soft parts of mammals 70 or 80 Ma ago. However, it has been realized in recent years that many features in marsupial organization are special developments, not 'primitive' characteristics carried over from the ancestral condition.

The piecing together of all the evidence to give a reliable picture of the history of mammals is a valuable exercise in zoological and geological method, as well as a means of becoming familiar with a fascinating story. We may now return to the Carboniferous times to examine the evidence more in detail.

Mammal-like Reptiles, Synapsida

All our evidence about the origin of mammals must of course be based upon the study of hard parts, which can become fossilized, and in particular on the skull. The characteristic feature of the skull in the populations that led to the mammals is the development of a hole in the roof, low down on the side of the temporal region, the lower temporal fossa. This hole was at first bounded above by the post-orbital and squamosal bones, below by the jugal and squamosal. It was at one time supposed that this single fossa was formed by union of the two present in diapsid reptiles, and thus the whole group acquired the inappropriate name Synapsida. Animals having this characteristic appear in the rocks at the same time as the cotylosaurs, with no hole in the skull roof (Anapsida), from which they were presumably derived. We have only fragmentary knowledge of anapsids from the Carboniferous; most of our information comes from lower Permian forms, such as ***Solenodonsaurus***, yet there are quiet well preserved synapsids of Upper Carboniferous date. Therefore we have no complete series of successive types to show how the earliest mammal-like types arose; nevertheless we can see something of the probable stages of this progress within the Permian Anapsida.

One of the characteristics of the skull of amphibians was the presence of an 'oticnotch', in which lay the tympanic membrane, at the back of the skull. In *Captorhinus* and similar anapsid fossils from the Lower Permian this notch had disappeared. The tympanum apparently lay behind the quadrate, leaving the whose side of the skull as a rigid support for the jaw. A long series of subsequent evolutionary

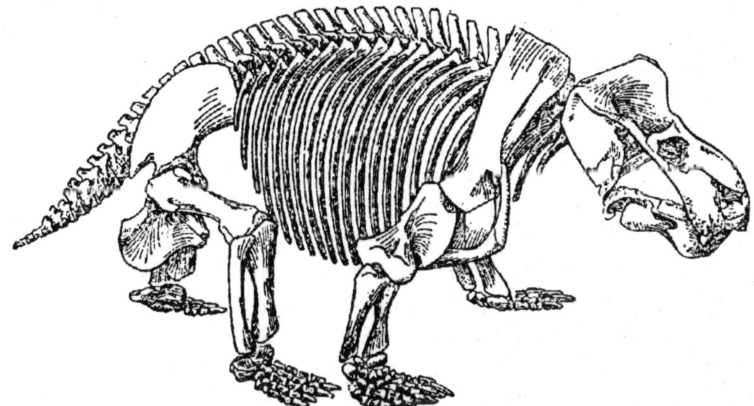

Fig. 17.6. The skeleton of Kannemeyeria, a Lower Triassic dicynodont.

changes led to one of the most characteristic features of the skeleton of mammals, namely, the conversion of the hinder jaw bones (quadrate and articular) into ossicles for the transmission of vibrations to the ear.

Captorhinus and its allies were 30 cm or more in length and showed some development of the limbs toward the mammalian condition, in that the bones were moderately elongated and slender and the limbs perhaps tended to be held vertically under the body. However, there was no great development of a backwardly directed elbow and forward knee. The teeth were considerably specialized, possibly for eating molluscs, and there were several rows of crushing teeth on the edge of the jaw, and long overhanging ones in the premaxillae.

Pelycosauria (=Theromorpha, Mammal-like)

The synapsid line began when an early offshoot from some such anapsid as *Captorhinus* developed a temporal fossa. This must have occurred in the middle of the Carboniferous, nearly 300 Ma ago, producing in the later part of that period, and in various Pennsylvanian and Lower Permian strata, especially in North America, there earliest synapsid populations, classified together as pelycosaurs or theromorphs. *Varanosaurus* from the Texas Red Beds, is a typical example, a carnivore, about 90 cm long, showing a lizard-like appearance little different from that of anapsids or primitive diapsids. Intercentra were found all along the vertebral column, as in anapsids, and there were abdominal ribs. The teeth showed the beginnings of the mammalian differentiations that one or more near the front of the series were elongated as 'canines'. The skull of pelycosaurs also showed tendencies in the mammalian direction in having a long anterior region and a relatively high posterior part, giving a large brain case and deep jaw.

Other pelycosaurs developed long neural spines, in more than one line *Edaphosaurus* and its allies being herbivores, some of them as much as 360 cm long. Whereas *Dimetrodon* was a carnivore. The spines presumably supported a web, perhaps used in temperature regulation to absorb or radiate heat. Which would explain that it was relatively bigger in some of the larger animals. These animals were important in the late Carboniferous and early Permian fauna, the carnivores preying on the order early reptiles and on rhachitomous amphibia. They did not survive long, however, being apparently replaced by their descendants. From some form similar to *Dimetrodon* arose a whole series of lines classed together as Therapsida and leading on to the mammals.

Therapsida (= Mammal-Arched)

In these animals there was never any trace of the bipedalism that developed in the Archosauria; the fore limbs were always at least as long as the hind, and tended to be turned under the body and to carry it off the ground. The skull became deeper and its brain case enlarged; a bony secondary palate developed from flanges of the premaxillae, maxillae, and pterygoids. The teeth became differentiated for various functions and the bones of the lower jaw, except the dentary, were gradually reduced.

These features seem to have developed in several different lines descended from theromorph ancestors; the sorting out of the various genealogies is not yet complete. It is therefore still difficult to decide for certain the interesting question whether parallel evolution occurred, and especially whether similar mammalian features appeared independently in animals of different habits. The fossils are nearly all found in South Africa and have been studied in great detail by Broom (1932).

Much of the surface of the southern part of Africa is covered by rock known as the Karroo system. These consist partly of shales and mud-stones formed by the matter brought down by a large Mesozoic river, and the remainder are sandstones composed of blown sand. Both

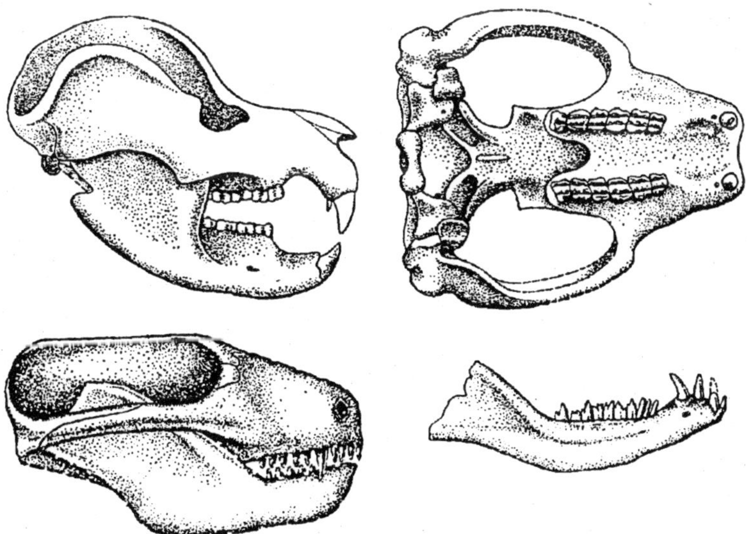

Fig. 17.7. Ictidosaurians. Above, lateral and palatal views of the skull of Bienotherium, a tritylodontid from the Triassic of China.

sorts of rock were particularly favourable for the preservation of the remains of terrestrial animals. Unfortunately the absence of marine fossils makes it difficult to give dates for these rocks. Altogether the strata present a thickness of some 4500 m, laid down over a period corresponding probably to that from the Middle Permian to the Upper Triassic in Europe, that is to say, from 250 to 180 Ma ago.

The therapsids fall into two groups, the mainly herbivorous Anomodontia and the carnivorous Theriodontia, probably preying upon the former. *Galepus* shows a stage of evolution of anomodonts. The temporal opening was small. The teeth all alike and the bones at the hind end of the jaw large. The early anomodonts include the Dinocephalia (=terrible-heads), such as *Moschops*, which retained many primitive features, but the legs were turned under the body and the phalanges became reduced to the mammalian formula of 2.3.3.3.3 The roof of the head was expanded into a large dome, giving the name to the group. The later anomodonts were a still more specialized and successful group, the first of the many great tetrapod herbivores. They became the commonest of all reptiles in the later Permian. They were large creatures (*Kannemeyeria* and *Dicynodon*), some probably living in marshes. The margins of the jaws were covered worth a horny beak and the teeth reduced to a single pair of upper 'canine' tusks (from which they get their name two-tuskers), and even these were absent from some.

The most interesting of the therapsid reptiles are, however, those placed in the suborder *Theriodontia*, which, by the early Triassic, had produced a very mammal-like type of organization, apparently in several independent lines. The mammals themselves very likely arose from one of the, though we do not know which. The temporal opening became progressively wider, presumably for the accommodation of larger jaw muscles, so that the parietal bone entered into the margin of the fossa and post-orbital and squamosal bones no longer met above it. Eventually the post-orbital bar itself became incomplete, leading to the typical mammalian condition, with the orbit and temporal fossa confluent. The jaw elements other than the dentary became reduced, but never wholly lost. There was a large columella, articulating broadly with the quadrate and perhaps serving to brace the latter as well as to transmit vibrations to the inner ear. A bone of this size could act as an efficient vibrator. The brain case was high and large and the cerebellum and cerebral hemispheres better developed than in modern reptiles. The occipital condyle became double as in mammals. A

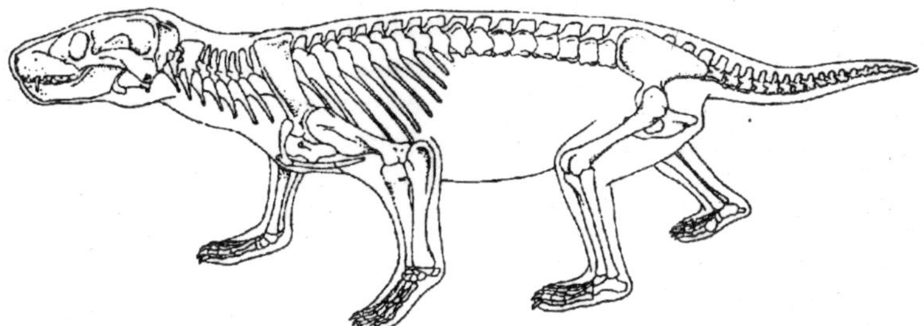

Fig. 17.8. Skeleton of Thrinaxodon, a mammal-like reptile.

secondary palate became developed, allowing breathing to continue while holding the prey.

The ribs were well developed and seem to have formed a cage, which may have been used, with a diaphragm, for respiration. The limbs came to support the body off the ground, and the dorsal parts of the girdles were developed accordingly, the scapula becoming large and bearing an acromial spine for muscle attachments, the coracoids being reduced. The anterior portion of the ilium became large and a hole appeared between the pubis and ischium. The head of the femur lay at the side, and a special knob, the great trochanter, appeared on it for the attachment of gluteal muscles running from the front part of the ileum and producing a backward thrust. In the hand and feet the phalanges show a reduction in several distinct evolutionary lines to 2.3.3.3.3. The teeth became differentiated into incisors, canines, and cheek teeth, the latter having several cusps in place of the single cones of a typical reptile tooth. In most forms the tooth replacement was serial as in reptiles but in some it was more limited.

These features were little developed in the earliest therio-donts, such as *Scymnognathus* and other Permian forms. The temporal fossa still resembles that of pelycosaurs, there was a single condyle, no secondary palate, and a phalangeal formula of 2.3.4.5.3 *Cynognathus* and other Triassic cynodonts (=dog-toothed) were typical theriodonts, showing the above 'mammalian' features. They were carnivores, of distinctly dog-fish appearance, although they still retained many signs of a heavy reptilian build. *Bauria* was of another type, still more advanced than *Cynognathus* in that the orbit and temporal fossa were confluent. The foramina of the maxilla of some of these animals suggest the passage of nerves and blood-vessels for a facial musculature, which is characteristic of mammals but absent in other vertebrates. Other types of theriodont were more specialized, with rodent-like features (*Tritylodon*, *Oligokyphus*).

These *Theriodontia* were the dominant carnivores of the early Triassic, but by the end of that period that had almost disappeared. The latest synapsids include in some classifications the Ictidosauria, such as *Diarthrognathus* of the Upper Triassic. They are rare fossils, showing almost completely mammal-like structure. There was no post-orbital bar and no prefrontal, post-frontal, or post-orbital bones. A well developed secondary palate was present. The bones at the hind end of the lower jaw had become very small. In at least one form the dentary probably articulated with the squamosal, though there was also a quadrate-articular joint. This fossil has been appropriately named *Diarthrognathus*. The double articulation by more medial quadrate-articular and lateral squamosal-dentary may have served to resist the forces produced by a shearing dentition. It is found again in the triconodonts. In the fully mammalian condition the medial articulation is lost.

Study of the synapsids, mainly from the Karroo system, shows us, therefore, a series of types, of which the earliest were very like the first reptiles and the latest very like true mammals. There can be no doubt of the general tendency, but the series is not complete enough to enable us to follow the details of the evolution of the populations. These fossils are revealed by denudation and can be given only approximate dates. It is certain that some of the mammal-like features appeared independently in lines whose evolution proceeded separately from a common ancestor. Thus the dicynodonts and later theriodonts all had the mammalian phalangeal numbers, but each of these lines had certainly evolved independently from pelycosaur-like ancestors having a greater number of phalanges.

The influences that produced the evolution of these populations must have been quite complex, since they did not affect all parts of the body at once. For instance, some early therapsids, in spite of their mammalian phalangeal formula, still showed pelycosaur features in the absence of a secondary palate, and presence of a single occipital condyle and small dentary. If the presence of a squamodentary articulation is taken as the criterion of a 'mammal' this condition was almost certainly reached independently by several different lines. We still know too little to be able to specify clearly the conditions controlling such evolutionary chances, but it seems possible that the gradual appearance of terrestrial life and of large herbivores led various animals of a suitable structure and disposition to a carnivorous life. For this purpose certain changes of the ancestral structure would be

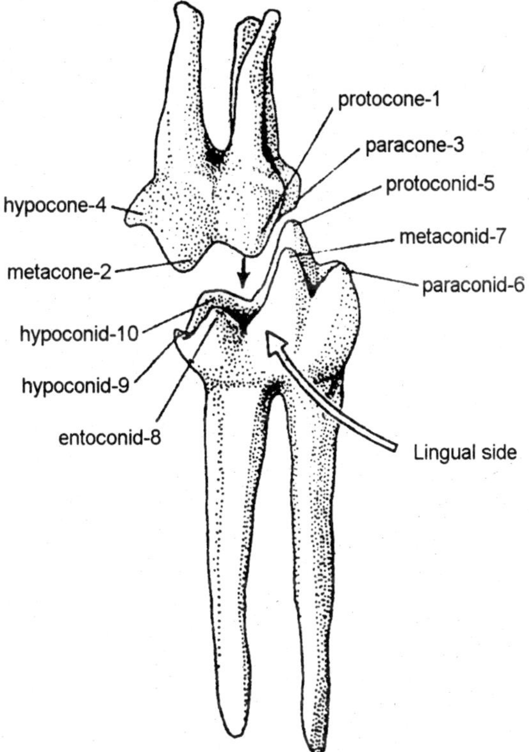

Fig. 17.9. Tribosphenic molar morphology.

needed, leading to parallel evolution in related stocks. However, at present we can hardly do more than pose questions about such matters and resolve to be rigorous in interpretation of the available evidence.

Mammals from the Triassic to the Cretaceous

The types classified as synapsid reptiles, which we have been considering, are not found later than the early Jurassic, 170Ma ago; mammals of approximately the modern type appear in the late Cretaceous. For the enormous time of more than 90 Ma between these dates the mammalian organization maintained itself in the form mostly of rather small animals, many probably arboreal and nocturnal. The fossil material of these early mammals is still scarce, and authors disagree as to the best interpretation and classification of them. It is often held that they include several different lines, which independently crossed the border between reptiles and mammals. As more is found out about the skull structure it seems likely that they are in fact all related and that we can recognize the Mammalia as a monophyletic class with two subclass Prototheria and Theria. The subclass Prototheria includes the order Multituberculata, which were the most abundant forms and are quite well known and the surviving order Monotremata. We shall also put here the orders Triconodonta and Docodonta about which there is much controversy. The subclass Theria includes all the modern mammals, with which we put the infraclass Symmetrodonta and the other especially controversial group the infraclass *Pantotheria* (=Trituberculata).

The differences between the two subclass Prototheria and Theria are largely in the structure of the teeth and wall of the skull. In therian mammals this is formed by a large squamosal and alisphenoid, which are small in Prototheria. The two subclasses have diverged from Upper Triassic forms, the triconodonts and docodonts, which are put in one group or the other rather arbitrarily. This simplified classification will not be generally agreed and may be upset by further discoveries, but seems to give a synthesis of recent views.

Recent Mammals

Vertebrates with elevated, constant body temperatures have high energy requirements. Much of mammalians evolution and the diversity of living forms can be attributed to trophic specializations that favour the evolution of increased efficiency in energy acquisition and the utilization of new source of energy. The classification of modern mammals into order is based primarily on trophic adaptations.

18

Fossils of Mammals

When the Tertiary period began, 65 million years ago, all the modem groups of animals were established. Evolution of the invertebrates, fishes, amphibians and reptiles still continued, although some members of these groups later became extinct. But the greatest progress took place in birds and mammals.

While the dinosaurs lived, mammals made little progress. But, in the Tertiary period, they had the chance to spread. All the habitats and food supplies that had once been the 'property' of the dinosaurs were available. One major food supply was the increasing variety of flowering plants. By the mid- Tertiary period, flowering plants were the dominant form of vegetation. The shrew like insect-eating mammals of the Cretaceous period evolved rapidly. By the end of the Palaeocene epoch (55 million years ago), 28 different orders of mammals had appeared. Each occupied its own habitat or relied on different foods.

In the Eocenean Oligocene epochs, between 55 to 22 ½ million years ago, mammals increased greatly in numbers. Some were unsuccessful. The Condylarths, such as *Phenacodus*, were extinct by the end of the Eocene (38 million years ago). The weird-looking Uintatheres and Baluchitheres died out by the end of the Oligocene epoch. At the start of the Miocene epoch, only 20 orders remained.

During the early part of the Tertiary period, dense forests covered much of the land. But in the Miocene epoch, between $2^1/_2$ and 6 million years ago, the forests grew smaller. In the Pliocene epoch, grassy plains covered vast areas. Enormous herds of plant-eating, hooved Mammals, such as three-toed horses, mastodonts and antelopes, roamed the plains.

During the Pliocene epoch, between 6 and 1.8 million years ago, two more mammal groups, the *Litopterns* and the *Notoungulates*, died out. Thus, 18 orders were left. These still exist today.

During the Tertiary period, the continents were steadily approaching their present positions. Most Tertiary deposits are found on the edges of the continents. But important marine deposits were formed in inland depressions, such as in the Mississipi River, London and Paris basins. Land deposits occur in central Asia.

Establishment of Mammalian Characters

There can now be little doubt that the mammals are an off-shoot of the reptilian sub-class *Synapsida*, and more particularly of some type or types of Triassic *Therapsida*, the true mammal like *reptiles*. The later were probably derived from primitive *Pelycosaurs*; and these in turn from the stem reptiles, *Cotylosaurs*.

Following characters were established during evolution of mammal:

1. Hairy skin.
2. Double occipital condyle.
3. Secondary hard palate.
4. Squamosal dentary joint.
5. Three ear ossicles from single (stapes) reptilian ossicle.
6. Large brain case.
7. Increased intelligence.
8. Differentiate dentition like incisor, canine, molar *etc*.
9. Double headed ribs.
10. Fused bones of pelvic girdle.
11. Reduced toe bones.
12. Constant body temperature.
13. Development of diaphragm.
14. Limited growth.
15. True copulatory organs with definite placentation except prototherians.

Development of Viviparity and Parental Care in Mammals

The dependence of the mammals upon intelligence as well as instinct for survival apparently grew concomitantly with the evolution of viviparity and continued postnatal association between the female and her young. Retention of the fertilized egg within the body of the mother may have been selected for initially as an adaptation for proper maintenance of the temperature of the developing embryo and as a way of protecting the embryo without reducing the mobility of either parent, but prolonged foetal-life and provision of sustenance for the new born also afforded a long period of time in which the intricate mammalian brain could establish itself and store its first impressions. If the condition in monotremes and marsupials-living mammals supposedly more primitive in their reproductive processes than placental forms- gives an accurate clue to the evolution of viviparity and maternal care, it seems that the suckling of the young was an early development and the extension of the time spent *in utero* a gradual process. Since the mammary glands are simply modified sweat glands, the secretion of milk might have followed soon upon the appearance of the latter structures. Holding an actively growing embryo within the body longer than it can sustain itself upon yolk material, however, required the evolution of an arrangement which would allow service of the embryonic tissues by the body of the mother. The uterine wall in both monotremes and marsupials affords nutriment to the resident embryo, but since the monotreme embryo is still surrounded by a shell and that of the marsupial associate itself only in a rudimentary way with the uterine wall, the period each spends within the mother is relatively short. Not until the advent of the choria-allantoic placenta, a structure in which the capillaries of embryonic membranes are brought into intimate association with those of the uterine lining, did extended development *in utero* become possible. The prolongation of the developmental period led, in the most advanced mammals, to the establishment of familial and social relationships far more complicated than those in any other group of vertebrates. In man, of course, the lengthening of the maturation process arid the supervision of the young have reached an extreme point, but other kinds of modern mammals also exhibit helplessness at birth, dependence upon learning for survival, and considerable ability to adjust their behaviour in response to changed conditions.

Traissic Mammals

Cynognathus

Cynognathus was a large, heavily built carnivore that is diagnostic of the *Cynognathus* zone in the upper part of the Lower Triassic of Southern Africa and is known from South America as well. With a skull that was as long as 40 centimeters, *Cynognathus* must have been one of the most formidable carnivores in the early Triassic. The snout is long and constricted behind the large canines. The adductor chamber is expanded laterally and posteriorly; the surrounding bones are wider and thicker than in *Thrinaxodon*, which suggests a very strong bite.

The chamber does not extend as far anteriorly relative to the total length of the skull so that the gape would be wider to accommodate very large prey. The cheek teeth are laterally compressed and coarsely serrated. The occiput was widely expanded, probably for insertion of very massive trunk musculature, and confluent laterally with the ends of the very high zygomatic arches.

The dentary bone makes up a substantially larger contribution to the lower jaw than does that of the galesaurids. Superficially, it

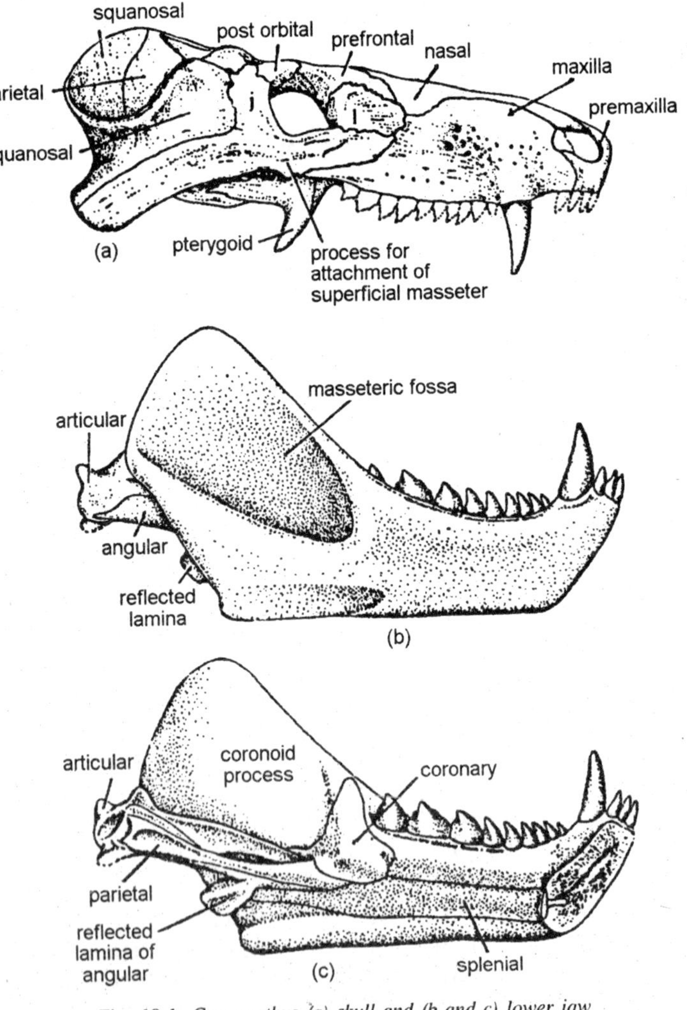

Fig. 18.1. Cynognathus (a) skull and (b and c) lower jaw.

resembles the dentary or large mammalian carnivores, with a high coronoid process. The *masseteric fossa* is deep and extends far forward. The ventral margin of the dentary extends posteriorly as the angular process. In modern mammals, this process serves as the area of insertion for the *superficial masseter* a muscle that originates on the lateral surface of the anterior portion of zygomatic arch. In *Cynognathus*, a masseteric process associated with the orgin of this muscle extends ventrally from the arch. The superficial masseter split off from the anterior margin of the earlier developed *deep masseter*, which is present in primitive cynodonts. The fibres of the superficial masseter are oriented posteroventrally, at nearly right angles to those of the deep masseter. Elaboration of the superficial masseter completes the major changes in the elaboration of the mammalian jaw musculature.

Although the lower jaw has a superficially mammalian appearance, closer examination shows that the dentary is accompanied by a number of smaller bones that are united in a narrow bar, which fit in to a groove on its medial surface. This bar is made up of the articular, prearticular, angular, and surangular. The coronoid remains a flat plate of the bone that overlaps the anterior end of the rod. The jaw articulation is formed in primitive reptilian fashion by the articular and quadrate.

The postcranial skeleton was very similar to that of *Thrinaxodon*, except for proportional differences that were associated with the greater weight of the body.

Gomphodonts

Accompanying *Cynognathus* in the early Triassic were primitive members of two families of herbivorous cynodonts, the Diademodontidae and the Traversodontidae, which are together referred to as *gomphodonts*. The primitive genus *Diademodon* closely resembled *Cynognathus* in most cranial and postcranial features, but the dentition was markedly different in the presence of transversely expanded cheek teeth.

Like *Cynognathus*, but in contrast with galesaurids, the posterior end of the squamosal is flared laterally to form a groove that extends down toward the area of the jaw articulation. This structure remains prominent throughout the gomphodonts but is less conspicuously developed in other cynodonts.

In Diademodontids, the snout is narrow behind the large canines, as in *Cynognathus*. The most anterior of the cheek teeth are conical and the most posterior are laterally compressed suctorial teeth, as in primitive cynodonts. The remaining teeth are transversely expanded

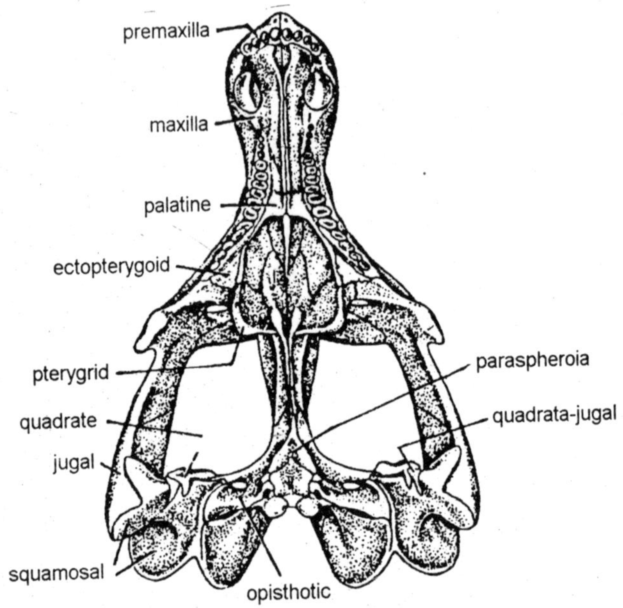

Fig. 18.2. Plate of the lower Triassic gomphodant–Didemodon.

and relatively flat at the tip. The upper teeth are much wider than the lower.

Diademodontids are known from China and east Africa as well as southern Africa, and they extend into the Middle Triassic. The Middle Triassic genus *Massetognathus* is typical of the group. The canines are greatly reduced and there is a short diastema between them and the cheek teeth. The snout is not constricted as in diademodontids but is expanded beyond the tooth row, which may indicate the presence of fleshy cheeks to retain the food as in ornithischian dinosaurs. The upper teeth are even more expanded than in diademodontids, with a strong transverse ridge on the posterior edge. This ridge would have

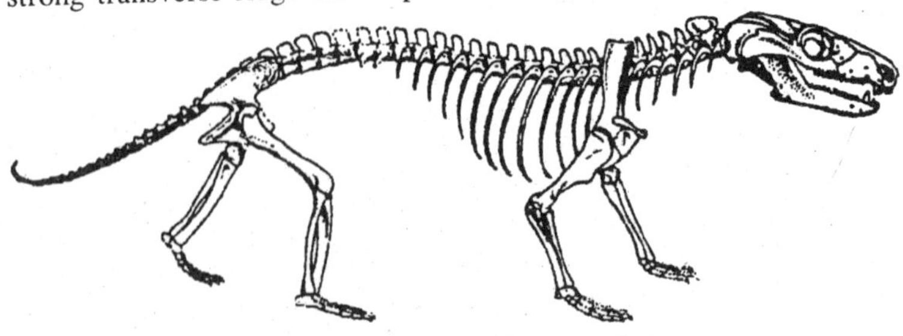

Fig. 18.3. Skeleton of Massetognathus.

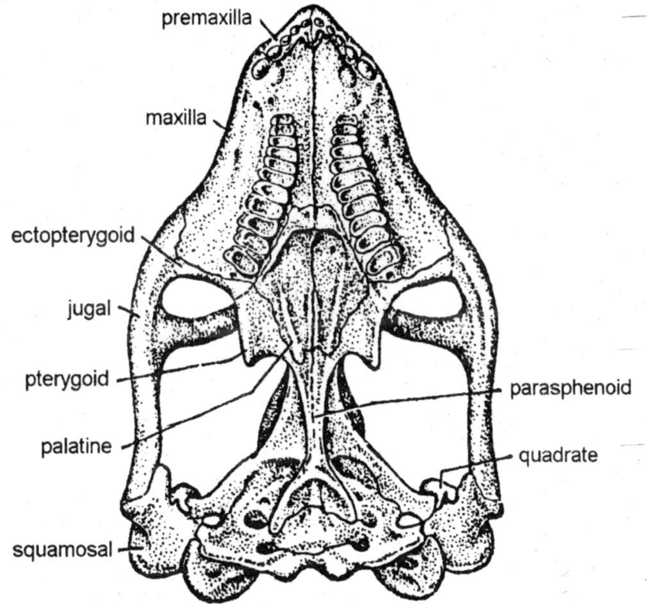

Fig. 18.4. Plate of the herbivorous Massetognathus.

sheared against an anterior ridge on the lower teeth. Both upper and lower teeth have an external ridge. Between the ridges is a basin where the food was crushed.

Traversodontids survive nearly to the end of the Triassic. We know them from the Lower to the Upper Triassic in South America, the Middle Triassic of East Africa, and the Upper Triassic of North America, India, and southern Africa.

Tritylodonts

The *Tritylodonts* are certainly the most specialized of the herbivorous cynodonts. They do not appear until the Upper Triassic but persist into the late Middle Jurassic as the last surviving therapsids. The skull is very mammalian in general appearances. The temporal opening is huge and confluent with the orbit. The postorbital and prefrontal bones are both lost. The dentition gives them a rodentlike appearance. One pair of incisors is greatly enlarged, but the canine teeth are not developed, which leaves a long diastema between the incisors (one to three pairs) and the molariform cheek teeth. In contrast with all other thereapsids, but like the mammals, the cheek teeth have multiple roots. However, the dentary retains a strictly reptilian patten, in which the condylar process does not reach the squamosal.

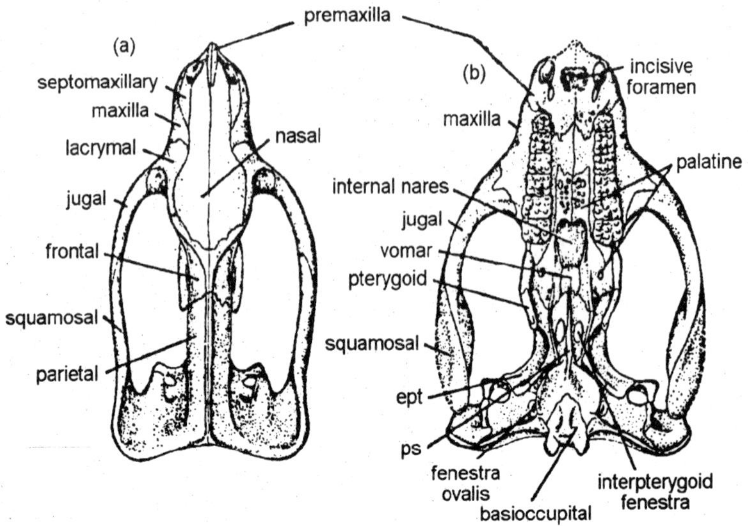

Fig. 18.5. Skull of lower Jurassic Tritylodont—Kayentatherium (a) dorsal view, and (b) palatal view.

The upper teeth have a squarish appearance in occlusal view, with three rows of cusps arranged longitudinally; the lower teeth have two rows of cusps. The individual cusps are crescentic; those of the upper teeth are concave anteriorly and those of the lower teeth are concave posteriorly. The lower teeth are drawn posteriorly along the grooves between the upper teeth as the jaw is closed. The pattern of tooth wear indicates that the similar jaw movements occurs in traversodontids. *Massetognathus* has a pattern of cusps that could have given rise to that of the tritylodonts, but the morphological gap between the two groups is significant. Only fragmentary remains of tritylodonts have been described from the latest Triassic in South America, but they are common in the early Jurassic of China, western Europe, southern Africa, and western North America. *Stereognathus* is known from the Middle Jurassic of England. The anterior portion of the postcranial skeleton is best known in *Kayentherium*. The limbs are short and there are no costal plates.

The generally mammalian appearance of the skull and the similarity of the dentition to that of gnawing mammals led in the nineteenth century to the identification of the tritylodonts as mammals. We now know that the postcranial skeleton is extremely mammalian in many features as well. *Kemp* (1983) suggested that they may be the most closely related of' any of the therapsids to the ancestry of mammals.

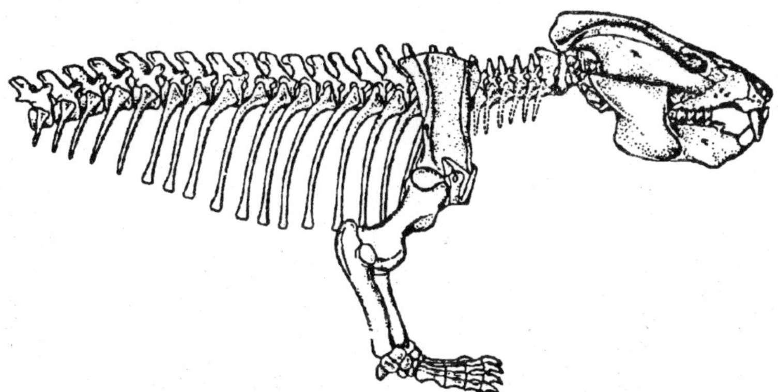

Fig. 18.6. Kayenatherium anterior portion of the skeleton.

The specialization of the dentition, with the loss bf canines arid the complex molarization of all cheek teeth, certainly precludes all known tritylodonts from the ancestry of mammals. *Kemp* proposed that the ancestors of mammals would lack all the dental spetializations of the tritylodonts but would have the derived postcranial features of this group. As yet, no such therapsids have been described. Alternatively, one must accept that a great deal of convergence in the structure of the girdles and limbs occurred between the ancestors of mammals and the ancestors of tritylodonts.

Chiniquodontidae

In addition to the large cynognathids and the *herbivorous gomphodonts* and *tritylodonts*, a more conservative line of small to medium sized cynodonts is known from the Middle and Upper Triassic, the Chiniquodontidae. We recognize five genera—*Aleodon* from the Middle Triassic of East Africa and the remainder from the Middle and Upper Triassic of South America.

In all members of this group, the temporal opening is very large, extending nearly half the length of the skull. Nevertheless, it is still separated from the orbit by a narrow postorbital and prefrontal bones are retained. However, the pineal opening is lost. The secondary palate extends back to the end of the tooth row with a major contribution from the palatine bones. The dentition closely resembles that of the galesaurids, with four upper and three lower incisors, moderately long canine teeth, and seven laterally compressed cheek teeth with a longitudinal arrangement of cusps. In most *chiniquodonts*, there is only a single row of cusps, but in *Probaingnathus*, a cingulum bearing cusps is present on the inner side of the teeth. The teeth show wear but

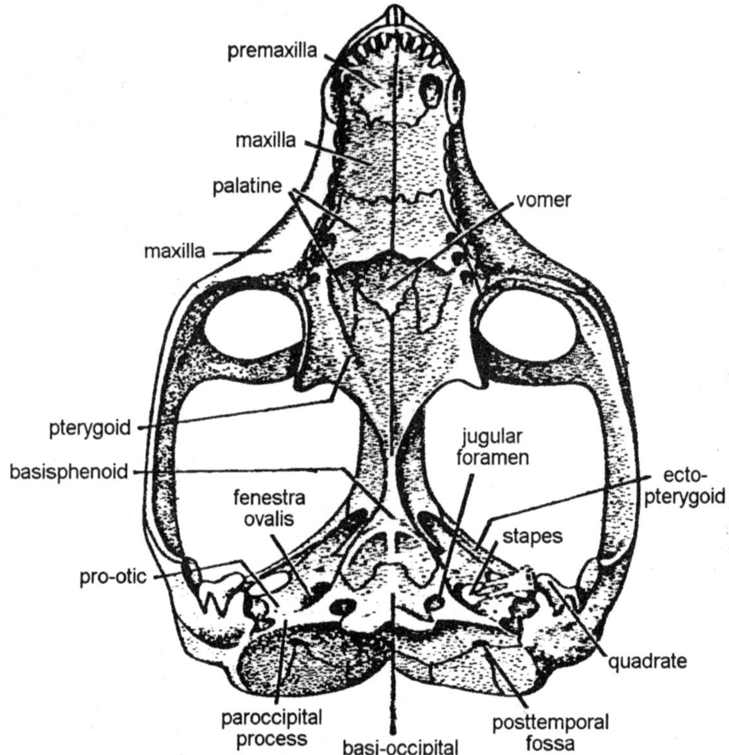

Fig. 18.7. Ventral view of skull of carnivorous cynodont—Probainognathus.

they did not have a regular pattern of occlusion. One of the most notable features of *Probainognathus* from the middle Triassic of Argentina is the posterior extension of the dentary. The adjacent surangular articulates with the squamosal, forming a new jaw joint lateral to the persistent reptilian joint between the articular and quadrate.

Tritheledonts (Ictidosaurs)

The assemblage of genera that is grouped as the trithelodonts or *ictidosaurs* are among the most tantalizing, or frustrating, cynodonts. Like the tritylodonts, they are known only from the very end of the Triassic and beginning of the Jurassic. Five genera have been assigned to the group, but all are known from fragmentary remains and few have been adequately described. The skull ranges from 3to 6 centimetres long, the temporal opening (as in tritylodonts) is confluent with the orbit, and the postorbital and prefrontal bones are lost. In

Diarthrognathus, the frontal extends ventrally anterior to the orbit to reach a dorsal process of the palatine that extends up from the paltal surface. The dentary appears to have made contact with the squamosal. As *Gow* (1980) described, the dentition is unlike, that of chiniquidontids and mammals in the relatively great width of the cheek teeth. In *Pachygenelus* there are only two pairs of upper and lower incisors, although *Chaliminia* has three uppers, behind which is a diastema. *Theioherpeton* from South America appears to be the most mammalian of them all with a small, low skull that has very narrow zygomatic arches and narrow, crowned cheek teeth with partially divided roots.

Because of the incomplete nature of most genera in this group, we are not certain that they are closely related. Nevertheless, they appear to show the closest approach to a mammalian morphology among small carnivorous cynodonts, which are otherwise the most appropriate ancestors of mammals.

Jurassic Mammals

Osteological evidence suggests that the Jurassic mammals were well above the evolutionary level of the egg laying monotremes thus properly to be classed with the marsupials and true placental mammals and the Theria. They include, presumably, the ancestors of these later groups and are thus of great phylogenetic importance. All, however were of modest size, and, in consequence, their remains are rare and consist mainly of isolated teeth, occasional jaws, and a few limb bones and skull fragments.

Docodonts

The *docodonts* are the shortest-lived and least diverse of all groups of Mesozoic mammals. They are known only from the Middle and Upper Jurassic and include only four genera from North America and Europe. Their specific ancestry has not been recognized and they apparently had no descendants.

The pattern of the molar teeth clearly distinguishes *docodonts* from all other Mesozoic. mammals. The lower molars are rectangular, with a row of high buccal cusps and somewhat lower lingual cusps that probably evolved by elaboration of the lingual cingulum of more primitive mammals. The principal lateral *cusp* is joined to the lingual row by a transverse ridge. The upper molars are much more expanded transversely, with a widely extended medial portion, that is distinguished from the more lateral surface by a waisted midsection. Wear produced a complex pattern of shearing surfaces. In older individuals, the cusps became blunt and they would have served to crush and grind the food.

If we judge from their jaws, docodonts were the size of small mice and had elongated snouts. The number of cheek teeth appears to increase over time, with Upper Jurassic species having as many as four premolars and eight molars. The jaws are grooved medially for the attachment of extradentary bones, which indicates that they still retained a reptilian jaw joint. An extension from the lower margin of the dentary is in the position of the pseudanglur process of Dinnetherium. The middle Jurassic docodont Borealestes and the late Jurassic Haldanodon are known from considerable postcranial material that has not yet been described.

Amphilestids

Amphilestids retained the cusp pattern of the morganucodontids, with a single row of linearly arranged cusps. The occlusal pattern more specifically resembles that of Megazostrodon and Dinnetherium, in which the upper and lower teeth more nearly alternate in position. Amphilestids are represented in the Middle Jurassic by the genera Amphilestes and Phascolotherium. Amphilestes has a tooth count of

$$\frac{?\qquad\quad ?\quad ?\quad ?}{3\text{-}4\qquad\quad 1\quad 4\quad 5}$$

They have neither an angular nor a pseudangular process.

Amphilestids also occur in the late Jurassic and early Cretaceous. Much of the skeleton is known of *Gobiconodon* from the early Cretaceous of the Mongolian People's Republic and the Western United States. The body was 35 centimeters long (excluding the tail) and more heavily built than the living opossum. An important difference from other primitive mammals is the development of a typically therian *supraspinous fossa* of the scapula. The alisphenoid is relatively larger than in *Morganucodon* and approaches the therian condition. *Freeman* (1979) suggested that the amphilestids be placed within the therian order Symmetrodonta.

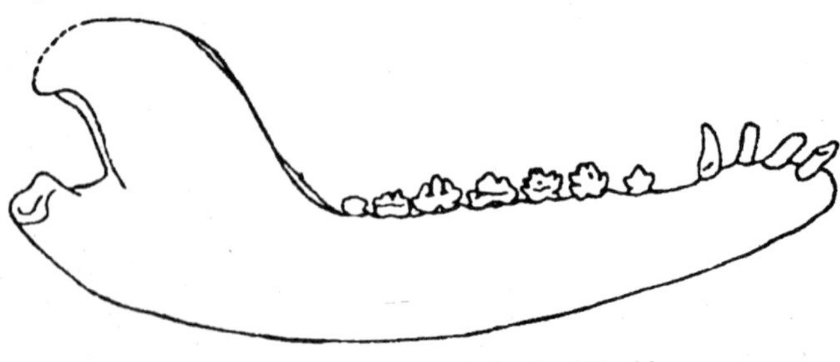

Fig. 18.8. Lower jaw of the Amphilestid.

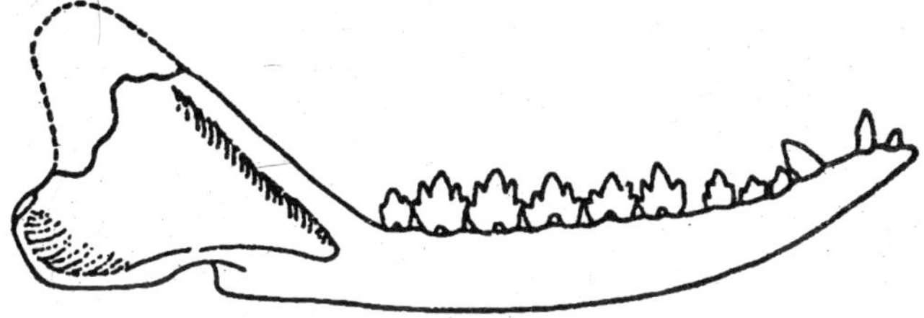

Fig. 18.9. Lower jaw of Dinnetherium.

Surprisingly, *Gobiconodon* shows replacement of the molar teeth, which may be related to the large size and prersumably longer life span of this genus. One of the reasons that tooth replacement was limited in the earliest mammals may have been because they were so small and, short-lived that there was no time for several generations of teeth to erupt. Other groups of mammals that appear first in the Middle Jurassic are clearly related to the ancestry of marsupials and placentals.

Triconodontids

Another long hiatus in the fossil record, which lasted approximately 25 million years, separates the Middle and Upper Jurassic faunas. Among the animals first known in the Upper Jurassic are the triconodontids, which extend into the Upper Cretaceous. Their closest affinities appear to lie with *Morganucodon*. This relationship is based primarily on dental similarities, since little of the skeleton of triconodontids has been described.

The fact that the principal cusps of triconodontids are arranged in a linear patter is certainly primitive. The particular pattern of occlusion shared by triconodontids and morganucodontids, in which the upper and lower teeth are aligned more or less one to one, may be primitive or derived relative to that "exemplified by Megazostrodon in which

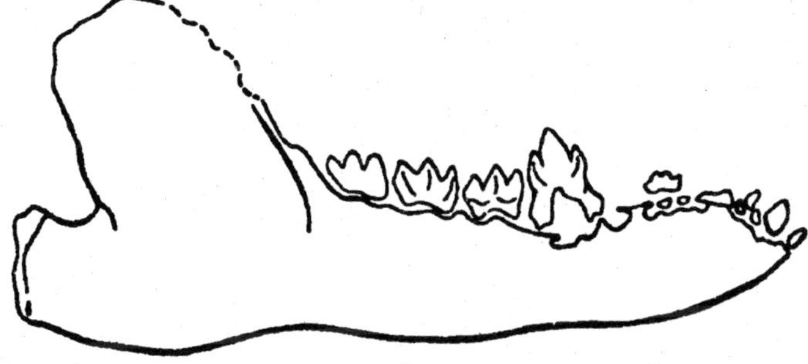

Fig. 18.10. Lower jaw of the Triconodont.

the upper and lower molars are opposed more or less two to one. The teeth of tiriconodontids are definitely specialized in having the major cusps of nearly uniform height, which gives the teeth the appearance of a saw blade.

A partial braincase described by *Kermack* (1963) demonstrates the presence of an uncoiled cochlea, as in early Jurassic mammals, and the cavum *epiptericum* is still open anteriorly. Newly discovered skeletal remains from the Lower Cretaceous being studied by *Jenkins* (1984) show the possible association of a scapula with a *supraspinous fossa*, as in the (mphilestids and therian mammals.

The only triconodontid in which the tooth count is known has a dental formula of

$$\frac{2 \quad 1 \quad 4 \quad 5}{1 \quad 1 \quad 4 \quad 5}$$

No clearly defined angular process is present.

Symmetrodonts

The order Symmetrodonta has been established for the inclusion of a number of late Jurassic forms, such as *Spalacotherium* and *Eurylambda*, also comparatively good size and comparatively predaceous habits. These forms also had three separate cusps in each molar tooth and hence were at one time included in the triconodonts, with which group they agree in the absence of an angular process on the jaw. But in the present types the three cusps were arranged in a symmetrical triangle, with base external above, internal beneath. Early advocates of the tritubercular theory believed that they had arisen from tricdnodonts by a rotation of the cusps. Bu there is no evidence to do with that group. Although possibly related, to the next order, the Symmetrodonts seem to have been somewhat off the main evolutionary line. The symmetry of the molar tooth in this group is on contrast to the molar shape of pantotheres, and the lower molars lack the "*heel*" characteristic of that order and of all later mammalian groups.

Pantotheres

Most important of Jurassic orders from an evolutionary point of view is that of the Pantotheria, frequently termed *Trituberculata*. Most 'representatives are from the late Jurassic *Amphitherium*, the only Middle Jurassic genus, is a generalized form but known only from the lower jaw. In *Amphitherium* there is long and slender, and, in contrast to other Jurassic orders, there is a well developed angular process. In addition to four incisors, a canine, and four premolars, there are seven molar teeth—an unusually high number but perhaps a rather primitive one. The molars show a three cusped, asymmetrical trignoid, comparable to that of later groups; and particularly significant is the presence of a "*heel*" or "*talonid*" absent in the orders so far mentioned.

Like the other Jurassic orders, pantotheres are unknown in later strata. But while our evidence is complete, it can at least be safely said that known characters of the pantotheres allow us to consider them as possible ancestors for all higher mammalian types.

Multituberculates

The most diverse and numerous of Mesozoic mammals, and the only group that were omnivorous rather than insectivorous or carnivorous, were the *multituberculates*. They are first known with certainty in the Upper Jurassic but have long been thought to show affinities with the haramiyids, which are recognized as early as the late Triassic.

Multituberculates range in size from the dimensions of a small mouse to those of a woodchuck. The skull has a superficially rodentlike appearance, which results from the presence of a single pair of long, procumbent, lower incisors followed by a diastema due to the loss of the canines. There may be one to three pairs of upper incisors, and the upper canine may be retained in primitive genera. In. most multituberculates, the anterior lower premolars are specialized as laterally compressed shearing blades that bear a series of vertical ridges. The molar teeth have two or three rows of linearly arranged cusps that may have been used to crush or grind food.

Tooth wear indicates that jaw motion was entirely in a sagittal plane, with no trace of the transverse component that was common to' the most primitive mammals. In most genera, the glenoid was open anteriorly, which permitted propalinal movement of the lower jaw. *Krause* (1982) demonstrated that the power stroke involved retraction of the mandible during closure. The symphysis is not solidly fused, which indicates the possibility of unilateral movement of the jaw rami. There is no angular process.

Like modem rodents, multituberculates may have been largely herbivorous, but the shearing lower premolars show a pattern that is particularly close to that of modern omnivorous marsupials, as *Clemens* and *Kielan-Jaworowska* (1979) pointed out.

The skull of multituberculates is low and broad. The frontal overhangs the orbit. The eyes faced primarily laterally, in contrast with their more anterior orientation in early therian mammals. The jugal is missing, and the zygomatic arcis is formed entirely by the squamosal and a long posterior process from the maxilla.

Cretaous Mammals

The multituberculates carried on the adaptations that had been established by their Jurassic fore bears, living and expanding through the complete span of the Cretaceous period and into the beginning of Cenozoic times. The Cretaceous and early Cenozoic evolution of the multituberculates was marked by certain refinements of the adaptation already outlined for these animals, especially by an increase in size. The culmination of evolution in these mammals was reached after the close of the Cretaceous period in the Paleocene genus *Taeniolabis*, an animal as large as a beaver, with a skull six inches in length and with large, chisel like teeth and in certain other forms that survived into early Eocene times.

A few triconodonts, symmetrodonts, and pantotheres survived into early Cretaceous times. The pantotheres were mentioned above as being probably ancestral to the marsupial and placental mammals that first appeared in the Cretaceous period. Marsupials closely related to the modern American opossum are found in upper Cretaceous sediments of North America; and placental insectivores, related to modern shrews and hedgehogs, have been discovered in upper Cretaceous beds of Mongolia and North America.

Archaic Mammals

Of the archaic mammals we will turn our attention to three groups, of which the first is the *Creodonta*. These forms resemble in many details the hoofed Condylarthra, but differ from them chiefly in the skull and teeth, in that they have more the aspect of a true carnivore than the condylarths which were largely of vegetarian diet. The terminal phalanges (unguals) are also moreclaw-like, although there are exceptions to this rule, notably in *Dromocyon*. The skull of a creodont differs from that of a carnivore, for while it is always large for the size of the animal, there is a much smaller brain-case, thus necessitating a high crest of bone along the mid-line of the cranium (sagittal crest) to obtain sufficient surface for muscular attachment. There are widely expanded temporal or zygomatic arches for the same purpose. The teeth also differ in not being so perfectly adapted for a flesh diet as in the true carnivores. In the latter, certain cheek teeth are almost always enlarged and modified to form a wonderful shearing device, and these so-called *carnassial teeth* (Lat. *carD,* flesh) are, when present,

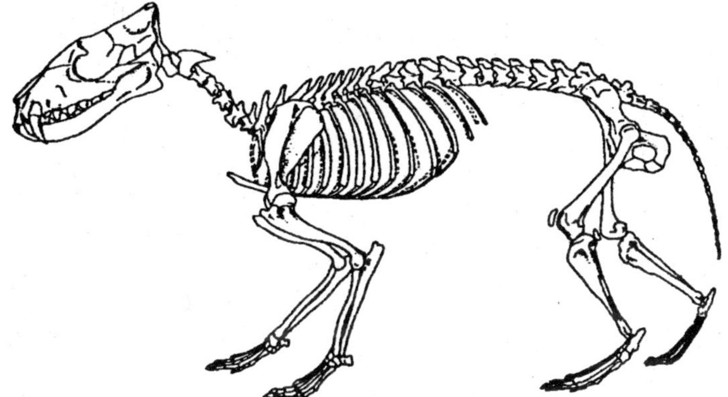

Fig. 18.11. Skeleton of Dimetrodon

invariably the fourth upper premolar and first lower molar, expressed thus: p^4/m^1 With the creodonts the carnassials may not be developed at all, and if they are, are variable and not necessarily, indeed rarely, p^4/m^1 and in addition they are rarely confined to a single pair of teeth but are two or more in number.

The creodonts have been divided into at least six distinct families, of which but one probably gave rise to true carnivores, the rest dying out one after another until by Upper Oligocene time none were in existence. The creodonts foreshadow the true carnivores in a number of ways, in that certain of them were bear-like (*Arctocyon*), others dog-like (*Dromocyon*) or otter-like (*Oxyxna*, *Patriofelis*), some like minks (*Sinopa*), others cat-like (*Dissacus*) or resembling hyaenas (*Hyaenodon*). The last genus is of especial interest because together with its Old World ally *pterodon* it is the last Ptreodon survivor, existing until the Middle Oligocene. One form, *Andrewsarcus*, from Mongolia, attained gigantic size, the skull alone measuring nearly a yard in length.

Fig. 18.12. Phenacodus.

The *Condylarthra* were a group of very primitive ungulates which, aside from the implied differences diet, paralleled the Creodonta closely, for in both groups there was the same generalized type of body, with a long heavy tail and rather stocky, more or less cursorial limbs. There were, however, relatively few of the condylarths, but four families being recognized as, against six for the creodonts. They range in time from Paleocene and Lower Eocene, but very little is known as yet of their geographical extent. *Scott* says of them: "They maybe looked upon as the connecting link between the clawed and hoofed mammals and therefore ancestral to most, or perhaps all of the ungulate orders." One of the condylarths, *Phenacodus*, was hailed by its discoverer, Professor *Cope,* as the five-toed ancestor of the horse, but this is now known to be impossible as it is too large and too highly specialized in certain directions, although very primitive in others, and also too late in time to be the founder of the great equine lineage. This genus, from the Wasatch beds (Lower Eocene), ranged in size from a fox to a small sheep. While the canines were tusk-like, they were not large, and the grinding teeth were low-crowned and of simple pattern, suited undoubtedly to a rather succulent herbage. The skull was long and low, with a well developed sagittal crest and, while that portion of the cranium behind the orbits was relatively long as with most primitive skulls, the brain case was of very small capacity. The feet are five-toed, semi-plantigrade, and built on a very primitive plan *Phenacodus* and the earlier *Euprotogonia*, represent the family Phenacodontida—while the other family, Meniscotheriidae, embraces but a single know genus, Meniscotherium. These forms, while contemporaneous with the phenacodonts, were, more advanced in tooth structure, for the cusps other grinders nave begun to assume a crescent shape such as one often finds in the higher odd- and even-toed ungulates. The body and tail were long and the limbs, while long resemble so much those of the Hyracoidea of Africa, as to cause the inclusion of *Meniscotherium* in that group by certain authorities. Others have considered the Hyracoidea 'to be surviving condylarths. There are, however, no very good grounds for such an assumption. The *Condylarthra* are of interest, however, in that they represent or were very similar to what was probably a very widespread group of primitive ungulates out of which, possibly, all of the other orders of ungulates arose. The genera which we know could not have been the direct ancestors, but they show us the nature of the ungulate ancestry.

The *Amblypoda*, or short-footed ungulates, are another group of hoofed forms, among which were some that attained a huge, almost elephantine size and in spite of a basic primitiveness developed a superficial specialization of a very remarkable sort. Their geological range is throughout the Eocene period, when they in their turn suffered extinction. Four families are recognized, of which the *two* most primitive are both Paleocene in distribution. *Pantolambda*, the type of the second family, while undoubtedly an ungulate, shows many points of similarity with the creodonts. It is described as having a head and body somewhat smaller than those of a sheep, and much shorter legs. The body and tail had somewhat the proportions of the larger cats, and the skull, as with the condylarths, was long and low, with small brain capacity and prominent sapittal crest. The limbs were very short and relatively heavy, with five spreading toes on each foot.

Coryphodon represents the third family and is in many ways a remarkable beast. The different species vary in size from a tapir to an ox, and thus are the largest forms we have so far considered. They were heavy, unwieldy animals whose short powerful limbs and spreading feet point to swamp-dwelling if not aquatic habits. The skull was large and flattened in such a way that no median crest is visible, nor are there any indications of horns such as the next genus possessed. The canine teeth were developed into huge flaring tusks suggesting those of the swine. Altogether it was a heavy sluggish brute whose very small brain gives evidence of great stupidity.

Uintatherium (*Dinoceras*) represents the last family of amblypod and in many ways size up to seven feet in-height; dentition, and horns was by far the most specialized, in fact-grotesquely so. Its limbs were graviportal, quite like those of the Proboscidea and like them an adaptation to carry the creature's great weight. The elephant-like characteristics extended also to the body, but there the resemblance ceased, for the skull was totally dissimilar in that it was extended upward into a series of horn-like prominences. These consisted of a pair upon the nose, which from their appearance may have borne dermal horns like those of rhinoceroses. The second pair were higher, with bluntly rounded ends, and were probably not sheathed with horn but covered with skin as in the giraffe. There was also a third pair, massive structures, 8 to 10 inches high, which again could not have borne horny sheaths. There was a high transverse occipital crest at the hinder end of the skull connecting the posterior pair of horns and giving, together with the prominences, a unique basin-shaped character to the top of the skull. Another remark able feature lay in the greatly developed canine teeth, which were curved sabers in some genera and spear-shaped in others, and were doubtless important weapons. Both the tusks and horn prominences were apparently better developed in the male than in the female, for their variation constitutes about the only difference seen in certain skulls. There is no indication of a proboscis, as the nasal bones, which are long and prominent in *Uintatherium*, are invariably shortened whenever that useful organ develops. The molar teeth *Uintathetium* were very conservative, for while one may trace a very marked evolution in the skull and tusks these important organ hardly change at a all. The brain also was- absurdly small- for so large a creature. The armament of *Dinoceras* may have served a useful purpose but one is constrained to believe that, together with the relatively great size, it indicates racial senility —the extreme of over-specialization attained by a primitive stock.

Fate of the Archaic Mammals

The archaic mammals as such have long since vanished from the earth, and were it not for their remains entomoed in the Eocene rocks would be unaware that they ever existed. Theirs was a brief span compared with that of the reptilian hordes and also with that of their mammalian successors; but for a while they throve mightly until competition with creatures of a better sort became too great for them to bear. That they strove to meet this competition is evident, for certain of the later creodonts, notably *Patriofelis* and the powerful *Harpagolestes* and especially *Andrewsarcus* of Mongolia, increased materially in bodily size, while the hyaenodonts actually increased the bulk of the brain, which .aided in making them the sole survivors of the group after the close of the Eocene. Competition was doubtless, therefore, a prime cause which led to toe extinction of these forms. We have argued racial old age in *Uintatherium*, but if that be deemed in sufficient in itself, we have the noteworthy fact that where evolution of an animal runs to the development of tusks and horns,- possibly favoured by sexual selection, the grinding teeth are apparently neglected and are apt to show arrested development. And bulk is fatal where correlated with inadequate feeding mechanism, and with brain power not adequate to enable the females to defend and care for the young as well as to meet new conditions of life.

Thus the fate of the archaic mammals was, first, extinction, and, secondly, transmutation of a few—a very few—into higher types. There remains yet a third possibility, and that is emigration, not of the late but of the earlier sorts, across the southern land-bridge into South America, where together with a certain and admixture of other stock they may have given origin to part of the remarkable South American fauna which rose and flourished during the long period of Neogaean isolation. Others, passing beyond the limits of South America, may have crossed an Antarctic land-bridge into Australia, where as marsupials they still persist. If, on the contrary, the entire marsupial order is to be considered archaic, the conclusion that they may still be surviving in these remote forms and in the American opossums is tenable. There is a further possibility that the American Edentata, sloths, armadillos, and their allies, should be included in the group. they also have survived because they found asylum isolated South America.

19

VERTEBRATE PALAEONTOLOGY

The history of vertebrates covers a span of 500 million years or more, from their appearance in the Cambrian Seas to the present rich and varied fauna. Their history is recorded in a sequence of fossils that documents their skeletal anatomy, distribution, and evolutionary change. From the fossil record we can determine the inter-relationships of the modern species and trace the origin of the skeletal features that characterize each major group. Fossil documents the rate of evolution and may enable us to establish the degree of which environmental, developmental or other factors influence its direction. We think of the human as the most highly evolved vertebrates - specialized in many structures - hands, feet, vertebral column, cerebrum but the structure and organization of the human body have been determined by a long and complex course of evolution. When we strip away all the special features of "humans and compare the result with other vertebrates we can Identify a "basic body plan", presumably the ancestral plan, which consists of a bilateral, tubular organization, possessing such characteristic features as notochord, pharyngeal gill slits, dorsal hollow nerve cord, vertebrae, and cranium, as well as other essential systems. One of the protochordates, Amphioxus, and the ammocoete larva of lamprey offer suggestions of what the earliest vertebrates were like.

DIVERSITY OF VERTEBRATE

What about the different kinds of vertebrates? Everyone knows that humans are vertebrates, that dogs and cats and cows and chickens are all vertebrates, but few people realize that there are some 40,000 living species that share this distinction, not to mention the many extinct, fossil forms. Most biology students know that there are different major groups of vertebrates-jawless fishes, cartilaginous fishes, bony fishes amphibians, reptiles, birds and mammals – each possessing certain distinctive features that set it apart from the others. The differences relate of adaptations to different environmental conditions or opportunities. Each species has an ecological niche that is different from all others and that is expressed at least in part, by altered body form and functions, and the diversity of species in the higher taxa, at the level of genera, families, order; and even classes, gives an indication of the genetic plasticity or responsiveness of that group to environmental differences.

JAWLESS FISHES

The earliest known vertebrates are the Ostracoderms, primitive jaw less fishes related to a lamprey and hag fishes and first appear in the late Cambrian and Ordovician periods about 500 million years ago. They were completely encased in heavy dermal bony armor. The name Ostracoderm (shell skinned) refers to an outer covering of dermal bony armor, which is especially elaborate on the head and around the thoracic region. Prehistoric species ranged from creatures a few centimeters long to a 14 metre long relative of man eating great white Shark alive today.

Fossils fishes have been found in every continent. Appropriate rocks yield isolated bony plates, spines, scales, vertebrae, and teeth. But whole skeletons are rare and many early fishes had gristly skeletons that seldom formed good fossils. The radiation of vertebrates began in the late Pre-Cambrian or early Cambrian periods. One or more lines not represented in the fossil record until the early Silurian developed jaws, came effective predators.

One of the most important changes we see among the Ostracoderms, is the evolution of different ways in which an exoskeleton forms early in development to allow for continued growth. We may conclude from the destructive nature of the armor that bone probably evolved separately in each of the major groups of Ostracoderms. Some Ostracoderms with a light flexible covering of small scales and a fusiform body may have lived in the open water. Others, with heavier armor and ventrally flattened body, were presumably benthonic. Ostracoderms fossils are plentiful in late Silurian and early Devonian rocks of Europe and North America.

E.R. Lankester divided the group into the Osteostraci and the Heter Thirty years later, RH. Traquair reported the discovery of two additional kinds of Ostracoderms, the Anapsida and the Coelolepida;

Modern workers however, have recognized two groups of Paleozoic: Agnatha: Cephalaspidomorphi a Monorhina and Pteraspidomorphi, or Diplorhina.

Peraspidomorphi

The members of this group retain a more primitive condition—in having the nasal sacs and, apparently, the external narial openings are . They are the first group of vertebrates to appear in the fossil record. The oldest known fossils that may be remains of vertebrates are phosphatic fragments widely scattered in North America, Greenland and spits bergen. These fragments are associated with benthonic trilobites, conodonts, and brachiopods. The surface shows scale like ornamentation but gives no evidence of the overall shape of the animal fragmentary remains are known as Anatoloepis.

The group pteraspidomorphi includes several genus.

The remains of the genes Arandaspis have been obtained from the of lower Middle Ordovician period in Australia. Their remains were in the form of molds in the rocks showing detailed impressions of internal as well as external surfaces. Large dorsal and ventral cover the head and pharyngeal region and smaller lateral plates, the area of the gills. Small scales covered the anterior trunk region probably extended posterioly to cover the remainder of the body and tail. Ritchie and Gilbert Tomlinson (1977) suggested that Arandaspis lived just above the bottom, feeding on organic debris and micro-organisms carried by currents immediately above the surface of the sediments.

The specimens do not show nasal capsule, a pineal opening, or the inner ear. Open lateral line canal grooves extend longitudinally along both the dorsal and ventral plates. They appear not to have opened to the exterior. The external plates are less than 1/10 millimeter thick. The external surface is ornamental with tiny scales like those of Anatolepis. The bone appears to have almost completely superficial to the soft anatomy of the body and would have served primarily as a protective covering.

The genus Porophoraspis accompanies Arandaspis but is represented only by fragments of dermal armor showing a distinct pattern of ornamentation. Thousands of bony fragments of two other Ostracoderm genera, Astraspis and Eriptychius were found in a series of localities within a narrow band of sediments termed the Harding Sandstone, which extends along the foothills of the Rocky, Mountains in the Western United States and Canada. These sediments were almost certainly deposited at the margins of the sea and contain marine invertebrates as well as fish fragments. Only two specimens of Astraspis have been found that show a significant portion of the carapace. The most striking difference from Arandaspis is the presence of many distinct plates called tesserae that make up the shield. Large tubercles at the center of each plate represent the point at which ossification was initiated in small individuals of Astrapis. The genus Eriptychius has calcified cartilage in the endoskeleton, which is not found in other pteraspidomorphs. Unfortunately, the fragments are too incomplete to provide information of the overall structure of this fish.

Cyathaspididae

Of the diverse agnathan families present in the middle and late Silurian, the Cyathaspids are the most closely related to the known Ordovician genera. The early specimens are entirely from marine deposits. The entire body is enclosed by bony plates and scales. The head and thorax are covered by extensive dorsal and ventral plates. The upper margin of the long narrow branchial plates is notched by a single pair of excurrent gill openings. Smaller plates border the orbits ventrally. A series of plates making up the ventral surface of the mouth were probably connected by soft tissues to form a flexible scoop.

The body is oval dorso-ventrally flattened with laterally compressed caudal fin. There are no dorsal, anal, or paired fins. The superficial ornamentation of the dermal armor of cyathaspids typically consists of series of rounded ridges of dentine running parallel with the margin of the plates.

Amphiaspididae, Pteraspididae, and Psammosteidae

The cyathaspids apparently include the ancestors of a wide variety of more specialized Ostracoderms that radiated throughout the Devonian. A family of highly modified descendants the Amphiaspididae are known from the Lower and early Middle Devonian of central Siberia and arctic Canada. All the forms show complete fusion of the carapace and reduction a complete loss of eyes. Pteraspids are among the best known, most or numerous, and taxonomically diverse groups of ostracoderms. The dermal shield shows more individual plates than are recognized in the cyathaspids. But like the cyathespids, these animals elaborated their armor only when of nearly adult size. The pteraspids did evolve laterally and dorsally projecting spines that may have served as protection, increased their stability in the water, and fixed .their position on the mud. The trunk and candal region is covered with smaller, more numerous scales than the cyathaspids giving than more flexibility for swimming. The ventral lobe of the tail is considerably

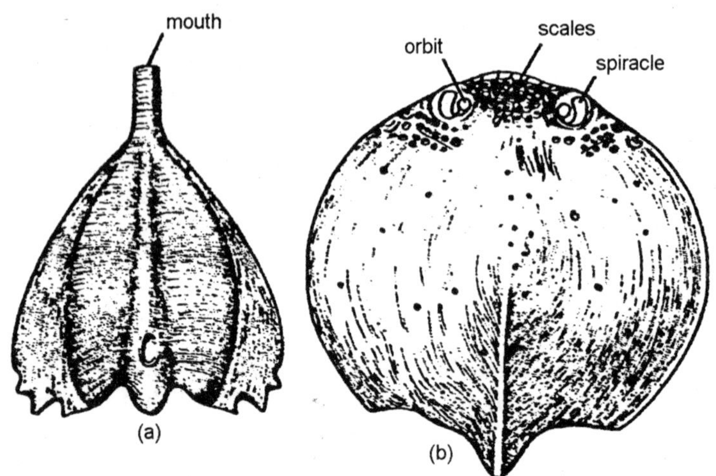

Fig. 19.1. Dorsal view of head shield of Ostracoderms: (a) eglonaspis, (b) Gabreyaspis.

longer them the dorsal lobe. Many pteraspids developed an anterior rostral plate, extending far in front 9f the mouth. The genus Doryaspis resembled that of a saw fish, and developed greatly extended lateral cornual plates.

The pteraspids are succeeded by the family Psammosteidae. This group has the same basic pattern of dermal armor, with the addition of a post orbital plate. The best known genus Drepanaspis is found in strictly marine sediments. They become more conspicuously dorso-ventrally flattened, with both the bronchial ferestral and the mouth opening dorsally. Most genera are known only from fragments, but isolated plates suggest a length of upto 2 meter.

Thelodonti

The coelolepid a thelodonts are frequently allied with the heterostracaus. Palaeontologists know little more about them than they did exist. Isolated scales have turned up in rocks of late Silurian and early Devonian age. They were small, apparently flattened, and covered with tiny spined nonimbricated scales. Presumably they swamin company with other forms of ostracoderms in bays where fresh water from rivers diluted the salty sea. Since their structure is poorly known, it is almost impossible to speculate upon their relationship to the other jawless fishes. Ritchie (1968) made the most complete recent study of bodv form. According to him the body is fusiform and somewhat dorso-ventrally flattened. The mouth is nearly terminal bordered by several, rows of transversely oriented scales. The orbits are laterally

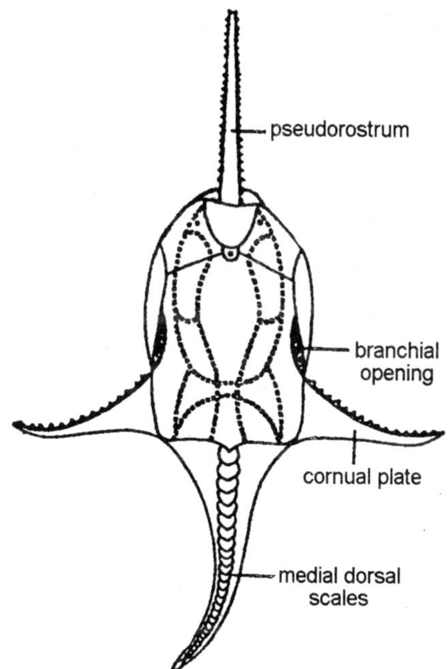

Fig. 19.2. Doryaspis, a pecullar pteraspidomorph.

placed, each surrounded by two crescentic scales. Exceptionally well preserved specimens show a straight row of approximately eight external gill openings, in contrast to the single pair of gill openings in all adequately known heterostracans. The gill openings are located beneath fragile, laterally oriented flaps of tissue identified as pectoral fins. In addition there are small rounded dorsal and oval fins; the candal fin is hypocercal, that is, the main axis is angled ventrally.

Cephalaspidomorphi

In most cephalaspidomorphs, both the narial opening and the nasal sac are single and medial, rather than pairea and lateral, as in other vertebrate groups. A detailed study of the brain and cranial nerves in fossil cephalaspidomorphs with heavily ossified endocranis demon startes a striking similarity to the living lamprey, In most cephalospidomorphs and the lamprey, both hypophyscal duct and narial openings emerge from the top of the head in a common nasopharygeal opening, The group caphalaspidomorphi includes genera:

Osteostraci

The Osteoslraci have the best known anatomy of all Palaeozoic Agnatha and are frequently used to initiate discussion of this

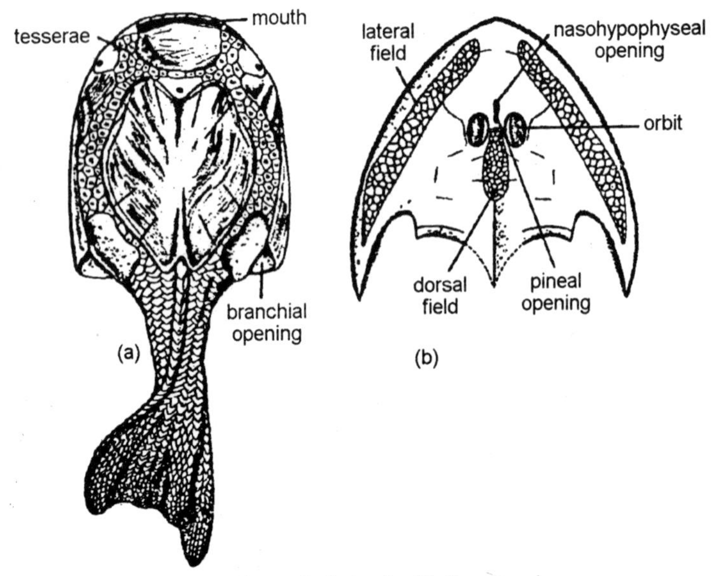

Fig. 19.3. (a) Cephalaspis, (b) Drepanaspis.

assemblage. In all the Osterstraci the head is covered dorsally with a shield of bone ornamented over—its surface by denticles, The shield shows a characteristic pattern of openings and depressed areas: close to the mid-line in the centre of the head a somewhat forward there lie a pair of circular holes through which in life, the eyes must have stared directly upward. Between these eyes a small pineal opening is invariably present, and in front of it a single aperture for the naso-hypophyseal canal. The head shield lapped over to protect the edges - of the yentral surface, the remainder of the area being covered by little scales fitted tightly together. Between the ventral tip of the head shield and the front of the scaly region, lay a very small mouth surely useless for ingesting any but the most minute particles. On each side of the mouth a curved row of external gill openings stretched to the posterior margin of the shield. The Osteostraci exhibit three advanced features that were independently evolved in jawed vertebrates: paired pectoral fins, an ossified endoskeleton, and lacunae for bone cells.

The Tremataspids are the most primitive osteostracans. They are known primarily from the Upper Silurian, are small, and have it solid carapace covering the body as far back as the base of the tail. The body is dorsa ventrally compressed, but the ventral surface is rounded rather than flattened. The gills (approximate number is 10) open separately to the exterior along the antera-ventral margin of the carapace. The head and pharynx are ossified as a unit, incorporating the brain case and the gill supports. Small scales cover the caudal region. There is no evidence of paired, anal, or dorsal fins.

The tremataspids, are succeeded by the better-known cephalaspids in late Silurian and Devonian. Janvier (1985) suggested them as an unnatural, polyphyletic, assemblage. They are more completely committed to a benthonic way of life to judge by their broad, flat, ventral surface. Nevertheless, they alone among the Agnatha evolved muscular paired fins. In later genera, the bony carapace is reduced and the scale-covered portion of the body extends further anteriorly. Pelvic fins never developed in the Agnatha.

The high degree of ossification of the endoskeleton in later forms among the Osteostraci provides detailed evidence of the structure of the brain, cranial nerves and mouth and gill region. This evidence is valuable in evaluating the origin of jaws and the ancestory of higher vertebrates. The Osteostracons emerged in the late Silurian, diversified in the early Devonian, and were extinct by the end of this period.

Galeaspida

The entirely new order of jawless vertebrates, the Galeaspida discovered in Southern China from the Lower Devonian. Members of this order have recently been reviewed by Halstead (1979), Janvier (1984), and Pan Jiang (1984). This group evolved in isolation, but it paralleled other ostracoderm orders in many ways. They resemble the Osteostracans in having a heavy carapace and a well-ossified endochondral brain case. The ventral surface of the carapace consists of a mosaic of small plates that lie between the multiple gill openings, similar to osteostracans. The axis of the tail in geleaspids is inclined ventrally, in contrast with that of osteostracans, and no specimens show any evidence of paired fins.

Anaspida

The anaspids are entirely different from the Osteostraci in general body form but have long been considered closely related because the have median nasopharyngeal opening. Typically, the body covered with small overlapping scales composed of laminar bone, resembling the basal layer of the dermal plates in other ostracoderms. The scale rows appear to correspond with the body segments. Dorsal and lateral scale rows are in the pattern of an anteriorly facing V, suggesting the outline of myotomes in Amphioxus, in contrast with the W-shaped pattern in living Agnatha and jawed fish. The bony plates covering the head are small but distinctly shaped. The orbits are more laterally placed than in the Osteostraci, and the pineal opening pierces the sales-just behind the nasohypophyshil duct. The mouth is terminal or early so. There are 6 to 15 gill openings running diagonally down the flank. Unlike any normal vertebrate of later times, the tail tilted down wards, rather than upward, to give a "reversed heterocercal" type of candal fin, found among animals only in the larval lamprey. An anal fin was present. Of paired fins there were more in the typical Silurian genera, but there was a prominent, projecting spine in the position of a pectoral fin, and other spines more posteriorly placed.

In genera Birkenia, Paryngolepis, Pterygolepis, and Rhyncholepis from the upper Silurian all have substantial bony scales.

Anaspids are known primarily from freshwater deposits, except for the oldest known genus jaymaytius, which is from near shore marine or brackish water deposits.

Living jawless vertebrates are lamprey and hag fish. Both these groups of modem agnathus appear in the Carboniferous. They completely lack bones, scales, and paired fins. The lampreys may have evolved from anaspid like jaymaytius and Endeiolepis that appear to have reduced or lost their body scales. There is no specific evidence to establish the origin of the hag fish. Much remains to be learned concerning the relationships among the early jawless and vertebrates.

INDEX